KRAKATOA

PRAISE FOR
Krakatoa

"A trove of wonderfully arcane information. . . . The rich and fascinating *Krakatoa* confirms [Mr. Winchester's] pre-eminence, not only for its inspired choice of subject, but also for its enthralling trivia." —*New York Times*

"In *Krakatoa,* Simon Winchester takes an event that happened in a white-hot second and expands it in both directions, filling in backstory and aftershocks to create a mesmerizing page-turner. . . . Winchester is an extraordinarily graceful writer. He may be the world's greatest crafter of smooth transitions, and he has the good sense never to resist an irresistible digression." —*Time*

"*Krakatoa* is deserving of superlatives: It is thrilling, comprehensive, literate, meticulously researched, and scientifically accurate; it is one of the best books ever written about the history and significance of a natural disaster." —*New York Times Book Review*

"Valuable in its resourceful analysis of what might have been one of the most startling events of the last 125 years." —*Rocky Mountain News*

"In a marvelous work of erudite scholarship and taut storytelling, Simon Winchester wraps a perfectly paced drama around intriguing history." —*Boston Herald*

"Winchester writes with thoroughness and authority."
— *Milwaukee Journal Sentinel*

"Winchester certainly knows how to make geology an alluring and wild place." — *Fort Lauderdale Sun-Sentinel*

"Vivid. . . . Dotting his narrative with learned asides and digressions, Winchester carefully builds a dramatic tale that begins with a few rumblings and ends with the end of the world as the Spice Islanders knew it. . . . Supremely well told." — *Kirkus Reviews* (starred review)

"Part history, scientific detective story, and travelogue, with all the storytelling zeal of his bestselling *The Map That Changed the World*, Winchester's new book complements the more scholarly approach of earlier volumes on the subject. With an eye for the smallest detail and a solid understanding of geology, Winchester's narrative culminates in an hour-by-hour account from the viewpoints of ship captains, a telegraph operator, a British consul, and a Dutch colonial official. . . . A good read." — *Washington Post Book World*

"Whether he is tracing the evolution of the *Oxford English Dictionary*, as he did in *The Professor and the Madman,* or detailing how England's geological foundations were first charted, as in *The Map That Changed the World*, Winchester's specialty is putting important historical events into wider context." — *BookPage*

"As Winchester explores the human terrain within the larger context of unprecedented disaster, the story reaches rapturous heights." —*Chicago Sun-Times*

"A rich blend of science and history." —*Library Journal*

"Winchester's exceptional attention to detail never falters. . . . Perhaps his greatest strength, exhibited remarkably here, is his ability to make a mystery out of that which is already known, compelling the reader forward to a tremendously satisfying conclusion." —*San Francisco Chronicle*

"An erudite, fascinating account by one of the foremost purveyors of contemporary nonfiction. . . . Winchester once again demonstrates a keen knack for balancing rich and often rigorous historical detail with dramatic tension and storytelling." —*Publishers Weekly* (starred review)

"Skillfully detailed. . . . In telling the tale of one disastrous day in 1883, Winchester has successfully blended history and science to create a concoction that is sure to please." —*Denver Post*

"*Krakatoa* is a fascinating adventure concerning not only a remarkable event, but a remarkable period of human history. As prose, Winchester's new book reads like the build up to a volcanic eruption; it is slow at first, then quickens to a bright conclusion." —*Tennessean*

KRAKATOA

KRAKATOA

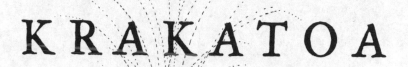

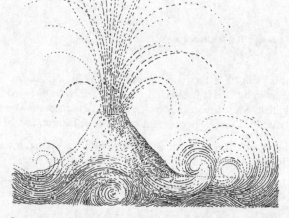

The Day the World Exploded:
August 27, 1883

SIMON WINCHESTER

HARPER ◉ PERENNIAL

NEW YORK • LONDON • TORONTO • SYDNEY

HARPER ● PERENNIAL

Originally published in Great Britain in 2003 by Viking, a division of Penguin Books, Ltd.

A hardcover edition of this book was published in 2003 by Harper-Collins Publishers.

P.S.™ is a trademark of HarperCollins Publishers.

First Perennial edition published 2004.
First Harper Perennial edition published 2005.

Library of Congress Cataloging-in-Publication Data is available.

ISBN-10: 0-06-083859-0 (pbk.)
ISBN-13: 978-0-06-083859-1 (pbk.)

20 ❖/LSC 30 29 28 27 26 25 24 23

I DEDICATE THIS BOOK, WITH PLEASURE AND WITH THANKS, TO MY MOTHER AND FATHER.

At any given instant
All solids dissolve, no wheels revolve,
And facts have no endurance—
And who knows if it is by design or pure inadvertence
That the Present destroys its inherited self-importance?

—W. H. AUDEN
"For the Time Being" (1944)

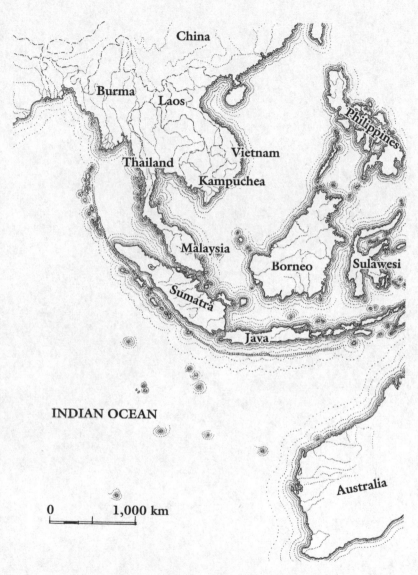

Southeast Asia.

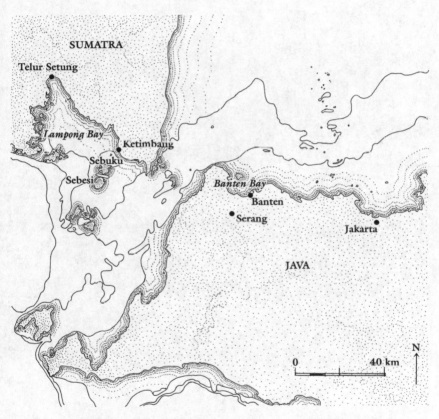

Southeast Asia, with the western islands of the immense archipelago of what is now Indonesia, formerly the Dutch East Indies.

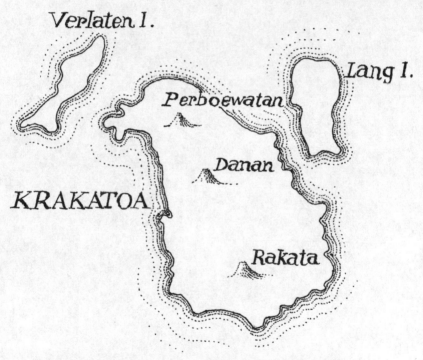

The islands of the Krakatoa group, as known before the 1883 eruption.

Contents

	List of Illustrations and Maps	xix
	Prelude	1
One	"An Island with a Pointed Mountain"	9
Two	The Crocodile in the Canal	37
Three	Close Encounters on the Wallace Line	51
Four	The Moments When the Mountain Moved	115
Five	The Unchaining of the Gates of Hell	149
Six	A League from the Last of the Sun	177
Seven	The Curious Case of the Terrified Elephant	197
Eight	The Paroxysm, the Flood, and the Crack of Doom	209
Nine	Rebellion of a Ruined People	317
Ten	The Rising of the Son	339
	Epilogue: The Place the World Exploded	367
	Recommendations for (and, in One Case, Against) Further Reading and Viewing	385
	Acknowledgments, Erkenningen, Terima Kasih	397
	Index	403

Illustrations and Maps

Southeast Asia	*xi*
Southeast Asia/Indonesia	*xiii*
Islands of the Krakatoa Group	*xv*
Cloves	*10*
Nutmeg and Mace	*11*
Pepper	*12*
Portugal's and Spain's Colonies	*14*
1595 Map of the Far East	*25*
Jan Coen	*33*
Logo of Vereenigde Oost-Indische Compagnie	*37*
Milliner Weaving Topis and Bonnets	*44*
Alfred Russel Wallace	*55*
Charles Darwin	*59*
The Wallace Line	*63*
Alfred Lothar Wegener	*70*
Pangaea	*72*
Greenland	*79*
Crystal of Magnetite	*83*
Process of Convection	*89*
Magnetic "Zebra Stripes"	*93*
Wilson's Transform Fault Structure	*104*
Subduction Zone	*108*
Geological Evolution of the Krakatoa Islands	*117*

Coastal Jungle of Krakatoa   120
Etching by Jan van Schley   137
Samuel Morse   143
Batavia City Scene   145
Frederik 's Jacob   146
Rogier Verbeek   169
Last Map of Krakatoa   175
The Agamemnon   185
Worldwide Telegraph Network   192
Ad for Anna Wilson's Circus   198
Grand Ballroom of Concordia Club   201
Pressure Wave of Explosion   216
Paddle Steamer Berouw *vs. Tsunami*   219
Brazilian Rubber   224
Charts of Sunda Strait Before/After Disaster   229
Krakatoa Dispatch by British Consul   235
Classic Tsunami   244
Tide Meter at Batavia   251
The Stranded Berouw   254
Chart Where Sound Was Heard   271
Chart of Oscillations in Tides   279
Frederic Church's Sunset over the Ice   281
Rafts of Pumice   296
The World's Tectonic Plates   304
Oceanic Plate Cross-section   312
Tectonic Structure of the Region   315
Title page of Max Havelaar   323
Muslim Imam   328
Anak Krakatoa   343
Chart of Anakrakata   344
Ballooning Spider   352
Life on the Beach   354
Casuarina Forest   363
Aerial photos of Anak Krakatoa and Surtsey   377
Varanus salvator   383

KRAKATOA

Prelude

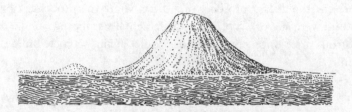

He also had one volcano that was extinct. But, as he said, "One never knows!" So he cleaned out the extinct volcano, too. If they are well cleaned out, volcanoes burn slowly and steadily, without any eruptions. Volcanic eruptions are like fires in a chimney. On our earth we are much too small to clean out our volcanoes. That is why they bring no end of trouble upon us.

—from *The Little Prince*,
Antoine de Saint-Exupéry, 1944

It was early on a warm summer's evening in the 1970s, as I stood in a palm plantation high on a green hillside in western Java, that I saw for the first time, silhouetted against the faint blue hills of faraway Sumatra, the small gathering of islands that is all that remains of what was once a mountain called Krakatoa.

There was a high peak to the left of the group, its pyramid shape cut off sharply by its vertical northern cliff. A couple of low islands hugged the horizon to the right. In between them was one small and perfectly formed, absolutely symmetrical low cone, from which rose a thin wisp of smoke. The smoke left a blackish, grayish trail that first rose vertically and then, as it caught the trade winds a few hundred feet above the darkening surface of the sea, was whisked off to the left, melting away until it became no more than a slow-fading stain against the salmon glow of the sunset.

I must have stood there, enraptured, until it got quite dark. And then I turned away for the drive back to Jakarta. This, I remembered thinking during the endless night of the flight back west, had been a scene of impeccable beauty. And all the more so because it presented a distant prospect of a place where the processes of the world were at work, a place of an elemental significance, and a disastrous place once—but these days quiet again, serenely biding its time.

It was almost another quarter of a century before I found myself back in Java. Most of the work I was doing kept me in the island's center, in towns like Jogjakarta, Solo, and Semarang. But just before I was due to fly away, and because I had a free evening until my plane left, I decided on a whim to make my way back across to the western edge of the island. I drove down to the coast road, just as I had back in the seventies.

I wanted to go back for no other reason than to take what I thought would be one final look at a place that, though few beyond the East Indies knew exactly where it was or what it looked like or just what had happened there, had a name—Krakatoa—that had for decades been firmly annealed into the world's collective consciousness. There was a famous film (which admittedly placed the island on the wrong side, the east side, of Java). There was a much loved children's book (which admittedly placed the island in a different ocean, the Pacific, rather than the Indian). The name had become part of the world's cultural lexicon—it had a vaguely exotic familiarity about it, a certain indefinable resonance. It was a word that people seemed to like, both to say and to have said to them. And now here I was, so close to the volcano that the chance of seeing it once more seemed an opportunity that I shouldn't pass up.

When I reached the best viewpoint on the corniche, it was evening, perhaps a little later, and so, rather darker, than when I had been there last. This time the enormous iron lighthouse near the port of Anjer—built by the Dutch to replace the one that had been torn away in the terrible waves caused by the great eruption—was sweeping its beam calmly across the unruffled waters of the Sunda Strait, beginning its night on sentry duty.

The group of islands was there, just as before, now black against the vivid deep pinks of the western sky. The giant peak to the left of the group was just as I remembered; the low islands, too, this time merging with the evening clouds. And in the middle of them all, its summit seemingly rimmed with a curious orange fire, rose the pyramid shape of the one active remnant of the great disaster. As I looked through my glasses, I could clearly see that the orange was indeed fire, and that from it rose smoke, just as before, only this time a tumble of black billows that towered straight up into the windless late-evening sky.

But there was one obvious difference. The pyramid—which I now knew to be called *Anak* locally, Malay for "the child" of what had once existed there before the great eruption—seemed this time somehow bigger, sturdier, and much taller than I had remembered.

I blinked hard, looked again. I measured it as best I could against the big peak to the left—trying to recall where the smaller mountain had stood in relation to that cliff wall. It was higher up now, surely. Yes, there was no doubt. Of course memory does play tricks in situations like this, but as I stared long and hard, I became ever more certain. The volcano, *the child of Krakatoa,* had grown very much larger during the twenty-five years that I had been away.

When I got back to the maps, I checked, and could see in short order that the modern surveys all agreed. The small island-mountain, which had been born out of the sea forty-odd years after the very explosion that destroyed and vaporized its parent, was now itself growing fast, thrusting upward at an extraordinary rate. By looking at the old charts and maps that had been published down the years since the last week of June 1927, when it was first seen above the waves, it was possible to calculate that it had been growing taller, fairly steadily, at an average rate of about five inches a week.

True, there had been some fits and starts—a lava flow here, a wild eruption there—but generally Anak Krakatoa had enlarged itself by twenty or so inches each month since 1927. Every year since its birth it had become higher by twenty feet, and somewhere near forty feet wider. And if this was indeed still so—I checked my figures once again—then it meant that my mountain was not simply taller: it was fully *five hundred feet* taller than when I had last seen it.

Which is why this sturdy stripling of a volcano has captivated me ever since. It is a volcano that absolutely and very visibly refuses to die. It is a volcano that seems to me to possess a wonderfully seductive combination of qualities, being beautiful and dangerous, unpredictable and unforgettable. And more: Though what happened in its former life was unutterably dreadful, the realities of geology, seismicity, and the peculiar tectonics of Java and Sumatra will make sure that what occurred back then will without a doubt one day repeat itself, and in precisely the same way.

No one can be sure exactly when. Probably it will be very many years—many years, that is, before anything will befall the world that could possibly be as terrible as what took place during the historic paroxysmal moment that reached its extraordinary climax at exactly two minutes past ten on the morning of Monday, August 27, 1883.

The explosion itself was terrific, a monstrous thing that still attracts an endless procession of superlatives. It was the greatest detonation, the loudest sound, the most devastating volcanic event in modern recorded human history, and it killed more than thirty-six thousand people.

Geological evidence from around the world admits of a number of bigger and more devastating volcanoes, true. Krakatoa is reckoned today to be only the fifth most explosive one in the planet's certain geological history—Mounts Toba and Tambora in the East Indies, Taupo in New Zealand, and Katmai in Alaska are all thought to have been very much larger, at least in terms of the amount of material they hurled into the sky and the height to which all that material is thought to have soared.

But these were all eruptions quite lost in antiquity, with rather little direct effect on human society. When Krakatoa exploded it was 1883, and the world was a profoundly different place. Sophisticated human beings were on hand to see this volcano's convulsions, they were able to investigate the event, and they were able to attempt to understand the processes that had caused such dreadful violence. And yet, as it happens, their observations, painstaking and precise as science demanded, collided head-on with a most discomfiting reality: that while in 1883 the world was becoming ever more scientifically advanced, it was in part because of these same advances that its people found themselves in a strangely febrile and delicately balanced condition, which an event like Krakatoa did much to unsettle.

The communications technology of the time, for example—the advances of telegraphy, the laying of undersea cables, the flourishing of news agencies—ensured that the world's more advanced peoples learned about the eruption within moments of its happening. But at the same time the limited, and only slowly unfolding, geological knowledge then to hand did not give the audience an explanation of the events that was sufficient to soothe their fears about all that they were learning. Hearing of the event baffled people thousands of miles away from where it happened, and left faraway populations bewildered and, in some cases, more than a little frightened.

Moreover, religious dogma still held a powerful grip on millions, even in societies whose scientific understanding was advancing so rapidly. Tribal ancestors might have explained away an eruption like Krakatoa's: It was simply a matter of the gods being angry. But the luxury of such a facile explanation was not available to the more modern people of 1883, whose growing sophistication meant both that they knew so much about an event like this and yet at the same time actually knew so little. They had sufficient knowledge of the facts—the news agencies saw to that—but it was knowledge coupled with an insufficiency of understanding. Many people worried in consequence that the eruption of Krakatoa meant something infinitely more dreadful than it did: that their world was being torn asunder, and perhaps even, as the Bible had foretold, was coming to an end.

Fretful and fascinated people around the world, in cities as distant from Java and from one another as Boston and Bombay and Brisbane, all came to know of the event in an instant—and they did so quite simply because this was the world's first major catastrophe to have taken place in the aftermath of the invention of the submarine telegraph. The newspapers were full of it, the descriptions of the happenings made all the more enthralling because they were so perfectly up to date. Words and phrases that had hitherto been utterly unfamiliar—*Java, Sumatra, Sunda Strait, Batavia*— became in one mighty flash of eruptive light part of the common currency of all.

And in learning of these places and of the terrible events that occurred there, so the world's people suddenly became part of a new brotherhood of knowledge—in a sense it was that day in August 1883 that the modern phenomenon known as "the global village" was born, in part through the agency of this enormous explosion. The word *Krakatoa,* despite being a word misspelled and mangled by the imperfect arts of Victorian telegraphy and journalism, became in one awful ear-splitting moment a synonym for cataclysm, paroxysm, death, and disaster. And the disaster left a trail of practical consequences—political, religious, social, economic, psychological, and scientific consequences among them. Even today their curious and chilling echoes are still faintly and worryingly discernible, both in Java and around the globe.

In all sorts of other observable ways the impact of Krakatoa on the world's consciousness was profound, and immensely more so than in the case of the eruption of the four technically greater volcanoes of earlier times. Krakatoa had an impact on climate, for example: The airborne debris hurled into the skies lowered the planet's temperature; it changed the appearance of the entire world's sky; it set barometers and tide meters flailing wildly thousand of miles away; it panicked American firefighters into battling what they thought were raging infernos, but that were in truth violent sunsets caused by the roiling clouds of Krakatoan dust.

Because of all the painstaking and precise work that was begun in the immediate aftermath of the eruption, we know today just why it was that Krakatoa erupted, and we understand only too well the nature of the forces that caused it. An entirely new science has come along to pare away the old and mythmaking mysteries of the event, and to render this volcano, and indeed all others, readily explicable.

Now, seen from a palm plantation high on a green hillside, Krakatoa looks peaceful and serene, with just a thin column of white or gray or on occasion black smoke easing up from its summit. But looks are deceptive: All the while the child-mountain is growing steadily and rapidly, as the elemental fires that created the world rage deep inside.

1

"An Island with a Pointed Mountain"

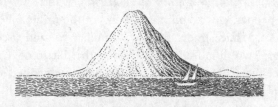

Volcanic Eruption Area: *Owing to volcanic eruption this area is considered unsafe for shipping . . .*

Pulau Anakrakata: (6° 06' S, 105° 25' E), *an islet, appearing midway between Pulau Sertung and Pulau Rakata Kecil in 1928, where formerly a bank with a depth of 27m was charted. In 1929, this islet disappeared again, but resurfaced by eruptions in 1930, and after heavy eruptions in February 1933, appeared to have increased in size. In 1935 this islet was almost circular in shape, with a diameter of about 1200m, and a height of 63m, and in 1940 it was 125m high. In 1948 there were a few casuarina trees on the N extremity of the islet; in 1955 the islet had an elevation of 155m and, viewed from the S, was devoid of vegetation. In 1959 it was in eruption and emitted thick black smoke to a height of 600m. Volcanic activity on Pulau Anakrakata was last observed in 1993.*

Danger Signal: *In the event of threatened eruption within the Krakatau area, Jakarta Radio will broadcast the necessary warning in Indonesian and English. (See Admiralty List of Radio Signals.)*

—from "Admiralty Sailing Directions,"
NP36, *Indonesia Pilot*, vol. 1, London, 1999

Though we think first of Java as an eponym for coffee (or, to some today, a computer language), it is in fact the trading of aromatic tropical spices on which the fortunes of the

great island's colonizers and Western discoverers were first
founded. And initially supreme among those spices was the one
rather ordinary variety that remains the most widely used today:
pepper.

   *Piper nigrum, Syzygium aromaticum,** and *Myristica fra-
grans*—pepper, clove, and nutmeg—were the original holy trinity
of the Asian spice trade. Each was familiar to, and used by, the
ancients. Two hundred years before the birth of Christ, for
instance, the Chinese of the Han Dynasty demanded that their
courtiers address their emperors only when their breath had been
sweetened with a mouthful of Javanese cloves, the "odiferous pis-
tils," as they were later more widely known. There is some vague
evidence that Roman priests may have
employed nutmeg as an incense;
it was definitely in use as a flavoring
in ninth-century Constantinople,
since the terrifyingly Orthodox
Saint Theodore the Stu-
dite—the scourge of the
image-smashing Icono-
clasts—famously
allowed his monks to sprin-
kle it on the pease pudding they
were obliged to eat on days when
monastery meat was forbidden.
And in Elizabethan times a nutmeg
pomander was an essential for keeping foul ailments at bay: The notion
that nutmeg could ward off the plague survived longer than many
another old wives' tale.

Syzygium aromaticum, the clove.

   Pepper, though, was of infinitely more moment to the ancients
than to be merely a topping, nostrum, or cachou. The Romans

---

*Some botanists regard the clove as more properly *Eugenia caryophyllata,* though both
agree it is part of the family *Myrtaceae,* of which the evergreen myrtle is the best known.

used it in abundance: Gibbon wrote of pepper being "a favourite ingredient of the most expensive Roman cookery," and added his authority to the widely held idea that Alaric, the rambunctious king of the Visigoths, had demanded more than a ton of it from the Romans as ransom when he laid siege to the city in A.D. 410. The *aureus* and the *denarius,* the gold and silver coins of the empire, became the preferred currency of the Spice Route, and the Indian pepper merchants of Cochin and Malacca and the ports of southern Ceylon were said to be impressed that the denomination of coins was indicated by the number engraved upon them, not by their size.

Nutmeg and mace.

However they may have been denominated, the coins must have been paid out in enormous numbers. Pepper was so precious and costly and so much in demand that the cost of it all had Pliny the Elder fulminating. "There was no year in which India"—and by this he meant *the Indies,* since pepper traded came both from the Malabar Coast and from western Java—"does not drain the Roman empire of fifty million sesterces." So dearly, he added drily, "do we pay for our luxury and our women."

(There is a pleasing symmetry about Pliny's involvement in this part of the story of Krakatoa, even if he appears in only a walk-on role. Although this rich and well-connected former soldier—he was a cavalry officer in Roman Germany—happily took on a variety of official duties on behalf of his emperors, Pliny was above all else a naturalist. He was a savant, or a student, as he once famously put it, of "the nature of things, that is, life." His reputation is based largely on his thirty-seven-volume *Natural History,* an immense masterpiece in which, among countless other delights, is the first use of the word from which we derive today's *encyclopedia*.

It was during the late summer of A.D. 79, while pursuing his official task of investigating piracy in the Bay of Naples, that Pliny was persuaded to explore a peculiar cloud formation that appeared to be coming from the summit of the local mountain, Vesuvius. He was duly rowed ashore, visited a local village to calm the panicked inhabitants—and was promptly caught up in a massive eruption. He died of asphyxiation by volcanic gases on August 24, leaving behind him a vast reputation and, as memorial, a single word in the lexicon of modern vulcanology, *Plinian*. A Plinian eruption is now defined as an almighty, explosive eruption that all but destroys the entire volcano from which it emanates. And the most devastating Plinian event of the modern era occurred 1,804 years, almost to the day, after Pliny the Elder's death: at Krakatoa.)

Pepper has a confused reputation. There is no truth, for example, in the widely held belief that it was once used to hide the taste of putrefying meat; this charming thought perhaps derives from the equally delightful notion, still recognized by pharmacists today, that pepper can be used as a carminative, a potion that expels flatulence. But it was very much used as a preservative, and more commonly still as a seasoning. By the tenth century it was being imported into England; the Guild of Pepperers, one of the most ancient of London's city guilds, was established at least before 1180, which

Piper nigrum, pepper.   was when the body was first recorded (they were in court for some minor infraction); by 1328 the guild had been formally registered as an importer of spices in large, or *gross*, amounts: its members were called *grossarii*, from which comes the modern word *grocer*.

Joseph Conrad caught the obsession, in *Lord Jim:*

The seventeenth-century traders went there for pepper, because the passion for pepper seemed to burn like a flame of love in the breast of Dutch and English adventurers about the time of James the First.* Where wouldn't they go for pepper! For a bag of pepper they would cut each other's throats without hesitation, and would forswear their souls, of which they were so careful otherwise: the bizarre obstinacy of that desire made them defy death in a thousand shapes; the unknown seas, the loathsome and strange diseases; wounds, captivity, hunger, pestilence, and despair. It made them great! By heavens! it made them heroic . . .

The Western appetite for the trinity of flavorings increased almost exponentially during the fourteenth and fifteenth centuries—the trade being dominated, at least after the Papal Donation of 1493, by the only serious maritime power of the day in the Orient, the Portuguese.† Vasco da Gama, who opened up the East and made it as far as Calicut, was said to be exultant at finding out that the pepper he knew would sell for eighty ducats a hundredweight back in Venice (which was the European center for the trade) could be bought in India for only three. A steady stream of Portuguese merchantmen and explorers promptly left the Tagus for the Orient—one of them, Pedro Alvares Cabral, discovering

---

*1603–25.

†The Papal Donation in essence "gave" the exploitation of the Western world to Spain and the Eastern to Portugal. The Spaniards, who were seamen and navigators of equal skill, had under papal supervision agreed with the Portuguese on the division of the conquerable planet—drawing Pope Julius II's so-called Tordesillas Line along the meridian 370 leagues to the west of the Cape Verde Islands (approximately 48 degrees west of Greenwich). To the west of the line, Spain had a free hand—hence Mexico, Chile, California; to its east—which crucially included the coast of Brazil—Portugal could freely operate its caravels. And since Africa, Asia, and the islands of the Spice Route lay similarly to the east of the Tordesillas Line, so Portugal dominated the exploration of the East and, for a while, the European pepper trade too. The antimeridian of the Tordesillas Line appears in the East too, of course, at around 129 degrees east of Greenwich. Spain colonized the Philippines as a consequence, and Portugal won parts of New Guinea and Timor. The Papal Donation, which had its origins in a ruling from Pope Alexander VI in 1493, cast a very long shadow indeed.

and then claiming Brazil on the way—and for a while the Portuguese entirely dominated the business. The ancient overland route, ships to Arabia, camels to the Mediterranean, was utterly changed: Now it was massive sailing ships all the way, via West Africa and the Cape. And in just the same way as Roman currency became the common coin of the old route, so the Portuguese language became the lingua franca of the new.

But slowly time and technology intervened: By the sixteenth century the Dutch and the English, now with all their shipbuilding skills finely tuned, with all the oak they needed for their hulls and all the flax they wanted for their sails and all the cannonry their foundries could produce and the navigating wherewithal for making long journeys fast and safely, found they could outrun and outgun the fine vessels from Lisbon. More than a few Dutch ships,

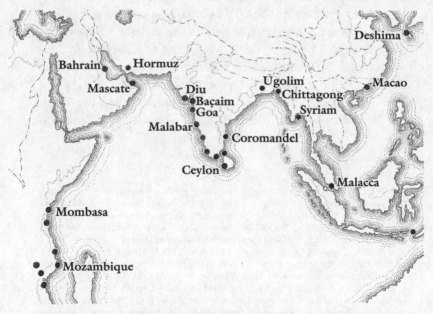

Portugal and Spain also exercised their rival colonial claims to the territories of the East, dividing their claims around the antimeridian of the Atlantic Ocean's well-established Tordesillas Line of demarcation.

flagged with the Portuguese *bandhera* to get around the royal pro-
hibition on non-Iberians trading in the Orient, were now coming
home and whetting the appetites of the Amsterdam merchants for
pepper and for profit. And so, slowly, under the force of these various imperatives, the balance of maritime power in the East started
to change. The Portuguese from the warm and lazy south were
slowly driven out and replaced by doughty Europeans from the
cold and more ruthless north.

The change began somewhat inauspiciously in late June of 1596,
when a ragged flotilla of four Dutch vessels dropped anchor in the
roads off the northwestern Javanese pepper port of Banten and
invited the Portuguese spice merchants, whose warehouses had
long lined the shore, to come aboard. The voyage had been spon-
sored by the nine merchant-adventurers of the Compagnie van
Verre of Amsterdam—in translation simply "The Long Distance
Company"—who had been inspired by the idea of blazing a spice
trade–route to the Indies. It had not gone at all well.

Cornelis de Houtman, who commanded the venture, turned
out to be both an inept navigator and a cantankerous martinet. Not
that he had lacked preparation: along with his brother Frederik he
had already spent two years in Lisbon gathering intelligence on the
Portuguese operations in the East.* His expedition was grandly
titled *Eerste Scheepvart,* "The First Ship Sailing." It excited much
attention as, with 249 men aboard, it swept out into the Zuider Zee
and, after provisioning at the merchant docks inshore of Texel, fell
away from the roads on the morning of 2 April 1595. It promptly
ran into a whole sea of troubles.

The provisioning had been too hasty. After only a few weeks
scurvy broke out, with sailors suffering such rending stomach

---

*Both had been imprisoned for their pains. They were alleged to have stolen a number of
Portuguese *portolanos,* the secret charts and sailing directions carried on all expeditions. No
doubt, given the furtive nature of their employment as "commercial representatives" of the
van Verre group, they were guilty as charged.

pains among their other symptoms that the Dutch still have a word for it, *scheurbuik,* "tearing-belly." Disputes raged between the on-board merchants and the ships' masters: One merchant was locked in his cabin in chains for the entire voyage, another was poisoned in India, a master fell victim to a mutiny. De Houtman proved himself to be no more than "a boaster and a ruffian." A short stopover in Madagascar for the convoy to catch its breath turned into a six-month deathwatch, during which so many crewmen died that there is still a Madagascan bay called the Dutch Cemetery. By the time the venture reached Banten* only a hundred Dutchmen were still alive.

The surprised Portuguese at first made them welcome, probably assuming that so wretched a gaggle of starving men could hardly pose much of a challenge to the might of Lisbon. The head of the entrepôt sent a message to his immediate superior in Goa, on the Indian coast, though more for reasons of diplomatic propriety than of disquiet. He then introduced the visitors to the local Banten sultan, who was sufficiently impressed with them to enter into a treaty: the first formal document between Dutchmen and Javanese, whose three-and-a-half-century relationship would prove to be one of repression, exploitation, and too often most cruel colonization. "We are well content," the sultan wrote, "to have a permanent league of alliance and friendship with His Highness the Prince[†] and with you, gentlemen."

This contentedness would not long survive. Some indication of what would evolve into a deeply unhappy relationship between the Dutch and their soon-to-be subject peoples came good and early, during the subsequent sojourn of the de Houtman expedition.

Much of Sumatra and Java had already become widely Islami-

---

*In many accounts this small northwest Javan port city appears with its original Portuguese-given spelling, "Bantam," which suggests, probably not wholly accurately, that the small and eponymous chickens that are actually believed to have first come from Japan originated there.

†Maurice of Nassau, after whom another Dutch-settled island, Mauritius, was named.

cized (the earliest-known Muslim grave on Java dates from 1419, after which the creed took hold rapidly), and the local people and their leaders were acutely sensitive to the strange ways of the European infidels. The Portuguese had seemingly shown a measure of tact; the Dutch, on the other hand, had a reputation for being crass and insensitive in their dealings with the "primitives" they met. Cornelis de Houtman himself insulted the Banten sultan—contemporary accounts are not specific, speaking only of de Houtman's "rude behavior"—and was ordered to leave port.

De Houtman's flotilla sailed east. It was then attacked by pirates off Surabaya, a dozen more of his crew were killed, and one of the ships had to be abandoned and set ablaze because there were too few hands to work it. In retribution for the piracy he ordered a brief campaign of systematic rape and pillage off the coast of Madura. But by the time he reached Bali he had calmed down, just as many others have been similarly soothed.

He found that the king of Bali was "a good-natured fat man who had two hundred wives, drove a chariot pulled by two white buffalos, and owned fifty dwarves." He was, perhaps understandably given his location, not the best informed of monarchs. When de Houtman, keen to impress him with Holland's importance, drew him a map of Europe showing the Dutch borders lying improbably close to Venice and Moscow, the old man did not for a moment even raise a skeptical eyebrow.

The crew liked the place hugely; only one sailor deserted. But there were yet more disturbances, and before they left an ill-tempered de Houtman almost had to be forced to accept a cargo of a few pots of peppercorns, which the expedition took back to Amsterdam. Their return voyage, by way of Java's south coast, was every bit as miserable, and almost as time-consuming, as the outbound one. And seven more sailors dropped dead almost as soon as they reached home, their stomachs reacting violently to the change back to civilized rations.

But for all the blundering tragedy of that first Dutch expedition, the achievement could hardly be gainsaid, for it did bring back those few pots of precious black Java pepper. And that, it seemed to the nine sponsoring merchants of Amsterdam, was all that really counted. The Portuguese stranglehold on Oriental spice trading could in theory at long last be broken. Other, better-furnished expeditions could now venture out and obtain more pepper, clove, and nutmeg (and its attendant aril, called mace, as well as the cinnamon that grew close by) and thus help make the mercantile barons of Holland richer than could be imagined.

One might have expected a spirited reaction from the Portuguese. They did indeed react—but then managed to blunder even more calamitously than had the Dutch. The Portuguese ran their *Estado da India* from Goa, on the western coast of India. Their viceroy there had received news of the de Houtman expedition from his agents and promptly decided he would see to it that no such impertinence happened again. He dispatched a specially built fleet* south to Banten, but the admiral he chose for the mission, a still-infamous-in-Lisbon figure named Dom Lourenço de Brito, took such a long time to get there that the Dutch, whom he had planned to confront, had already upped and sailed back to (and at the time were in fact comfortably ensconced in) Amsterdam.

So the admiral, apoplectic, turned his fury on the Bantenese—despite his chief in Goa having ordered him specifically to do no such thing. He tried to teach them a lesson for having dared to be hospitable to his masters' rivals—and was promptly outfoxed by the Javanese sailors' battling techniques. He lost two of his four ships and withdrew, humbled, to Malacca to lick his wounds.

No one—not the admiral or the viceroy or King Philip back home—recognized the import of the moment: that the skirmish between the Bantenese and the tiny fleet from Goa marked the beginning of the end for Portugal's imperial role in the East. It took a while for it to be wholly extinguished (the reversion of Macao from Portuguese to Chinese rule in 1999 marked the final drawing

---

*Included were a pair of Mediterranean-style rowing galleys, surely inconveniently lumbering beasts to handle in the open waters of the Bay of Bengal.

down of blinds); but the lights began to go out from that time. "Look at the Portuguese," noted the British ambassador to India, Sir Thomas Roe. "In spite of all their fine settlements they are beggared by the maintenance of military forces; and even their garrisons are only mediocre." Within half a century only Goa, Macao, Mombasa, the ports of Mozambique and, in the Indies, Flores and Timor remained. The power of the old Iberian empires was suddenly waning; new trading empires, directed by men from the chilly northern European capitals of Amsterdam and London, were about to be born.

From the moment of the return of that first unhappy excursion to Java and beyond, shoals of Dutch fleets, each backed by packs of excited adventurers, began coursing across the ocean, bound for Banten and the fast-opening Oriental universe beyond. In May 1598 a fleet twice the size of de Houtman's, under the command of the more competent and distinguished sea captain Jacob van Neck, left Texel for Banten, reached there in half the time, and sailed away with enough pepper for its backers to realize no less than a 400 percent return on their investment. The floodgates then opened, once and for all.

Before the end of 1601 fourteen fleets had gone East—a total of sixty-five vessels. Most had passed along the conventional route via the Cape of Good Hope. Others, keen to get first to the spice-rich islands on the eastern side of the archipelago, took the much riskier path southwestward across the open Atlantic, through the newly discovered but windy and reef-strewn Strait of Magellan and then across the entire Pacific Ocean. It was an extraordinary, boisterous, devil-may-care time, and the docks in Hoorn and Enkhuizen and Amsterdam were frantic with shipping, always leaving, leaving, leaving. Some of the more sobersided Hollanders sniffed over their glasses of genever and accused the adventurers of indulging what they termed "the wild navigation." And wild it may well have been—but the commercial results were obvious. The Hollanders were on to something big.

*   *   *

Banten itself—though it was once the largest city in Southeast Asia and one of the most famous of the world's seventeenth-century ports—is these days anything but big. The garrison that the local sultan permitted the Dutch to build, Fort Speelwijk, still stands: With its ten-foot-thick walls, its worn machicolations and breast-works, its muddy tunnels and embrasures now greasy with tropical mold, it is a melancholy reminder of how mighty and ambitious the Dutch had once been. The sea that their vigilant sentries scanned for approaching enemy sails now laps a full mile away beyond the fort's curtain walls: Silt oozing from the river started to choke the estuary in the nineteenth century, and before long had rendered Banten port unusable for vessels much larger than whalers and large canoes. Its rise was spectacular; its fall—like the fall of empires more generally—slow and inevitable. It has long since been abandoned as a port; nowadays Banten is little more than a collection of shanties and ruins, with a lane of shops selling Muslim caps and boxes of locally grown dates. There are no local pepper plantations: Western Java has turned its plantations to tapi-oca and coffee, and it is Sumatra that now produces about a sixth of the world's two hundred thousand yearly tons of *Piper nigrum*.

Yet Banten still makes its own contribution to the economic well-being of the islands. Just beside the old Dutch fort, at the entrance to the maze of lanes that lead to the former sultan's ruined palace, stands a curious trio of immense and very ugly cement tow-ers. They look mysterious and rather sinister. They are formidably well guarded by razor wire, attack dogs, and civilian guards (one of them a young woman who contentedly suckled her infant when I inquired—in vain—if I might go inside). They look as if they might be kin to Fort Speelwijk, fortresses of a kind too, a protection for a population fretful about some nameless and more contemporary disaster.

In fact they are nothing of the kind. The towers are manmade homes for the local birds called swiftlets, members of the genus *Collocalia*, whose saliva is the basis for that most celebrated Cantonese aphrodisiac, bird's nest soup. The towers' owner is a Chinese restaurateur from Jakarta. He thinks it most unlikely that any client would ever realize that the soup he serves comes not from nests plucked in the traditional manner from dangerous clifftops in Thailand, but from those farmed weekly inside a cement blockhouse in a former Dutch pepper port. He is reputed to be one of the wealthiest Chinese in all of modern Java. He was blissfully unaware that he had made his fortune in the town that had once made other outsiders, the entirely dissimilar burghers of seventeenth-century Amsterdam, extremely wealthy too.

With the advent of the Dutch, whose maritime traditions had already spawned legions of cartographers, came maps: good, beautiful, accurate, and before too long utterly magnificent maps. With the making of maps came the observed and calculated details of place. With details of place came names—and among them, quite early in the process, the first naming of the tiny island that is the subject of this story.

For fifteen hundred years the notion of the great Greek astronomer Ptolemy, published in his definitive *Geography,* was that Africa and Asia were one, connected by a land bridge across the southern Indian Ocean. It took a very long while to discredit this idea. But by the fourteenth century, on the basis of reports from early explorers (such as Marco Polo) and intelligence gathered from Arab travelers and traders, cartographers in Europe began to suppose the existence of a number of peninsulas and discrete islands littering the ocean between China and Africa. The dangling appendages of India and the Malay Peninsula began to appear on maps first; and then three of the largest supposed islands, today's Borneo, Sumatra, and Java, started to be depicted on charts printed in the sixteenth century (though there was a long period when lozenge-shaped Sumatra was confused with its not-too-distant neighbor, pearl drop–shaped Ceylon).

Martin Behaim's globe of 1492—his *Erdapfel,* or "earth apple," as this solid wooden sphere was charmingly called—shows

painted on its surface the islands of Java and Sumatra, part of a
ragged chain of land between a quite distinct Malay Peninsula and
a body of vaguely determinate shape that is just recognizable as
Borneo. And Martin Waldseemüller, the German mapmaker who,
in 1507, was the first to put the name *America* on to a world map,
clearly indicates on one of the same woodcut's dozen sheets an
island, to which he gives the name *Java Minor,* quite separate from
another island to its west.

And even before these, there were prescient maps made elsewhere. Early Thai, Indian, and Korean maps, often delightfully
fanciful, indicate bodies of land that could well be the two huge
islands. There is also a copy of a fourth-century Roman road map
from England to India (with the distances variously measured out
in Roman miles, Gallic leagues, and Persian parasangs) showing
Ceylon as a decidedly insular *insula Taprobane* at its bottom right-
hand corner, but which could, from its size and familiar shape, easily be mistaken for Sumatra.

Once the Italian, Spanish, and Portuguese sailor-adventurers
began to penetrate to the east of Malacca, this imprecision vanished. Sumatra and Java were by then named and shown to exist,
most assuredly, separated by a body of water called the Sunda
Strait. Far to their east the nutmeg island of Banda* finds its way
on to a map of 1516, as well as a proper (and more or less properly
spelled) Borneo, located five years before Magellan's expeditions
reached it during history's first (and for Magellan, tragic and fatal)
circumnavigation. The Moluccas, Timor, the Celebes, and the
Philippines likewise all begin to appear on maps, *portolanos,* and
charts as well, their positions and their dimensions depicted ever
more accurately—such that by the end of the sixteenth century,
when the Dutch were poised to sail in from Texel to try to wrest
the spice monopoly from the stranglehold of the Mediterranean
powers, every major island between the Maldives and the coast of

---

*Known also because one of its neighbor islands, Pulau Ran, was seized by the British and
later swapped for a Dutch-held North American island, Pulau Manhattan.

China was illustrated, reasonably correct in outline, place, and size.

And then, with the turn of the century, and as those first commercial fleets began their laborious voyages out from Texel, so along with the merchants went the Dutch cartographic masters. These men possessed formidable naval knowledge, great artistic brilliance, and still unrivaled mapmaking skills, who would push back the boundaries of world cartographic knowledge farther and ever farther, and also add to their newly developed picture of the planet a wonderful measure of beauty and style. Their maps are treasures, awesome in both utility and aspect. And it was one of these mapmakers, Jan Huyghen van Linschoten, who was to make the first positive identification of the volcano-island that would cause such mischief in the years ahead.

Van Linschoten was a curious and energetic traveler, a man afflicted by wanderlust centuries before the ailment was properly known. "My heart is longing day and night for voyages to faraway lands," his journals note in an entry for 1587. His travels were prodigious. He was keenly interested in the Arctic, and his numerous journeys there—to the lonely island of Novaya Zemlya, north of Russia, for example, to which he went with the Dutchman for whom the Barents Sea was named—were recounted in a hugely popular book that inspired both the English and the Dutch to try to forge a sea-route through the ice to India (they never did).

But it was for the maps he created as a by-product of his six years in service to the Portuguese that he remains best known today. He had traveled east via Africa and the Cape across to the administrative capital of Goa, in western India, working as bookkeeper to the Portuguese Catholic archbishop. He spent the six years from 1583 in the immense province, traveling with the divine, noting furiously details of every place and peoples he visited (though we do not know if he actually visited Java and the islands to its east). In 1595 he published his account of it all—the *Itinerario,* one of the most detailed travel guides ever known. Bound into the volume were a number of remarkable maps—some by Portuguese mapmakers, some by Spaniards, and one exquisite map of the Far East drawn by van Linschoten himself.

The map itself has a curious history: It was executed by a Dutchman, based on information that van Linschoten admits was "from the most correct charts that the Portuguese pilots nowadays make use of" and later handed over and republished for the English. The custom was for these four powers jealously to guard such information; yet in the case of this one beautiful creation—with great cartouches, sea monsters, rhumb lines, the arms of Portugal and a flourish of compass roses—it was allowed to be seen by the navigating hierarchs of all four competing nations.

On the map much is recognizable—China, the Mekong, Malaya, Luzon. The islands of Java and Sumatra are also named and shaped more or less as they should be (though the island of Sulawesi nearby is a strange "paramecium" shape, according to one critic). Between Sumatra and what van Linschoten called *Java Major* is a narrow strait, and within it a tangled mess of unnamed islands. But though they are anonymous, there is, and crucially for this story, the text from the *Itinerario,* which acts as vade mecum for any Java-bound navigators using this map, and whom it duly advises:

> to reach the mouth of the Sunda Strait stay close to the mainland of Sumatra, always keeping a good eye out for the mountains and cliffs, of which there are many ahead, for one does not know where one will find the mouth of the Strait except only by the knowledge of the islands; look for a high island located straight across from the land tip on the north side of Sumatra, which with the island of Java Major forms the Strait that ends here. On the northwest side of the coast are two or three small islands about one mile away from land. On the island closest to land a ship with Frenchmen was run aground once. Its guns went to the king of the Island Bantam, and the one from Calapa. And a mile from land toward the south is an island with a high top or pointed mountain.

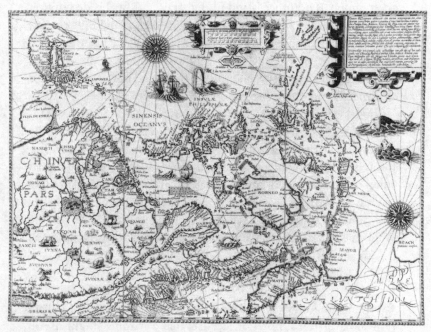

Jan Huyghen van Linschoten's 1595 map of the Far East, which shows a dusting of islets and skerries in the strait between *Svmatra* and *Java Mayor*. Krakatoa—seen for the first time ever on a European chart—is probably the large island marked opposite *Palimbam*.

There can be no doubt at all, despite van Linschoten's failure to name the island on either his chart or in the *Itinerario,* that this is Krakatoa.

He probably never saw it; though he traveled widely from Goa, there is no certainty that he visited the archipelago itself; all his reports come from his immense network of Portuguese pilots. But there simply is no other island in the strait that fits his description (or, rather, until 1883 there *was* no other such island in the strait, for since that August morning the original island has essentially not existed at all). There is no "pointed mountain" in van Linschoten's reported position at the northern end of the Strait or, for that matter, anywhere else nearby.

Other early Dutch cartographers are revered along with van Linschoten: Blaeu, Lodewijcksz, Hondius, Visscher, and, equally important in this one context, Lucas Janszoon Waghenaer. For though van Linschoten writes briefly of his island with its pointed summit in 1595, and though a year later Willem Lodewijcksz records so many small islands in the strait that one crew said they had difficulty finding a channel along its western side, it was the Waghenaer map that first depicted the island and properly captioned it with an approximation of a soon-to-be-familiar name.

Lucas Janszoon Waghenaer was a Dutch pilot who almost single-handedly revolutionized the entire business of sea-borne maps. Until his publication in 1583 of a *rutter,* a mass-produced collection of navigational notes and charts that a mariner might use to help plan a voyage in unfamiliar waters, such maps as a ship's captain had taken on board were handmade, singly produced, and often ultrasecret. Waghenaer decided instead that ships' charts should be printed, using wooden or engraved metal blocks. They should be made by the hundreds, the better to encourage men to sail, to explore and to discover.

His maps are many, and among the finest ever made. They are colorful, filled, like van Linschoten's, with elaborate cartouches, ships in full sail, fabulous sea monsters and a host of devices—compasses, scales, indications of deeps and shallows—that are still in use on charts today. His name was anglicized to *Waggoner,* and so popular and respected did his works become that the word is still heard in some of the older ship's chandleries and appears in recent dictionaries of the sea; Dalrymple's famous chart book was once known colloquially as *The English Waggoner.*

His exquisite chart of the Sunda Strait was published in 1602, seven years after the *Itinerario.* Engraved on the map, halfway between the immense body of land called *Suma. Pars.* and its equally immense eastern neighbor, *Javae Pars.*, is a group of what appear to be four small islands—a trinity of larger bodies partially enclosing a fourth. The inscription says simply, *Pulo Carcata.*

It was half a century before the island was named in a prose work. In October 1658 a Dutch doctor, Wouter Schouten, noted in his travelogue that he passed the "high tree-covered island Krakatau"—the first mention of the island by its currently accepted name.

As to the origin of this name, it sometimes seems that there are almost as many theories as there are geographers. The word *Pulau* that appeared on the Waghenaer chart remains, since it is the modern *bahasa Indonesia* for "island." But Carcata? Or Krakatau? Or something in between? Both the idea of what should be the spelling and the etymologies of its various names are an enduring mystery.

There was an early and linguistically alluring report by a French Jesuit priest, Guy Tachard, suggesting that it is an onomatopoeia. Tachard passed the island eighty years after the Dutch cartographers had, and wrote in his log to the effect that "we made many Tacks to double the island of Cacatoua, so-called because of the white Parrots that are upon that Isle, and which incessantly repeat the name." It sounds improbable, not least because of the difficulty that any mariner might experience trying to hear the call of land-based birds from high on the windy deck of a passing ship.

Others subsequently thought that Krakatoa, or the more common local form Krakatau, derives essentially from one of three words, *karta-karkata, karkataka,* or *rakata,* which are the Sanskrit and, according to some, the Old Javan words meaning "lobster" or "crab." Then there is a Malay word, *kelakatoe,* which means "flying white ant." Since crabs and parrots belong on the island—or since they did, at least, until that dire August morning in 1883—either of the two last lexical explanations seems reasonably acceptable.* White ants only occur in the eastern part of the archipelago, rendering this theory rather less credible; though perhaps rather more credible than the notion, briefly popular in Batavia, that an Indian ship's captain had asked a local boatman what name was given to the pointed mountain he could see, prompting the local to reply, *Kaga tau,* meaning "I don't know."[†]

The effect of the eastward onrush of armadas of Dutch sailing vessels, their captains now well equipped with their bound collections, their *rutters,* of the region's new-printed charts, was immediate and profound. Such Portuguese as remained were soon overwhelmed and dismissed. They were eventually and officially sent packing from the Banten pepper port in 1601. In 1605 the Dutch seized all the eastern islands of the Moluccas, which included the nutmeg and clove islands of the Banda Sea. The fort at Solor fell in 1613, and Malacca, the great trading station on the western side of the Malay Peninsula, passed from the control of Goa to that of Amsterdam in 1641. The Dutch speedily assumed power across fully forty-five degrees of eastern longitude, and before long effective control of what were to be called the East

---

*Modern navigational charts call what remains of the island Pulau Anakrakata. The official details of its ever-changing condition can be found in this chapter's epigraph.

[†]As to why *Krakatoa,* and not the more properly Javanese *Krakatau,* it is said—but not proved—that this was a spelling mistake made in an early telegraphic cable to London—a spelling error that, thanks to British domination of so much nineteenth-century science and geography, came to be accepted for many years after as the preferred (but technically incorrect) spelling. Robert Cribb, editor of *The Historical Atlas of Indonesia,* wonders if the ending *-oa* simply sounded, to the British ear, more euphonious and charmingly like an idyll in the South Pacific. I shall revisit what is to some a very vexing matter in chapter 6, when I look in more detail at those first telegraphic messages about the eruption.

Indies was almost (aside from tenacious Portuguese holdouts on the islands of Flores and Timor, and fast-fading competition from the British in a variety of tempting places) entirely theirs.*

In 1602 the Dutch took a step of a profound importance. It was a move that, as well as hugely improving their own merchants' fortunes in the East, helped to create a business model for the foundations of all of modern capitalism. The government formally chartered a company, the Vereenigde Oost-Indische Compagnie, known to this day as the VOC, to act in cooperative concert, and as a monopoly, in all trading matters to the east of the Cape of Good Hope. And not just trading: The VOC was given exclusive and quasi-sovereign rights to enter into treaties with local princes, to build forts, to maintain armed forces and to set up administrative systems of governments whose officials pledged loyalty to the government of the Netherlands.

The idea of officially sanctioned trading cooperatives was far from new. The British had had the Merchant Adventurers and the Merchant Staplers; they had created the Muscovy Company in 1555 and the Turkey Company in 1583. The Plymouth Company and the Massachusetts Bay Company were established at the beginning of the seventeenth century, to settle colonists. The Hudson's Bay Company, set up half a century later solely to trade, remains today: The Bay, its flagship department stores, can be found in all of Canada's cities (and in not a few more isolated Arctic settlements),

---

*I have no wish to belabor a historical nicety that will be familiar to most, but it seems worthwhile pausing to underline the fine irony of the coming change of power centers in the East. Since the early sixteenth century the Netherlands were in fact under Spanish control, as a province of the Spanish branch of the Hapsburgs. A revolt led by William of Orange ensured that the seven most northerly districts—including the best known, Holland, Zeeland, and Friesland—became independent in 1579. In 1648, under the terms of the Treaty of Westphalia, Spain recognized the independence of these seven and an additional southern group of provinces as what was then called the Republic of the Netherlands, headed by members of the Orange-Nassau family as *stadhouders,* or governors. The Netherlands became a monarchy—as it remains today—only in 1815. What is now the Kingdom of Belgium—those additions that were to be known colloquially as the "Spanish Netherlands"—split back away in 1830. The sudden rush of Dutch colonial energies and anti-Iberian zeal in the 1600s came just after their first emergence from under the yoke of Spain.

and its owner, a cheerfully eccentric peer called Ken Thomson, lives modestly and happily in a suburb of Toronto.

But there was a difference about the VOC. Right from its beginnings, it was cleverly constructed as a joint-stock company. The good burghers of Holland who had initially sent out their own small fleets decided to band together to back a much larger, much more ambitious company, with each backer owning a "share" of this company's value, with the value of each share depending on the amount by which each shareholder backed it. This new concept, of a joint stock company, with in this case a start-up capital of six and a half million guilders, was to become the model for all the thousands upon thousands of publicly traded firms that are listed on the world's bourses and stock exchanges today, and whose raison d'être, the sharing of risk and sharing of reward, lies now at the beating heart of the modern capitalist system.

The "Gentlemen Seventeen" who were the directors of the VOC may have been ultimately responsible to the Dutch parliament; but they were also financially responsible to those who bought shares in the company, to the merchants and the bankers who had taken the bold and revolutionary view that such a costly undertaking—impossibly expensive for any one Dutch company to afford—might, in time, win vast profits for them if they supported it collectively and backed it in concert. The VOC may be best known to historians for having ruled most of the East Indies for two centuries, from 1602 until its collapse in a welter of ignominy and corruption in 1799; but students of finance know it best for its seminal effect on the making of the model for an institution that underpins much of the prosperous West's current way of economic life.

The first steps were modest enough: a renewal of the pepper treaty with the sultan of Banten; the takeover of spice factories on the Banda Islands, in Aceh, in central Sumatra, and at a number of small ports on the Malay Peninsula; the rebuilding of a conquered Portuguese fort in Ambon in the Spice Islands that was to become the first permanent Dutch military base in the region and that would protect (from the newly predatory English, mainly) their fleets, which were now carrying cinnamon, cloves, and nutmeg home in enormous tonnages.

Before long, however, it became evident to the stockholders back in the Netherlands that there might be much more money to be made if at least some of the ships stationed out east began actually to trade *within the East*—taking goods not just from Java to Holland but from Java to Sumatra, say, or from Galle in Ceylon to Macassar in the Celebes. The local people weren't fully exploiting their potential for trading, so why shouldn't the Dutch—who had the ships and the growing navigational knowledge, confidence, and skill—trade for themselves? And so the practice known as "country trading" began, with skippers from faraway Holland sailing along the immense coastlines of an archipelago that is now known to embrace more than seventeen thousand islands, bringing cargoes from merchant to merchant and carrying not a small tonnage of their own goods, by the sale of which some became extremely wealthy men.

Just one warning was sounded during those early, braggartly, optimistic days. It was one that would be heard intermittently through all the succeeding years of Dutch rule in the East, that would find echoes in the rule of other European powers beyond Europe, that would extend past and become surprisingly important locally immediately after the explosion of Krakatoa, and that would reverberate across the world, with varying intensity and degree, right through to modern times.

And this tocsin note was the suspicious and on occasion the openly hostile relationship that at times developed between the Dutch and those local people who were, and already for the two past centuries had been, as the papers of the day called them, *Muhammadans.*

From the very moment the Dutch admirals met with the sultans of Ambon and Banda, the easygoing arrogance of the visiting Christian Dutch sat uneasily with the rigid formalisms of Islamic belief. An entire spectrum of antipathy developed: Almost from the beginning suspicion, disdain, and contempt were felt and expressed on both sides. Relations between Muslim and Christian, in the East Indies and throughout the Eastern world, present a discordant *continuo* in the telling of the tale of Krakatoa.

Surviving portraits usually show Jan Pieterszoon Coen hatless and crop-headed, with a menacingly thin moustache and goatee, and invariably dressed in the frills and furbelows of the time: a soft lace millstone ruff, black and heavily embroidered doublet, an elaborate system of belts and buckles holding up a silver-handled Toledo sword. He never smiles or looks in any way genial. To those who painted him he seemed always stern, forbidding, ruthless. Whether he did indeed look that way, or whether the portraitists simply reflected the abiding perception of the man, we cannot know. But everyone in the Netherlands and in the East seemed to know only too well that Coen, the founding father of the Dutch empire in the Orient, was not by any stretch of the imagination a kindly man.

No doubt his Calvinist roots—he was born in the small, religious, and conservative fishing town of Hoorn on the Zuider Zee—had much to do with his attitude. The fact that he witnessed an early slaughter of Dutchmen at the hands of Spice Islanders—the admiral in command of his first voyage to the East and fifty of his men were killed in the Banda group in a massacre that Coen long vowed he would avenge—may have been of even greater influence. Whatever the roots of his disdain, by the time the "Gentlemen Seventeen" saw fit, in 1618, to promote him as the fourth and most famous of their governors-general of the East Indies, Coen was a man in a mood to go out East only to enforce, expand, discipline, and punish. And to lay the foundations for what would in due course become a world-class trading empire—an empire that would have at its epicenter a world-class capital city, as Rome and Athens and Venice had in their turn been before.

The directors back in Amsterdam recognized early on the need for a regional headquarters—a place from where they might

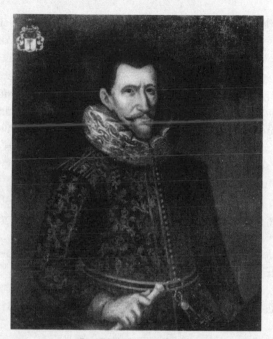

Jan Pieterszoon Coen.

administer their Indies,* somewhere where they might victual and water and rest their imperial adventurers, somewhere where they might repair their warships or build new trading sloops, somewhere from where they might pause to reflect on the vast territory they were fast acquiring and learn about its subtleties and plan its future course. The existing major centers, Malacca and Banten, had problems, Malacca still being run by the Portuguese and Banten ruled by a sultan whose congeniality was capricious, to say the least.

The only site that tempted Coen, when he became governor-general in 1618, was the tiny fort that the Banten sultan had

---

*And well beyond—Dutch outposts in Japan, Formosa, India, Burma, Laos, Thailand, Cambodia, Vietnam, Mauritius, Ceylon, even the Cape Colony in southern Africa were all, at various times, run from this Javanese HQ.

allowed the Dutch to build on the right bank of a greasy and slow-flowing river called the Ciliwung, on Java's north coast, opposite a village called Jayakarta. The only obvious disadvantage was the local presence of a fair number of British adventurers and would-be colonizers. Ever since Sir Francis Drake, on his great circum-navigation, had arrived in the Moluccas in 1579, there had been a British presence, and men like Sir Thomas Cavendish and James Lancaster had enjoyed some success with the local chiefs, present-ing themselves as potential enemies of the Portuguese. The British colonial intentions for Java and Sumatra were formless and lacked direction, but British engineers had built a fort on the left bank of the Ciliwung River, beside Jayakarta, at least to protect their traders. Coen, aware of their potential rivalry and sure that they meant, in time, to become a local imperial power, decided to toss them out.

It looked likely to be a tougher task than he at first supposed. The British army garrison was larger than his, and the Royal Navy squadron had fourteen vessels, while Coen had only eight. He wrote beseechingly to the directors in Amsterdam, asking for rein-forcements. His request was ignored. He promptly left town in high dudgeon—"I swear to you that . . . the Company has no enemies who more hinder and harm it here than the ignorance and thoughtlessness which reigns among Your Excellencies and defies understanding"—and sailed off to the Dutch fort in Ambon, hop-ing that he might find reinforcements there.

By doing so he missed such action as then took place—which wasn't much, and most of which descended rapidly into farce. The British laid halfhearted siege to the tiny Dutch fort, irritating the Dutch storekeeper, a Mr. van den Broecke, by forcing him to build barricades out of the very costly bolts of silk and batik cloth he had waiting for export. Then a row broke out between the Britons and a local junior sultan over how the spoils of the coming fight, if won, would be divided. Next the sultan of Banten brought in his fleet, to make sure that neither the British nor the junior sultan should benefit from Dutch weakness. The upshot of all this noise and tumult was the opening of four-way negotiations, the sudden abandonment of the scene by the British, and the overthrow of the junior sultan by the sultan of Banten.

All of a sudden, by dint less of savage and courageous fighting than of unseemly argument among their many foes, the Dutch found that they had won. Their fort was unscathed and secure. Mr. van den Broecke could take down his precious bolts and bales and send them back to the godown. And when Jan Pieterszoon Coen returned home from the Moluccas, shipless and much delayed—but that is quite another story— he found to his surprise and (considering that he was such a bellicose man, always spoiling for a fight) some small dismay that his men and his little fortress were all safe and sound. He also discovered that in a moment of relief, celebration, and sentimental rapture the men he had left behind him had on March 12, 1619, realized they were now free to transform their fortress into their planned regional headquarters, and so they had given their tiny settlement what they considered a properly dignified Dutch name: Batavia.

Jan Pieterszoon Coen, the founder of the Dutch East Indies, can take credit for much. But the naming of the capital that he founded is an honor that belongs to an unsung and quite forgotten soldier, and decidedly not to Coen. And yet to whomever the glory of foundation belongs, the coming role of this city is undeniable: A great Oriental conurbation was soon to be in the making here, a headquarters for a gathering of Europeans that would all be key to the enginework of a great colonial capital. There would soon be small armies of Dutchmen working in Batavia as merchants, traders, bankers, surveyors, soldiers, farmers, engineers, tax collectors, teachers, accountants, spies, philosophers, historians, and scientists, and in a whole host more of professions and trades.

It was the particular concentration in the city of the scientists, however, that is most important to this story. For there would soon be men, institutes, and laboratories dotted around Jan Pieterszoon Coen's capital specifically charged with keeping under close scrutiny the East's manifest natural peculiarities—peculiarities that included on more than one occasion the spectacularly unpredictable nature of its geology, especially that which was readying itself to erupt so terrifyingly in the narrow body of water that lay quietly off to the west, less than a hundred miles away.

2

The Crocodile in the Canal

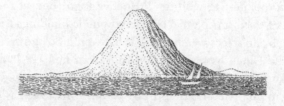

Dust lies hot on streets
Clearly empty of love and pity;
It's not like my green village
Here.

—Ebiet G. Ade,
from his song "Jakarta 1,"
the album *Camellia 1,* 1979

The name *Batavia* had a kind of easy, silky poetry to it. The Dutch, who were exceptionally proud of having created their great administrative supercity for the Orient from scratch—a somewhat less than altogether accurate claim, as the equally proud Javanese are still eager to point out—liked to think of it as their "Queen of the East." The choice of the name was a nicely senti-mental notion. *Batavia* was the old name for Holland, later the Nether-lands more generally, the Batavi having been a tribe, first recognized by the

One of the world's first corporate logos, that of
Holland's Vereenigde Oost-Indische Compagnie.

Romans, who inhabited a muddily fertile peninsula between the Rhine and the Waal, a few miles south of what is now the city of Utrecht.

There had in fact been a village beside the muddy seepings of the Ciliwung River long before the men of the VOC had planted their corporate flag with its distinctive logo (one of the world's first ever, used on colonial coins and public buildings too) beside their silk and spices warehouses. What would be known as Batavia until the Dutch were compelled to leave it in 1949 had hitherto been known by its more appropriately Javanese name, *Jayakarta,* which meant "victorious and prosperous." In 1949, as capital of the newly independent Indonesia, the city reverted to what its new leaders thought of as its happily suitable old name, though they modernized it a little to today's *Jakarta.* There are many, and not simply elderly Dutchmen of nostalgic temperament, who still think *Batavia* a sweeter sound.

And generally speaking old Batavia was, outwardly at least, and for much of its history, a very much sweeter place than now. Today more than seventeen million people cluster around the Ciliwung's long-cemented and mostly vanished riverbanks, clambering and jostling and polluting in the kind of cheerful jumbled mayhem that marks many a modern Asian city. Jakarta is not at all a pretty place, and for a visitor stuck in an interminable traffic jam, between gaudy hotels, gimcrack office blocks and tarpaper shanties, it is difficult to imagine it ever enjoying a queenly status, or ever having been a favored place for posting or employment.

But though Batavia has had more than its share of urban wretchedness, there was a golden era, most especially so around the time that Krakatoa exploded, when it was very much a queen among cities, and a place for which many felt great fondness.

Less so, however, during those formative, early company years, when the VOC was feeling its way into the East. The first settlers were by and large rather frightened men, understandably bewildered by the environment they found themselves in, and not at all sure whether they were in Batavia to lay the foundations of a great city or simply to build a hasty confection of a town that could serve as company headquarters, while somewhere more congenial was waiting to be discovered.

They did their best, however, to turn the sultry and fetid estuary that Jan Pieterszoon Coen had chosen as his base into somewhere that might remind them, at least a little, of their home. They built their fortress, a prison, an armory, a treasury, a Protestant church, and a modest palace for their governor-general* on a sandspit out in the roads. With quiet and deliberate speed all of this complex became landlocked, as more and more of the lagoon was dredged, more and more real estate was reclaimed, and more and more houses were thrown up around it.

A network of narrow streets (Amsterdam-straat, Utrecht-straat) and sixteen canals (the evocatively named DeLeeuwinengracht, Bacharachtsgracht, and Stadsbinnengracht among them) was then built in the jungle. The canals in particular, lined with flowering tamarind trees, were also intended to remind the settlers of home; but in fact, since local crocodiles got into the disagreeable habit of sauntering carelessly along them and poking their noses into the residents' doorways, the gesture had for a while rather the opposite effect.

Across the Ciliwung River, which was straightened and given high earthen embankments, an engineer constructed a classic Dutch drawbridge: the kind of double-sided bridge with struts and wires and T-shaped wooden beams that is still found spanning the canals in Amsterdam today, but of which the most famous is actually at Arles in southern France, because a homesick Vincent van Gogh had painted the specimen there. The very first Batavia bridge still exists: Called Hoenderpasarbrug—the "Chicken Market Bridge"—it is one of the more powerful reminders, of only a very few that have survived, that the Dutch did ever hold sway over this bustling modern city. In some evening lights, if one can

---

*The governor-general soon took on airs: He used to walk around town with an umbrella carrier, a dozen halberdiers, and a detachment of sentries armed with muskets.

forget the sight and sound of the choking traffic, there is a touch
of Rembrandt to the scene, a hint of van Gogh among the diesel
fumes.

Because the early Dutch were petrified of attacks from the
Javanese—from the often hostile sultan of Banten nearby and his
opposite number at Mataram in the island's center—they also built
themselves a wall. They did so at least in part because there was a
particular fear abroad, especially among some of those Dutchmen
who knew a little of the ways of India, Malaya, and Arabia, that
they were at risk of being murdered in their beds by fanatical Muslims. But, as it turned out, the Dutch in Java had precious little
cause to be so apprehensive.

The Javanese sultans were indeed Muslims, as were their followers, and if Islamic orthodoxy was being strictly obeyed, then they
were men who were perhaps in theory likely to be ill-disposed
toward the pale-skinned infidel invaders. But here the theory simply did not apply. In the seventeenth and eighteenth centuries, at
the very time the Dutch were arriving and setting up colonial camp,
the old orthodoxies of Islam were hardly being obeyed at all. The
faraway mullahs of Araby were not being heeded. Nor were their
teachings: A home-grown, locally brewed version of the creed was
proving immensely more popular.

The Islam that had swept so furiously through Sumatra and Java
in the fifteenth century had evolved rather swiftly into a gentler
amalgam of beliefs and passions, and become very different from
the rigorous pursuits of the desert-dried Arabs. Here in Java in particular, on an island of lush and fecund tropics—a place where there
was color, gaiety, and a tradition of vibrant animist religions and
curious and long-revered local gods and where sex was *fun* and girls
went half naked and were unlikely ever to wish to veil themselves—
Islam took on a very different form. One of the great scholars of the
period, Snouck Hurgronje, observed in an essay written in 1906
that the Javanese "render in a purely formal manner due homage to
the institutions ordained of Allah, which are everywhere as sincerely

received in theory as they are ill-observed in practice." Java, in short, was a place where Islam was amiably syncretic, more or less well disposed to anyone—whether a Christian from Europe, a Hindu from the Malabar Coast, or a Buddhist Chinese from Amoy—who happened by.

Very much later—indeed, at about the exact time of the Krakatoa eruption in the late nineteenth century—this was all to change. Orthodox Islam, its revival in part triggered by tragic events such as the great cataclysm, was totally transformed in Java during the nineteenth century, with fundamentalism, militancy, and profound hostility to non-Muslims its watchwords. But that was later. At the time of the building of early Batavia, the Dutch had little reason to be fearful of the Javanese *as Muslims*. They had other reasons to be uncertain, true, but it was not the thought of a *fatwa* or a *jihad* that made most of them jumpy. They built their fortress Batavia for more mundane reasons, bowing to the kind of unspoken fears anyone camped in an unfamiliar jungle might experience, and that might prompt him to mount a picket, light a bonfire or be ready with a gun. Or to construct a wall.

At first the Dutch created a series of high wooden palisades around their little town; but after thirty years of a growing insecurity the governor-general agreed to raise funds to enclose an area of about a mile square within a formidable *laager* of stone. In some places this was simply provided by the massive outer walls of the dockside spice warehouses; elsewhere sappers built a twelve-foot-high masonry structure, with embrasures, barbicans, donjons, battlements, a moat, and a sentry walk. Beyond it stretched the jungles, hot, dense, soggy, ever hostile, and alive with animals: the tiger and the panther, the tapir and the one-horned rhino, black apes and giant rats, a range of giant pythons and venomous cobras together with a gaudy wealth of cockatoos, parrots, and birds of paradise.

Inside the walls grew up the curiously compounded population of this quintessentially company town. Dutchmen were at first reluctant to come—only the "scum of the earth," complained Coen, wanted to settle there—and in the early years only a vanishingly small number of Dutchwomen appeared on the scene. In fact there were so few females that Coen was forced to appeal to Holland: "Everyone knows that the male sex cannot exist without women . . . if your Excellencies cannot get any honest married people, do not neglect to send underage young girls: thus do we hope to do better than with older women."

At first only company servants from the other Asian outposts of the VOC would deign to work in Batavia: Company employees, their slaves (for slavery, usually of men from faraway islands or from elsewhere in Asia, was permitted, and was extraordinarily widespread in the early years of Dutch rule in the East), a motley garrison of soldiers (with troops from as far away as Japan and Philippines brought to do guard duty under their bewildered Dutch officers), and, on occasion outnumbering everyone else, a very large number of Chinese.

There had been Chinese in Java long before the Dutch, long before the Portuguese. Along with squadrons of coolies hired to perform the hard work, scores of Chinese merchants had sailed down from the southern ports of Fujian Province and set up a prosperous agriculture on the Javan shore. They cultivated sugar and made gallons of the coco-palm toddies and rice-and-jaggery arracks that so beguiled (and rendered happily insensible) generations of visiting Western sailors.

Coen immediately spotted their usefulness. He insisted that they stay and become part of his new community—offering them (unlike his fellow Dutchmen) the right to trade privately, and to take pepper and birds' nests and sea cucumbers, all of which were readily available in Java, back to their homes across the South China Sea without interference from the monopoly of the company. "They are an uncommonly clever, courteous, industrious and obliging people," wrote one of Coen's colleagues. "There is nothing you can imagine that they do not undertake and practice. . . . Many keep eating-houses or tea-houses . . . or earn money fishing or carrying or conveying people." It has been four hundred years since this was written. So far as the impression offered by the diaspora of overseas Chinese is concerned, very little seems to have changed.

Slowly the community was born and struggled to its feet, matured, and began to grow. At first no Javanese were allowed to live within the city walls—and no Javanese were employed as slaves, lest they band together and conspire against the Dutch. But by the middle of the seventeenth century the ban had been relaxed somewhat, and a census in 1673 records the presence within the walls of twenty-seven thousand inhabitants, of whom thirteen hundred were classed as "Moors and Javanese." Two thousand were Dutch, nearly three thousand Chinese, and five thousand were members of a curious group called *Mardijkers,* who were Portuguese-speaking Asians, most of them freed slaves from Malacca and India who had been converted to Protestant Christianity.* The result was that Batavia had a cosmopolitan air of the most exotic stripe: There were turbaned Macasserese and long-haired Ambonese, Chinese with festoons of black queues, Balinese Hindus, "black Portuguese" vegetable hawkers, Moors from Kerala, Tamils, Burmese, and a few soldiers from Japan. And overseeing them all, with the haughty disdain that is born of vague fear, were the stout and pale-skinned burghers from Holland, Zeeland, Friesland, and elsewhere in the flat and cold European north.

And then there were the others, now nearly sixteen thousand of this counted population of 1673, the slaves. Their use (which remained legal until abolition in 1860†) made life exquisitely comfortable for some. Since no one would risk having slaves from Java, they had to be brought in by ship from elsewhere—an inefficient process, one slave peddler complained, after noting that of a consignment of 250 slaves sent down to him from the Arakan Hills of Burma, only 114 had been delivered. And some slaves did flee over

---

*And who under the peculiar rules of the VOC were, as Christians, permitted to wear hats: the only nonwhites in Batavia allowed to do so.

†When slavery was abolished in the Dutch East Indies it was still widespread in the United States. Indeed, the final slave ship, the *Clothilde,* had arrived in Alabama's Mobile Bay with its grim cargo only a few months before.

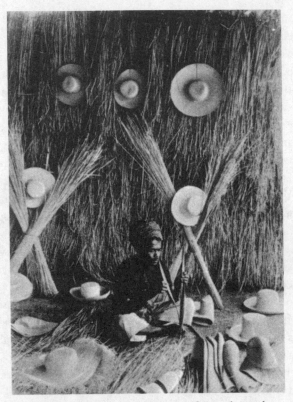

A milliner weaves topis and bonnets from *alang-alang* grass, the better to keep off the sun and the flies.

the city wall, gathered into gangs that lived in the jungle, and raided parties of wandering Dutch; one, a Balinese named Surapati, had a band of rogues so large and powerful that he formed his own fiefdom in eastern Java, which was ruled as an independent state for more than a century.

The wealthier Europeans in seventeenth-century Batavia might own a hundred or more slaves, and the town's main slave market was from the beginning a bustling, crowded place. These Malay, Indian, Burmese, and Balinese workers were trained to occupy the tiniest of niches in the household labor structure—advertisements spoke of a need for lamplighters, coachmen, pageboys, tea makers, bakers, seamstresses, and, most specialized of all, makers of a spicy side dish known as sambal.

The ladies'-maid slaves would be put to work as masseuses or hairdressers; these girls were skilled in fashioning hair into the bun-shaped style known as the *conde,* much favored in the salons of the time. Since they were so plentiful and so cheap, slaves frequently had little to do and sat around gambling their days away. But if they ever tried to escape, or, worse, if they ran *amok*—the word is Malay for "a state of frenzy" and was used as a legal term in the VOC courts—punishment was severe: they could be whipped or imprisoned. A Dutchman who shot one of his slaves dead and injured three others was merely told to leave Batavia, and not to do any further business with the VOC for the rest of his days.

Though the townsfolk in those closing years of the seventeenth century did not yet know it, their nearby neighbor island of Krakatoa was itself also, and for the first time within their sight, about to run amok.

No one had yet noticed that the island in the Sunda Strait had any potential for trouble. None of the navigators who had passed northward into the Java Sea and gazed, as seamen do, at the island with the "pointed mountain" on their port beam had supposed that one day it might do something quite terrifyingly and world-shatteringly dramatic. They, like the citizens of their destination city, were blissful in their ignorance of the tectonic complications then beginning to unravel many miles beneath their feet. They carried on instead with the serious business of colonial life, with a magnificent insouciance that was to be their motif for the next two centuries—right up to the moment of the cataclysm that engulfed their lives when the tiny parrot-filled and palm-covered island finally went mad.

But on the eve of this first recorded volcanic throat clearing, life for the European men and women who lived eighty-three miles east in Batavia had assumed an atmosphere of near-settled urbanity. Perhaps the frantic gaiety that would characterize life in the nineteenth century was not yet apparent: seventeenth-century eve-of-eruption life tended toward the more formal, strict, luxurious, and, at times, appallingly cruel.

The buildings that were being constructed in midcentury were by now quite substantial affairs. The warehouses were massive with teak and mahogany. The grand mansions along the Jacatra Weg that were built for the pepper planters and ship dealers, with their ornate wrought-iron gates, their gilded carvings and Delft tiles, moved an otherwise forgotten Mr. Speenhoff to song:

> At long last I enjoyed myself
> Outside Batavia
> Along the green heather
> On the Jaketra road.

And the great Town Hall was first built during this rather sedate and pompous period:* a cupola, shutters, columns, and the porte cochère all part of the standard architecture of the East.

This building served a myriad of functions: The magistrates' bench was here, licenses were issued, slaves were freed, ships were sold. On the cobbled square outside was a set of stocks, with miscreants frequently seen locked into them. Inside and below ground there were dungeons, and many are the stories of how the VOC security officers, who ran their company town with a ferocious rectitude, resorted to torture to extract confessions. A visiting German soldier named Christopher Schweitzer wrote an account of the harshness he saw:

> The 29th. Four Seamen were publicly Beheaded at Batavia (which is here the common Death of Criminals) for having

---

*The third and present incarnation of the Batavia Stadhuis, built in 1707, remains intact, serving as the Jakarta History Museum.

killed a Chinese. At the same time, six Slaves that had Mur-
thered their master in the night were broke upon the Wheel.
A Mulatto (as they call those that are betwixt a Black-a-Moor
and a White) was Hang'd for Theft. Eight other Seamen were
Whipt for Stealing, and running away, and were besides this
Burnt on the Shoulder with the Arms of the East India Company. Two Dutch Soldiers that had absented themselves from
the Guard two days, ran the Gauntlet. A Dutch Schoolmas-
ter's Wife that was caught in Bed with another Man (it being
her frequent Practice) was put in the Pillory, and Condemn'd
to 12 years Imprisonment in the Spinhuys, the women's
prison.

Christopher Schweitzer's account is dated 1676. It is suggestive of
a certain public unhappiness, of a draconian degree of Dutch
response, of a current of distemper in the land.

And then, four years later, with the situation between rulers and
ruled still uneasy, Krakatoa very noisily awakened its slumbering
self. It was an event that astonished and perhaps even briefly terrified the new European arrivals. Yet most Javanese, long immersed
in a balm of myth and legend relating to their volcanoes, would
later say that, with all the evident unhappiness abroad, they could
have seen it coming.

Orang Alijeh, the Javan god and mountain ghost whose task is
to superintend the emissions of smoke and fire into the eastern
heavens, is said to breathe sulfur from his nostrils when all is less
than well on his earthly dominions. Krakatoa, which, along with
Tambora, Merapi, Merbapu, and Bromo, was one of his most
potent mountains, had been blessedly quiet, or comparatively so,
for at least the previous twelve hundred years. The only event that
some might say had taken place to tax his patience was that foreigners, white men from far away, had now come by sea to rule
over the people of Java. This, not a few Javan mystics liked to say,
was one of the reasons why volcanoes occasionally made their fire,

the more forcefully to display the degree of Orang Alijeh's grave displeasure.*

Yet, however displeased Alijeh might have been, what followed was by all accounts not the greatest of pyrotechnic performances. And no one who witnessed what happened ever came forward to write a first-person account. All we know comes essentially from one man, a Dutch silver assayer from the western Sumatran mining town of Salida, named Johan Vilhelm Vogel.

Vogel, who was said to be so "pious and studious a servant of the Compagnie" that he eventually became mayor of Salida, first passed Krakatoa in the usual way en route from Holland to Batavia—thus with the island to his port side—aboard the long-range packet *Hollandsche Thuyn* in June 1679. He waited ten weeks in town, then left Batavia for Sumatra in September aboard the yacht *Wapen van ter Gos,* this time passing Krakatoa to starboard. He saw nothing that struck him as remarkable.

But then in due course he fell ill. The company, sedulous in caring for its more valuable employees, ordered him to visit their doctors in Batavia, and he left the Sumatran port of Padang in January 1681 aboard the yacht *de Zijp.* This time the Krakatoa he saw presented a very different aspect.

> I saw with amazement that the island Cracketovv, on my first trip to Sumatra completely green and healthy with trees, lay completely burned and barren in front of our eyes and that at four locations was throwing up large chunks of fire.
>
>   . . . the ship's Captain told me this had happened in May 1680. That time also he had made the trip from Bengal, had run into a heavy storm, and about ten miles away from the island had experienced an earthquake. This was followed by a tremendous thundering crash which had made him think that

---

*Evidence both geological (fresh lava flows on the island) and anecdotal (references in Javanese oral histories) suggests that Krakatoa has erupted as many as ten times prior to whatever happened in 1680. Few of the dates of these previous eruptions can be pinned down with any certainty.

an island or otherwise a piece of land had split apart. . . . He and the whole ship's population had smelled a strong and very fresh sulphur odour. Also the sailors had retrieved with water pails from the sea some very lightweight rocks, very much resembling pumice stone, which had been thrown from the island. They were scooped up as a rarity. He showed me a piece of the island. He showed me a piece of it a little larger than a fist.

By checking the port records of vessels sailing in and out of Batavia, we can see that the captain of the *de Zijp* had indeed traveled between Batavia and the port of Bengalen* in May 1680 aboard the cutter *Aardenburgh*. The story, in this respect, does thus seem to tally. If it is true, then this long-forgotten and so far anonymous sea captain and his crew were the first Europeans ever either to see the volcano of Krakatoa erupt or to see the results of its recent eruption. The *Aardenburgh*'s log, however, has never been found; and the day register of Batavia Castle, an official journal that records all ship movements in and out of the harbor and any pertinent comments from the vessels' various masters, is silent.

A writer named Elias Hesse then wrote a vivid account of an eruption, suggesting that it was continuing in November 1681 when he and Mayor Vogel left together aboard a Sumatra-bound ship called the *Nieuw-Middelburgh*. He first mentions passing an island he calls Zibbesie (today's Sebesi, a couple of miles north of Krakatoa) and being unable to sleep because of the crying of ghosts (which the apparently rather more sober Vogel later reported were orangutans, "which produce a terrible howling, often when the weather is about to change"). He continues: "then still north of the island Cracatou, which erupted about a year ago and is also uninhabited. The rising smoke column of this island can be seen from miles away; we were with our ship very close to shore and

---

*Or he may have been on international duty, journeying to and from the British possessions in India: the record is not clear as to whether the *Aardenburgh*'s destination was Bengal or a Sumatran port named Bengalen.

could see the trees sticking out high on the mountain, and which looked completely burned, but we could not see the fire itself."

Later the *Nieuw-Middelburgh* and its crew of company servants and miners were forced to heave to in the Sunda Strait, where they experienced heavy sea quakes and learned of an earthquake that, Hesse reported, "did considerable damage to the buildings of the Company."

A close study of the records of other ships passing through the Strait at the time—and for a variety of reasons there were very many—shows no other suggestion of an eruption or earthquake in 1681. And further—the day register has no information even in May 1680 of anything of interest having taken place in the Sunda Strait. The register reports the most mundane occurrences of city life: a tiger found outside the walls, crocodiles captured in the city streets, a comet seen in the sky, servants running amok. Yet nowhere in 1680 or 1681 is there any mention at all of an eruption on an island that was passed by scores of company ships every week.

From this dearth of information it is perhaps fair to conclude only three things: First, that Elias Hesse was an inventive fantasist and probably made up his entire account of volcanic activity in November 1681. Second, that the silver assayer and sobersided mayor Johan Vogel was similarly afflicted, and that his suggestion of seeing "large chunks of fire" at "four locations" on Krakatoa in February 1681 is also fictional. He did, however, probably see evidence—burned trees, barren plains of ash—that some disaster had befallen Krakatoa a while before. And third, that the captain of the *de Zijp* and the *Aardenburgh* almost certainly did see something of an eruption in May 1680. But since no other passing vessel did, and since no mention was made of anything grave having occurred to the ever vigilant bureaucrats in the castle, whatever had happened was small beer indeed, and the captain had, like many seamen, made the story into a considerably better one the moment he stepped ashore.

3

Close Encounters on the Wallace Line

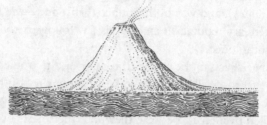

Southeast Asia is probably the finest natural geological laboratory in the world . . . It is a spectacular region in which the
manifestations and processes of plate collision can be observed
at present and in which their history is recorded. It is a region
which must be understood if we are to understand mountain
belts, arc development, marginal basin evolution and, more
generally, the behaviour of the lithosphere in collision set-
tings. . . . Furthermore the region is developing rapidly.

—from the introduction to
Tectonic Evolution of Southeast Asia,
R. Hall and D. J. Blundell, eds., London, 1996

DELINEATION

On a bleak December evening shortly before Christmas
1857, in the severely elegant lecture theater of what was
then biology's sanctum sanctorum, the Linnean Society
in Burlington House on London's Piccadilly, a young man named
Philip Lutley Sclater rose to present a paper that, though all but
overlooked by most biologists today, was to lead steadily to a revolution in scientific thinking about the history of the world.

Philip Sclater may have been only twenty-eight years old when he stood before the assembled worthies, but already he had won an awesome reputation. He had been to Winchester and Christ Church, Oxford, and was inordinately clever. He was unusually well traveled (Argentina, Malaya, India, Australia, and most of the United States), and was both a superb artist and—the basis of his near-legendary reputation among his few followers today—a brilliant ornithologist.

He specialized in the exotic, the colorful, the tropical, the odd. He came to know all there was to know about the South American finchlike birds called tanagers and their country neighbors that were called ovenbirds, because they built their nests out of clay in the shape of ovens. He would write definitive books about Passeriformes, that huge class of rather less unusual perching birds; he had a lark named after him; he founded the British Ornithologists' Club and was for many years secretary of the Zoological Society of London. And he gave the name *Lemuria** to a landmass he believed (wrongly) had in Eocene times connected the Malay Peninsula with Madagascar, and on the remaining peaks of which now sits the enormous mid–Indian Ocean, British-owned and American-leased military base of Diego Garcia.

Crucially for this Krakatoa story, Philip Sclater was also a specialist in a new science that was growing apace during late Victorian times, and that has been called variously either biogeography or zoogeography. As more and more specimens of animals, insects, plants, and birds were collected, classified, and cataloged, so it became steadily more apparent that geography had a major influence on zoology and botany—that certain living creatures were peculiar to certain parts of the world and not merely to certain cli-

---

*Helena Blavatsky, the creator of the genial Hindu-Buddhist religion called theosophy, seized upon Lemuria as the likely home for a people she insisted were the "Third Root of Mankind." As she described them, they lacked classical beauty: They were fifteen feet tall, brown-skinned hermaphrodites with four arms, some possessing a third eye in the back of the skull. Their feet had protruding heels enabling them to walk either forward or backward, and their eyes were sited so they could see sideways.

matic zones. One might find precisely the same climate in Uganda and Queensland, for example, but because Africa and Australia were so far away from each other, their native inhabitants had developed without ever making contact with one another. In the same way their flora and fauna were now entirely separate, different, and peculiar to each of the two places. The polar populations are another obvious example of zoogeographical reality: Polar bears and people inhabit the north, penguins and albatrosses the south, even though the climatic environment in each place is essentially identical.

The birth of the study of such regional differences in populations of plants and animals—of the *biota,* to use the proper if rather unattractive word—was a natural by-product of the frenetic pace of exploration of the time; and Dr. Sclater, an explorer of the old school, had by the middle of the nineteenth century a keenly developed interest and expertise in the topic.

The paper he presented to the Linneans that winter's night was on precisely this subject: "On the General Geographical Distribution of the Members of the Class Aves," with a particular interest "in the islands around New Guinea." And Sclater offered what to his listeners was a remarkable and startling discovery. He had observed what seemed to be a rigid zoogeographical distinction deep within the islands of the East Indies.

The kind of dramatic geographical variation that might be seen on the two sides of a vast body of water—the measure of difference between African birds and Brazilian birds, for example, separated by the whole of the South Atlantic—might not be expected within the continuum of an archipelago as closely knit as the East Indies. But here, astonishingly, there was. In months of research that took him right along the two-thousand-mile volcanic chain stretching from Sumatra to the Trobriand Islands, Sclater found the occurrence of a huge, unanticipated, and very sudden change: The jungles in the west of the archipelago, for example, sported birds that were akin to those found in India, while those to the east of a vaguely imagined line were alive with an avian fauna that was to be found only in Australia.

Sclater had not completed enough research to be able to draw a firm demarcation line between the two populations. All he could say for certain was what he had seen. Certain birds—say, parrots—found in wild abundance in the eastern, Australian end of the archipelago were, quite literally, *rarae aves* on the western, Indian side. There are barely any parrots on Java; but there is every imaginable kind in Sulawesi, Irian, and Timor. Not a single cockatoo is to be found in the west; but two entire families each of cockatoos and lories—this last a kind of parrot with a brushlike tongue—are found in the east.

Barely any thrushes are to be found on the islands to the east of Bali. In the western half of the archipelago there are woodpeckers, barbets, trogons, paradise flycatchers and paradise shrikes, minivets, blue drongoes, pheasants, and jungle fowl—but none of these, just like the humble thrush, is to be found in the east. The forests of this eastern end of the chain, on the other hand, are populated uniquely with legions of showy and exotic birds that one might just as easily find in Queensland and New South Wales: honeyeaters and birds of paradise, cassowaries and emus.

Any traveler with half an eye and half an ear open will realize simply from the colors and the birdsong that something profound is going on. In making an eastward journey from Sumatra to Irian, though he may not once pass out of sight of land, the traveler will have most decidedly left one world and entered another one that is utterly different.

It would be forty more years before Sclater, working then with his son, would draw a map with the formal delineation, as the pair saw it, that separated the two avian worlds. But his work, which culminated in the reading of his 1857 paper outlining but not quite delineating their meeting place, excited the interest of a much older, bolder and less well-educated Briton who was then living in the Indies. It was Alfred Russel Wallace who came swiftly to understand that it was not simply birds who inhabited two quite different worlds: plants and animals did also. And, just like the

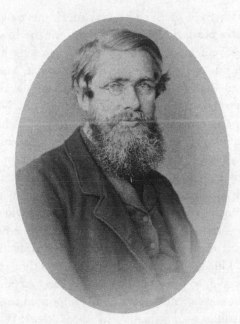

Alfred Russel Wallace.

birds, they all met—collided, even—somewhere among the maze of jungles of the myriad islands of the Dutch East Indies.

Alfred Russel Wallace, who at the time was collecting, studying, and living in a grass hut on the spice-rich island of Ternate, would take the observations of this young naturalist and, adding a vast amount of information from his own observations and collections, transform them into a theory and a grand cartographic creation that would survive to this very day. It has a name that remains its architect's most familiar memorial: the Wallace Line.

Philip Lutley Sclater, learned, patrician and well connected, might have thought he had some vague right to have this two thousand miles of tracery named after him, in recognition of his pioneering work on the region's bird geography. But the honor was eventually to be given to his very much more capable successor, the lowly born, hugely tall genius from the town of Usk in south Wales, remembered today mainly for having imagined and then drawn this vast and invisible line in the sea. Mainly, but not solely: Alfred Russel* Wallace has a trench off Java named after him too, as well as a 13,300-foot peak in the Sierra Nevada, a garden in Wales, an aviary in Bristol, a bird of paradise, biology prizes in both Kansas and Australia, countless lecture theaters and university halls, and craters on both Mars and the Moon.

Whoever named a lunar crater in his honor was a man with a mordant wit, or an ear for the labored pun. For many years Wallace had been known, perhaps more widely than anything else, as *Darwin's Moon*†—the lesser body that was bound by the Fates to dance in constant attendance around the orbit of the greater. For although the Wallace Line has yet some importance, and though it has a singular connection to this story of Krakatoa, any account of Wallace himself must make mention of the more important reasons for such fame as he still has—and that is that Alfred Russel Wallace, alongside Charles Darwin but always as his satellite, is the other true but largely unremembered pioneer of the science of evolution.

His birth in Usk in 1823, as the seventh child of a librarian, was into a threadbare but intellectually respectable family. He started his own life as a Leicestershire schoolteacher, but one whose life was marked from the start by passionate fascinations—in this first case, with the life and times of the *Coleoptera,* those insects with hard coverings for their flight wings that are generally and more prosaically known as beetles. He was also interested in spiritualism, and indeed rediscovered this fascination much later in his life; mercifully for the advance of science, his dabblings into what he called "phreno-mesmerism" were suddenly nudged back toward wor-

---

*The spelling was an error by Usk's registrar of births; Wallace also long thought, to compound the mistake, that he had been born on January 8, 1822—it was in fact 1823.

†This was the title of a 1966 biography of Wallace by Amabel Williams-Ellis: well known as a Strachey, the editor of a number of science-fiction anthologies for Victor Gollancz, and the wife of the Portmeirion architect and stylish eccentric Clough Williams-Ellis.

thier pursuits when he stumbled upon a similarly insect-captivated city lad, a hosier's son and sometime warehouse sweeper named Henry Walter Bates.

This pair, soon inseparable, hunted, collected and cataloged beetles in every corner of Leicestershire and points beyond—with Bates publishing his first paper "On Coleopterous Insects Frequenting Damp Places," the title telling us perhaps more about Leicestershire than about beetles.

In 1848, three years after meeting and having exhausted the coleopterous delights of the meadows and damp places of the Midlands, this pair of redoubtable young men combined what little savings they had to begin a remarkable scheme: to go off specimen collecting in the wilds of the beetle-rich, insect-rich Amazon rain forest.

For Bates it was the beginning of a lifelong love affair with what he called "the one uniform, lofty, impervious and humid forest," a love affair that started in the remote Brazilian jungle village of Ega, fourteen hundred miles upriver. During the six years he lived there he collected and named no fewer than 550 new and distinct species of butterfly (at the time only 66 were known throughout the entire British Isles). He traveled thousands of miles in the deepest and most remote corners of the forest, and over the years became a world authority on insect mimicry; and, together with his friend Wallace, ultimately came to be an instigator of the theory of natural selection and an unalloyed supporter of Charles Darwin once *On the Origin of Species* had been published in 1859.

For Wallace, however, the journey to Amazonia was just the start of a fascination that soon ranged around the entire world, but that is most closely associated today with the region he found the richest and most intellectually satisfying: the then Dutch East Indies. Yet his interest did not properly begin until he had been severely tried and tested on his way home from Brazil: the square-rigger *Helen,* in which he was bringing home his valuable collections of Amazon specimens, caught fire and sank in mid-Atlantic, and Wallace spent ten days in a longboat before being picked up near Bermuda. He wrote two books on his experiences. Darwin,

who trawled through both looking for evidence to bolster his own
fast-coalescing ideas on biological variation, natural selection, and
species origin, found them frustrating. "Not enough facts," he
harrumphed, cruelly unaware that Wallace had lost not only his
notes but his specimens, all gone down with the foundered ship.

In 1854 Wallace set off, alone, for the East Indies—the chain of
islands that he, as a geographer, preferred to call the Malay Archipelago, since the thousands of islands were indeed a classic example of an archipelago, and the islands' lingua franca was, in most
cases, and to a greater or lesser degree, based on Malay. His keen
conviction remained, for all the eight years that he spent there,
that evidence to be found in this archipelago would substantiate
his two growing beliefs: that geography was highly influential in
the development of biology, and that species originated by the natural selection of favored types from within the variations of any
population. He spent the better part of his life seeking to prove
both points—and by and large (and in the penumbra of Charles
Darwin) he succeeded magnificently.

His collecting zeal was prodigious. He sent back to London, or
eventually came back with, a meticulously organized collection of
no fewer than 125,660 specimens of the plants, animals, insects,
and birds of the islands. There were 310 mammals, 100 reptiles,
83,000 beetles (not surprisingly), 13,000 other insects, 8,000
birds, 13,000 butterflies and 7,500 shells. His study of this vast
number of living things brought him to both of the epiphanies
that he had sought, and at almost the same moment. He suddenly
realized evolution's existence and mechanism, and he immediately
recognized the profound difference between the two basic animal
and plant populations of his chosen archipelago, and he realized
and recognized these two profundities at essentially the same time:
during what for Wallace were the seminal and intellectually furious
years of 1858 and 1859.

Wallace's sudden understanding of evolution is one of the most
romantic tales of modern science, ranking with the sudden

achievements of Archimedes and Galileo, Becquerel and Newton, Fleming and Marie Curie. The vision came to him not in a bathtub, or in Pisa, or on a Paddington windowsill, or under an English apple tree—but in a grass hut on stilts, in a village on the island of Ternate, during a bout of jungle fever. There are not a few who believe that it is the Spice Islands, and not the Galápagos, that one day should come to be recognized as the true birthplace of the science of evolution.

At the time, thousands of miles away in Kent, Charles Darwin was working, with painstaking tardiness, on what he called his "big book." He knew he was on to something. He had the core of an idea as to how and why new species evolved. He had the facts at hand. He had collected his own beetles; he had bred pigeons; he had observed and measured and cataloged the tens of thousands of living creatures from his years aboard HMS *Beagle*. And he was well aware of Alfred Russel Wallace—not least because in 1855

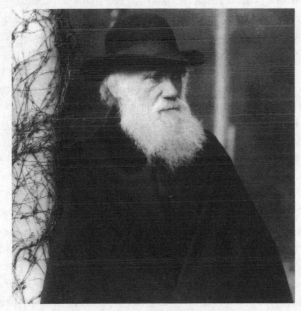

Charles Darwin.

Wallace, then in Sarawak, wrote a paper entitled "On the Law which has Regulated the Introduction of New Species," which argued that new species arise from within variations in a population that is somehow (as he was so often seeing in the East Indies) geographically isolated from other populations of the same creatures. The paper presented only an incomplete theory, and it presented no mechanism for this introduction of new species—but it set Darwin once again to pondering, ruminating, and (for which he is well known) procrastinating.

And then, just after New Year's Day in 1858, Wallace arrived in those eastern islands that the Dutch called the Moluccas (or Maluku, as they are once again today). During the intervening three years of wanderings between Sumatra and Irian he had had his full share of excitements: Headhunters had severely scared him on the island of Lombok, his arms and legs had been badly infected by leeches and biting insects, he had found a ten-foot snake in the ceiling of his hut on Ambon, he had suffered from unremitting dysenteric fever, and he had contracted case after case of malaria. Indeed, it was a particularly nasty attack of this, caught when he was on brief insect-collecting trip to Halmahera, that prompted him to go back to his little house on Ternate at the end of February. He sat and perspired and wrote and pondered and ruminated—and then, in one of those fascinating manic moments of fever-induced excitability, *he suddenly got it.*

While wondering yet again what it might be that caused the break between one species and the success of a specialized variety that would lead to the creation of a wholly new species

> there suddenly flashed upon me the idea of *the survival of the fittest*—that the individuals removed by these checks must be, on the whole, inferior to those that survived. Then, considering the variations continually occurring in every fresh generation of animals or plants, and the changes of climate, of food, of enemies always in progress, the whole method of specific

modification became clear to me, and in the two hours of my
fit I had thought the main points of the theory. [Italics added]

Wallace wrote this account in his book *The Wonderful Century* a
long while later, which might prompt some suspicions. But any
doubts about the legitimacy of his claim to be the first to recognize both the idea and the mechanism of evolution must be allayed
by his having immediately written a paper and dashed off a letter
to Down House—asking Darwin to forward it for possible publication to the geologist Charles Lyell, whom Wallace thought he
had better impress. It was Darwin who was impressed. The Wallace paper was, he later said, "a bolt from the blue." It deeply disturbed him. It was the very idea for which he was looking. It
provided the jolt he needed to get on and finish his "big book."
And when he did, and the book appeared under the now-famous
title *On the Origin of Species,* Wallace's crucial references to "the
struggle for existence" and the "the survival of the fittest" appear
as the keys to the entire mystery.

Charles Lyell and Joseph Hooker persuaded Darwin to share at
least some of the glory with Wallace—faraway, low-born, and an
upstart though he may have been. The formal announcement that
the puzzle at the root of the science of evolution had now been
cracked was made at the meeting of the Linnean Society on July 1,
1858: It was a joint announcement, made by the introduction of
papers by Darwin and Wallace—with Darwin admitting, somewhat
lamely, that though he had had the very same idea as Wallace some
days before, he had never committed it to paper. Procrastination
had done for him.

But not, it seems, for his reputation. *Darwinism* is the accepted
word, invented by Wallace and the title of one of his later books.
Wallace never displayed in public a scintilla of bitterness, and contentedly and generously gave Darwin all the credit. His classic, *The
Malay Archipelago* (in which Wallace tried to popularize the idea
of a zoogeographical division, which so neatly coincides with the

phenomenon that gives birth to Krakatoa), is dedicated to Darwin, "to express my deep admiration for his genius and his work." He remained loyal, almost servile—the ever revolving little moon around Darwin's glittering and far grander planet.

He did not receive a knighthood, as did so many of his more nobly born contemporaries, like Galton, Huxley, Lyell, and Hooker. He did, however, become a member of the Order of Merit, which many in Britain think more worthy. His public standing then went into a long and slow decline, and outside Indonesia, where he is still well known, he has until lately been little recognized in the world.

Perhaps his spell in history's wilderness is now coming to an end. In April 2000 his grave in Dorset was refurbished;* and in November 2001 a plaque was unveiled in what is now the Royal Society's Reynolds Room, noting that a century and a half earlier the papers of Charles Darwin and Alfred Russel Wallace had been read there, which formally set in motion a whole new science of evolutionary studies.

In recent years there have been kindly new biographies of Wallace, new studies of his contributions to science, and fresh examinations of the papers of Darwin, Lyell, Hooker, and all the others involved in the evolution of evolutionary thinking. There seems in consequence a growing sympathy these days for the idea that Darwin may have behaved less than fairly to the man who had the same idea as he had, and at the very same time—but who had the misfortune of writing everything down almost immediately and shouting the news from the rooftops, of being less gently born and less well connected than the master of Down House, and of spending his declining years more interested in eccentric sciences, out on the fringes of respectability. To some these days, the idea that Wallace's only memorial is merely a line that passes unseen across the sea seems a melancholy insufficiency for all of his many achievements and ideas.

---

*With a descendant of Charles Darwin on hand to observe.

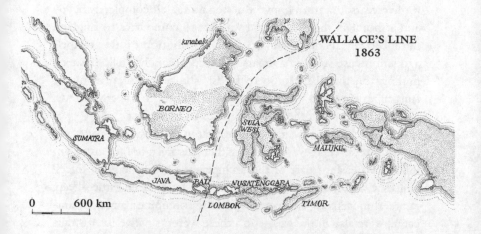

The Wallace Line—Australian fauna (cockatoos, kangaroos) to its east, Indo-European thrushes, monkeys, and deer to its west.

*　　*　　*

The Wallace Line, however invisible and however controversial (in later years, because of technical arguments among the world's community of zoogeographers), does at least have the merit of being directly relevant to both the makings and the violent unmakings of Krakatoa. It may have played only a marginal role in the development of the theory of evolution of life. But it does play, if unwittingly, a very significant part in the much newer theory of plate tectonics, the evolution of the earth.

Wallace's ideas about it were first made public at the Linnean Society on November 3, 1859—seventeen months after the famous Linnean Society presentation on evolution and, as then, while he was five thousand miles away from Burlington House. (He was in Ambon at the time, just back from a half year in Sulawesi spent indulging in what he incautiously regarded as the "capital sport" of shooting birds of paradise. He was coming under considerable pressure from his family, who had not seen him for five years, to return home. But he said he felt completely happy and at home in the islands, and was not to come back to England until 1862.) His observations of the distribution of the archipelago's immense stock of animals reflected precisely Philip Sclater's findings on the distribution of birds: The representatives of the Australian kingdom of animals lived in the more easterly islands, and the members of the Indian families to their west.

He was able to show, for example, that in the west there were apes and monkeys: There were none in either Australia, or the Australian-influenced eastern part of the archipelago. Flying lemurs, tigers, wolves, civets, mongooses, polecats, otters, bears, deer, cattle, sheep, tapirs, rhinoceroses, elephants, squirrels, porcupines, and scaly anteaters were also in the west—and these weren't native to Australia (or, it almost goes without saying, to New Zealand, which was so isolated for so long that it had no native mammals or snakes at all) or to the eastern islands.

Out in those eastern groups, though—where the cloves and nutmegs grow in wild profusion—there were all manner of animals wonderfully unfamiliar to the newcomer from Usk. These included kangaroos, opossums, wombats, and the duck-billed platypus. Not an ox or a squirrel, an elephant or a tapir, on any of these islands— but instead animals that raise their young in pouches, or hop, or live half in and half out of water, have webbed feet, lay eggs, and suckle their young, and flightless birds, and cockatoos.

Alfred Russel Wallace enjoys such fame as he has because he observed all of this bewildering profusion with great care, noted with exactitude where each animal or bird lived—or, just as important, *did not* live—and then drew a long and sinuous line that separated these two very distinct biological regions. The line, which was announced to the Linneans in 1859 and refined in a more substantial speech in 1863, wandered in an approximately northeast to southwest direction. It began to the south of Mindanao, the most southerly of the main Philippine Islands, and snaked to the north of the oddly shaped island now known as Sulawesi, but which was then called the Celebes.* It then swept south through the Strait of Macassar, leaving Borneo on its western, Indian side, and then headed across the Java Sea to the tightest-imaginable division: the 15-mile-wide and very deep strait between the islands of Bali and Lombok.

> The contrast is nowhere so abruptly exhibited as on passing from the island of Bali to that of Lombock, where the two regions are in closest proximity. In Bali we have barbets, fruit-thrushes, and woodpeckers; on passing over to Lombock these are seen no more, but we have an abundance of cockatoos, honeysuckers, and brush-turkeys, which are equally unknown in Bali and every island further west. The strait here is 15 miles wide, so that we may pass in two hours from one great division of the earth to another, differing as essentially in their animal life as Europe does from America. If we travel from Java or Borneo, to Celebes or the Moluccas, the difference is still more striking. In the first, the forests abound in monkeys of many kinds, wildcats, deer, civets and otters, and numerous varieties of squirrels are constantly met with. In the latter none of these occur; but the prehensile-tailed opossum is almost the only terrestrial animal seen, except wild pigs, which are found on all the islands, and deer (which have probably been recently introduced) in Celebes and the Moluccas.

What Wallace realized well at the time was that the reason these two biological regions had so nearly merged and yet had remained so distinct was entirely due to geology. "Facts such as these can only be explained by a bold acceptance of vast changes in the surface of the earth," he writes, and goes on to speak of a "great

---

*At my school we were more than familiar with at least the shape of the Celebes, since its long, fingerlike upper peninsula runs almost exactly along the equator, and thus provided us (as did Belém, at the mouth of the Amazon on the other side of the planet) with an easy way of drawing the line of zero latitude on any blank map of the world.

Pacific continent" that had "probably existed at a much earlier period." He talks vaguely of submergences and breakups of masses of land into islands, and other devices that may have caused animals to have existed in isolation and then to have become near, but separated, neighbors.

Four years later, when he presented a much longer paper on the same topic, Wallace was clearly excited, trembling on the brink of a bold new idea.

The nature of the contrast between these two great divisions of the Malay Archipelago could best be understood, he might have said to himself, by considering what would take place if any two of the primary divisions of the earth were brought into equally close contact. Africa and South America, for example, differ very greatly in all their animal forms. On the African side, we have baboons, lions, elephants, buffalo, and giraffes; on the other, American side, spider monkeys, pumas, tapirs, anteaters, and sloths; while among birds, the hornbills, turacos, orioles, and honeyeaters of Africa contrast strongly with the toucans, macaws, chatterers, and hummingbirds of America.

But, he continued to suppose, let us endeavor to imagine that a slow upheaval of the bed of the southern Atlantic takes place, and that earthquake shocks and volcanic activity on the landmasses on each side of the ocean cause increased volumes of sediment to be poured down the rivers, such that the two continents gradually spread out by the addition of newly formed lands. As a result of these two slow processes, he went on, the Atlantic, which now separates Africa and South America with its thousands of unbridgeable miles of water, would be reduced to an arm of the sea no more than a few hundred miles wide. At the same time we may suppose islands to be upheaved in midchannel; and, as the subterranean forces varied in intensity and shifted their points of greatest action, these islands would sometimes become connected with the land on one side or the other of the strait, and at other times be separated from it. The barrier of the ocean all of a sudden would cease to be a barrier at all. . . .

He was floundering here, becoming nervous and stutteringly prolix, knowing only that somehow he was on the verge of something, some explanation, some answer to the question posed by all he had seen. Yes, indeed—some geological process, some series of events that had something inexplicable to do with movement and submergence and upheaval and spreading and uplift and earthquakes and volcanoes (for he was a keen and sometimes very frightened observer of these in his beloved archipelago), had caused this curious avian and zoological division. But as to what: We can almost see Wallace dabbing the perspiration from his brow, as he omits to follow through, as his thinking draws to a halt, as he fails to come up with the solution, well aware nonetheless that the answer he seeks is out there, but has proved too elusive for him, this time around.

What Wallace was never to realize was that the mechanism driving all the geology was, in due course, going to be recognized as the then entirely unimaginable process of plate tectonics. And what he had not the least inkling of was that the tectonic collision that had brought his animals and birds together, brought the cockatoo so close to being snared by the squirrel, and the retromingent tapir so near to encountering the web-footed monotreme more familiarly known as the platypus, was also the selfsame collision that had brought about Indonesia's reputation as the volcanic cockpit of the world, with its most notoriously dangerous volcano, Krakatoa, a classic of the kind.

Alfred Russel Wallace knew none of this. But his papers, discoveries, and still-surviving line prompted others to begin thinking and pondering too, and to start asking why this encounter between the animals and birds of Asia and the animals and birds of Australia might have happened and, much more importantly, since it evidently *had* happened, just how?

The scientific world of the time was in the midst of a terrible ferment, with discoveries and realizations coming at an unseemly rate. To many in the ranks of the conservative and the devout, the new theories of geology and biology were delivering a series of hammer blows to mankind's own self-regard. Geologists in particular seemed to have gone berserk, to have thrown off all sense of proper obeisance to their Maker.

Men like James Hutton, Charles Lyell, and William Smith were fast beginning to suppose that man's existence was, in the grand scheme of the very things that they were delineating, of utter insignificance, his sojourn on the planet temporary and vanishingly brief. Darwin's discoveries (along with those of Wallace, naturally—these very parentheses serving to remind us how simple it still is to overlook the dyer's son from Usk) had overturned so much of man's own certainty about his own beginnings. And now, thanks to Wallace's other discoveries in the Celebes and Borneo, in Lombok and Bali, the certainties about the very immobile and unyielding solidity of the world itself were being challenged.

Such a series of hammer blows! Mankind, it seemed, was now suddenly really rather—dare one say it?—insignificant. He may not after all have been, as he had eternally supposed, specially created. The Book of Genesis, believed by so many to be Holy Writ, was perhaps no more than the stuff of myth and ancient legend. And now even the continents themselves, long supposed to be the most reliable and unshifting bedrock of our very existence, had become mobile.

The earth's surface, said this new breed of seers and iconoclasts, was shifting and unstable. The world was not rock solid at all. Such musings among the scientists were seen variously as dangerous, unsettling, ungodly, and evil—and small wonder, after so many centuries of man having been comforted by the certainties of simpler beliefs. It was a bold man indeed who would now pose further questions about such matters head-on.

One man who did take the great risk of placing his ponderings on paper, and who came up with what to many was a highly unwelcome answer (suffering mightily as a result) was Alfred Lothar Wegener.

DISPUTATION

Alfred Wegener was German, an Arctic explorer, a meteorologist,*
pipe-smoking, taciturn, tenacious—once described simply as "the
quiet man with a charming smile." But the theory he advanced in
a book published in 1915 made certain that he became famous—
though for the heresy for which he was famed he was vilified and,
most cruelly, denied his deserved academic reward. And when he
died, at the very early age of fifty, he was a figure of notoriety and
ridicule. Only in the last few decades has the wheel come full circle, and has Alfred Wegener come to be regarded as one of the
most prescient figures of twentieth-century science.

The problem that led to Wegener's personal trials was the very
virtue that gave him the insight. He was a generalist, interested in
everything, content to step outside the perimeters of his chosen
science—meteorology—and to dabble in the wide variety of other
unrelated sciences that fascinated him. Scientific specialists, who
still today guard jealously their own fields of research, attacked him
roundly for daring to invade their territories—and never more so
than in 1915, when he first published, in German, his now-famous
book *The Origin of the Continents and Oceans.* In it he used a
phrase that was to make him especially notorious: *die Verschiebung
der Kontinente,* which translates literally as "continental displace-
ment," but which by 1926, when it made its first recorded entry
into the English language, had transmogrified into what is today's
more familiar phrase: "continental drift."

Wegener's roving attention had first been attracted by a simple
Mercator map of the world. He noticed, in a cursory glance that
bred a sudden revelation, something that seems perfectly obvious
today. He wrote a hurried note to his fiancée: The coastlines of
Africa and South America are such—with the huge eastward con-

---

*He is memorialized by the names of two rare ice-crystal halo arcs and by the magnificent
eponym of the *Wegener-Bergeron-Findeisen* procedure, the mechanism that creates the peculiar shape of raindrops, which Wegener helped to discover.

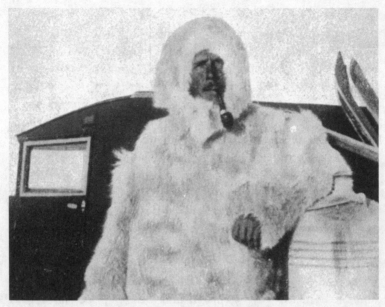

Alfred Lothar Wegener.

vexity of Brazil so alluringly similar to the immense eastward concavity between Nigeria and Angola—that *they seem to fit*. Might it not be, Wegener wondered, that once in some incalculable past the two continents had actually been joined together? And further— might it not be possible, since they had been joined and no longer were, that they had in subsequent years slipped, shifted—nay, *drifted*—apart?

Others had noticed the congruency of Africa and South America too—indeed, they would have been blind not to.* The differ-

---

*Francis Bacon, the philosopher, wrote about the "fit" as far back as 1620; the great French naturalist the Comte de Buffon, author of a majestic thirty-six-volume *Natural History*, a pioneering paleontologist who suggested that geological history had developed in a series of identifiable stages, speculated on the reason in 1778; and in 1858 a noted catastrophist named Antonio Snider-Pellagrini went so far as to propose that a single continent had once existed and then broken up, its parts torn away from one another to create today's arrangement. Eduard Seuss, an Austrian, posited the existence of a Gondwanaland in 1885—but to create

ence was that only Wegener, a trained planetary astronomer then working as a weather forecaster, had the temerity to take the notion and look further. He still had his work to do: He went on a series of expeditions to Greenland (an island so wonderfully complicated and instructive that it appears all too often in the stories of geology and geologists); he flew in balloons[†] to check upper-atmosphere jet streams; he set dozens of high-flying kites to study the polar weather phenomena.

But though the climate and the atmosphere fascinated him, the idea that the continents below might have somehow moved was fast becoming an obsession. He looked for support for the idea. He examined carefully the observations of other scientists and the conclusions of their fields—he looked at geology, at paleontology, at paleoclimatology and (most importantly for this story) at Sclater's and Wallace's new-fangled zoogeography and biogeography. He wanted to see if there was any hard evidence to back his idea that continents had somehow moved from their initial positions to where they are now.

And he found plenty, some of it hard and convincing, some of it circumstantial and tempting, much of it vague and alluring. The easier evidence comprised those existing mountain ranges, coal deposits, and fossil appearances that were to be found on the far sides of the oceans, right across from the obvious "fits": when maps of the continents were pushed together to fit properly. Then the ranges and the outcrops of exploitable minerals and the lines of ammonites, trilobites, and skeins of graptolitic shales themselves also slotted together perfectly, like pieces of a gigantic jigsaw puzzle.

To deal with drifting continents where there was no obvious

---

the oceans he imagined sinking and foundering, not creeping and drifting. An American geologist named Frank Taylor wrote in 1908 about the likelihood of continents inching toward the equator. But few of these men are now remembered for their geology; and Wegener was soon to be rudely rebuffed for his.

[†]For a while he and his brother Kurt held the world hot-air balloon endurance record, at 56 hours.

visible fit, instances of which were much more numerous, Wegener found it simpler to work backwards. He drew maps of the world's surface as he surmised it might once have been—and then looked to see if there was geological, climatic, or biological evidence of possible fits that would then back him up.

He created charts based on notions put about by earlier theorists. There had once been a great southern continent, Gondwanaland (first named by Eduard Seuss in 1885), and an equally great northern continent, Laurasia, and the pair had been separated by a great sea, the Tethyan ocean. According to these early theorists the Tethys had originated by way of continental sinking—an ever popular explanation for all manner of earthly mysteries, and which of course remained the basis for such enduring myths as the vanishing of Atlantis.

Wegener plotted onto images of these supercontinents and superseas the fossil evidence of a number of well-known and easily identifiable past events, like the great Permocarboniferous Ice Age. Amazingly, the fossil trails, which were so broken, discontinuous,

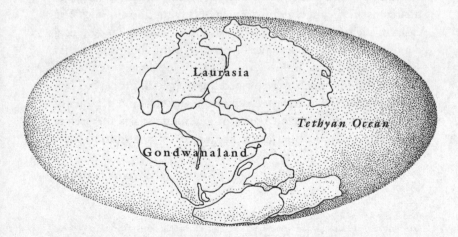

Pangaea is here beginning its division into Laurasia and Gondwanaland, with the Tethyan Ocean slowly opening up between the two giant super-continents, from which all the present-day smaller continental bodies were born.

and almost randomly scattered across the continents today, joined up when the continents were theoretically pushed together, and huddled around what looked like what had once been the world's three-hundred-million-year-old South Pole.

It all began to look very compelling, with more than enough evidence of later fossilized events to allow Wegener to follow the tracks that the chunks of continental crust must have taken across the surface of the globe to their present positions. Gondwanaland, for example, could be seen breaking up into the massive bodies that would in due course become Africa, Antarctica, South America, and Australia, with the peninsulas of India and Arabia floating off into their own separate existences from within it: Wegener was able to plot how the fragments moved, when they moved, and where they ended up at different times in the geological past.

Laurasia had exploded too, fragmenting into wandering lumps that had the vague appearance of North America, Greenland, Europe, and that part of Asia that today lies north of the Himalayas. Wegener was able also to suggest the paths of their various intercontinental excursions, and to nudge them into their present shapes and dispositions.

The more that this modest, tweedy weatherman thought and theorized, the more it seemed probable to him that the two early continents of Laurasia and Gondwanaland had themselves in fact been one. This single ur-continent came to be called Pangaea (though Wegener, who was credited with its invention, never actually used the word: the very Nordic term *Pangäa* appears in a later edition of his book, though it is not at all clear that he coined this either. The first use of the word comes in a 1924 translation of Wegener's book, by a man named Skerl). Pangaea had then broken apart in some giant mitosis, the Tethyan ocean had swept in between the superfragments that resulted—and then many tens of millions of years later these two smaller bodies had broken up as well, eventually giving us the world we know and recognize today.

"One day a man visited me whose fine features and penetrating blue-gray eyes I was unable to forget," the great German geologist

Hans Cloos later recalled of his first meeting with Wegener. "He spun out an extremely strange train of thought about the structure of the Earth and asked me whether I would be willing to help him with geological facts and concepts." Such was the beginning of Wegener's attempts to get his idea accepted. It was a battle he fought long, and ultimately in vain. The world was simply not ready to accept that its surface moved with such immense drama.

Hans Cloos, who was himself to become a grandee of fundamental geology, with papers on the physics of faulting and the deep-seated deformations of granites, was kindly and sympathetic—though never entirely convinced. Wegener's theory, he wrote, "placed an easily comprehensible, tremendously exciting structure of ideas upon a solid foundation. It released the continents from the Earth's core and transformed them into icebergs of gneiss on a sea of basalt. It let them float and drift, break apart and converge. Where they broke away, cracks, rifts, trenches remain; where they collided, ranges of folded mountains appear." It was tempting to believe it— but it was not a temptation to which Cloos, thus far inhabiting a comprehensively proofless world, was quite prepared to yield.

The rest of the academic community was implacably hostile, almost to a man. "Utter, damned rot!" said the president of the American Philosophical Society. "If we are to believe this hypothesis we must forget everything we learned in the last seventy years and start all over again," remarked Thomas Chamberlin, a towering figure in American geology, on hearing Wegener speak in New York in 1923. "Anyone who valued his reputation for scientific sanity," said a British geologist, at the same time, when Wegener's ideas were being given wide airing, "would never dare support such a theory."

Harold Jeffreys, a giant among early geophysicists, denounced Wegener with indignation and scorn: However powerful the forces that might be brought to bear beneath the earth's crust, none could be sufficiently strong to move it. And what of all the suggestive evidence of the Permocarboniferous Ice Age cluster, neatly arranged around Gondwanaland's South Pole—perhaps the strongest piece of evidence that Alfred Wegener had? Mere "geopoetry," they said, the stuff of little more than idle fantasy.

And so Wegener was pushed out into the cold, like a querulous customer in a barroom of rowdies. Not a single German university would give him the professorship that his otherwise impeccable pedigree deserved; it was left to the University of Graz, in Austria, to offer him only the chair in meteorology. He was obliged to stand away from the geology that was the business of others.

Alfred Wegener died young, quite convinced that he was right, but with the world beyond equally convinced that he was wrong. His ideas, it was almost universally agreed, were the results of bad science at best, wishful thinking at worst.

He died in his beloved Greenland, contentedly engaged with the problems of what the weather was doing above the earth, rather than worrying about the complex strangulations of what might have happened below. He had helped set up an observation camp 250 miles inland, high up on the icecap, to study yet again the Byzantine wonder of the polar climate. He was said by his companions at the time to be deeply happy, blissfully unworried by the firestorm he had ignited among the academic community back to the south.

It was shortly after his fiftieth birthday, on November 1, 1930, that he and his faithful Greenlander companion, Rasmus Villumsen, set off to return to the west coast. It was very cold—recorded temperatures of −58°F—and dark; the only blessing was that the howling gales were, at least when they left the ice camp, at their backs.

But neither man was ever seen alive again. The following May an expedition was sent out, which found Wegener's body, fully dressed, lying on a reindeer skin in a sleeping bag. His blue eyes were still open, and he seemed, the expedition report noted, to be smiling. It appeared that he had died of a heart attack; his companion had seemingly pressed on for the sea, but perished in the attempt. His body was never found.

The men who found him erected a 20-foot iron cross above the nameless spot on the glacier where he died. Some time in the fifties, when another dog team passed by, the cross had vanished. The glacier ice had moved on and torn itself to pieces, taking Wegener's body with it.

It would be stretching a point to suggest that the ice moved in quite the way the earth did. Indeed, the mechanism for the earth's postulated crustal movement was one of the factors that Wegener did not understand and could not fully imagine; his inability to do so was one reason why his skeptical foes were so active and effective. But in Greenland the ice moved, and yes, whether it was provable now or sometime in the future, the earth's crust surely moved too. One can almost hear Wegener, calm and pipe-smoking to the end, insisting to those around him who would not believe, and using the very words that Galileo Galilei had used to the churchmen who had made him recant nearly three centuries before: "*Eppur si muove.*" ["You may force me to say what you wish; you may revile me for saying what I do. *But it moves.*"]

And as with Galileo, so with Alfred Wegener. Aside from the barest murmurs of dissent from groups of present-day fundamentalists, creation scientists and flat-earthers, the entire scientific world now happily acknowledges that Wegener, whom all once thought a crank, was in essence, and in fact, quite right.

Moreover, in the particular context of this story, it turns out that the very processes that Wegener first and so bravely suggested do indeed happen to underlie, both literally and figuratively, the making and unmaking of all volcanoes—the formidably spectacular eruption of Krakatoa among them.

DIVINATION

Yet it was not anywhere near a volcano—and certainly not in the steamy heat of Java or Sumatra—where Wegener's theories first came to be confirmed properly. In the 1960s hard scientific evidence decisively proving the occurrence of continental drift flooded in from a great many sources—some of the most compelling being found fully ten thousand miles and two hemispheres away from Krakatoa, and eighty years after the catastrophe that happened there. Wegener would have savored this particular irony of geography, since it was early exploration work in the high Arctic snowfields of east Greenland where some of his ideas were first fully tested and shown to be sound.

As it happened (and by a series of strange coincidences that I was not to appreciate fully until the day many years later when I stood watching the sound and light show from Anak Krakatoa), I played a part—a very lowly part but an unforgettable one—in one of the Greenland expeditions where some of those first confirmations of continental drift were gathered in. I was lucky: I happened to be in just the right place at what science has now shown to be just the right time.

It was the summer of 1965, and I was a twenty-one-year-old geology student at Oxford. And though I did not attach any special significance to it then, much of the unraveling of the most profound enigma of the world's volcanoes—Why do some parts of the world erupt, while the rest do not?—started at the very same moment that I happened to win a place on a small expedition to a wild and unknown part of the east Greenland coast. While I was packing my steel crampons, shark-skinned skis, and moleskin salopettes for the journey north, I had not the vaguest thought in my head that this trip might have anything to do with a tropical mountain of which I knew very little, and that lay half a world away.

From the moment I first spotted the Greenland announcement, thumbtacked to a department bulletin board among offers of secondhand mineralogy textbooks and a former student's barely used Estwing hammer and Brunton compass, I was entranced by the thought of a summer in the cold of the far, far North. I positively yearned to go.

I had long felt a strange compulsion toward high latitudes. As a child I had been raised, quite conventionally for a Briton of my generation, on the heroic imperial stories of Scott and Shackleton, and, less conventionally, on the tall tales of even more heroic foreign figures like Fridtjof Nansen and Peter Freuchen. Much later, and thanks to the peculiar Arctic interests of my Everest-climbing Oxford professor, a tiny but physically and intellectually powerful man named Lawrence Wager, two of the most renowned of Greenland wanderers, Knud Rasmussen and Gino Watkins, became the greatest of my heroes too. Suddenly the chance of being able to take off to spend some time in their Arctic, in the wastes where they had made their names, seemed to me the most noble and romantic of ideas.

Since I had no obvious qualifications for making the team, I decided to teach myself a vaguely appropriate skill that might make me of some potential use. Morse code seemed as good a choice as any, and so I spent two weeks learning it, and became in short order reasonably competent and fast. I then put the word around Oxford that I might in consequence have something valuable to offer. The ploy evidently worked. Shortly before the expedition was due to set off, and just as I had hoped and schemed, the leader called me for an interview, and, on hearing of my self-professed ability to tell a dot from a dash (he had me tap out the highly rhythmic code for the word "essences" on his desktop), he signed me up. I was to go along as sled hauler, because I was reasonably fit and strong; and I was required to add to my duties that of radio operator. Only out on the icecap did I discover that the team's radio was in fact set up for voice transmissions alone: There was no Morse key, nothing that would allow me to show off.

What happened that glorious high Arctic summer still remains paramount as the purest of adventures, the dream of every schoolboy everywhere. Despite my having subsequently lived a life of fairly unremitting world wandering, that two-month expedition to the Blosseville Coast, south of the great high Arctic fjord-system (the world's biggest) known as Scoresby Sund, has never once been matched. The memories of those fifty days stay with me yet.

It started with embarkation in Copenhagen, the piles of expedition boxes among the coils of rope and crates of fish, the cold

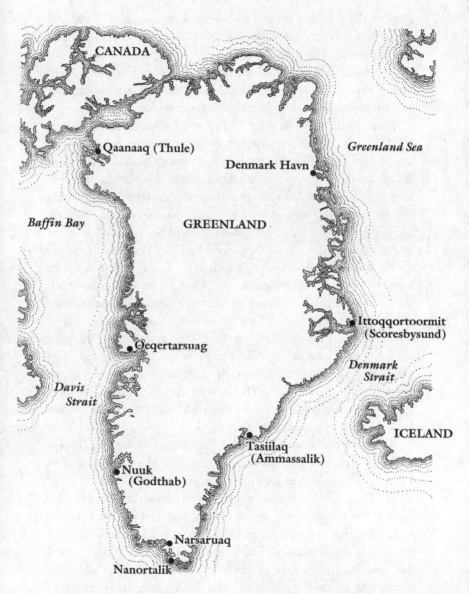

Greenland. The vast fjord system of Scoresbysund begins
halfway up the east coast.

smell of the northern sea and the sweet-sharp aroma of Stockholm tar. It got properly under way a few days later, at the moment when, from the icebreaker's chilly bridge-wing, we first picked out the luminous glow of the *isblink* on the north horizon of the Denmark Strait, and our little red ship began bucking and cracking her hull through the thick and wind-scoured pack ice.

From then on, as we went higher and higher above the Arctic Circle, every subsequent moment, every experience, became vivid, intense, unforgettable. We landed on a remote beach on the iron-bound coast of the immense, mysterious island. We climbed, in brilliant sunshine, the ice wall and then the crevassed length of a fast-moving, mile-wide glacier. We spent weeks camping high on the icecap. We rappelled down sheer walls of black basalt. We skied scores of miles over snow where none had ever been before.

We learned to speak the Danish-Inuit linguistic blend called Greenlandic, in which the country is called *Kalaallit Nunaat,* "our land," in which a snowflake is *qanik,* snow flurries *nittaalaq nalli-uttiqattaartuq,* and a good forty-seven other words besides speak of snow or ice and their many varieties. We grew beards, we grew strong, we became bronzed by the perpetual midnight sun. And when the season was ending, and the dark and the cold crept in, we would thaw our boots out each morning over the Primus stove and watch as our hot washing water, when we tossed it into the air, fell back as a mist of perfect snowflakes.

There were problems, naturally. We ran out of supplies (except for low-temperature margarine, of which we had a good quarter of a ton), and so I—no great shot at the best of times, but the caretaker of the expedition's only rifle—had to shoot a polar bear for food. It was an aged bear, not at all tasty, and its limbs were riddled with *Planaria,* flatworm parasites that we had to tease from between its thigh muscles. The following day, entirely by chance, I followed the shooting of the bear by bringing down a goose, in flight, with another single shot. From today's perspective, all horribly incorrect things to do—except that we were keenly hungry, and there was nothing else to eat.

We then found ourselves socked in by bad weather, were delayed by two weeks, and our Danish icebreaker had perforce to leave for Copenhagen without us. We had to risk walking for an endless day, sixty-pound boxes of precious rock samples on our backs, over a crazed sea carpet of thin and fast-shifting ice floes in order to reach an Eskimo settlement and relative safety. We still needed food: We hunted musk ox with the local men, and then dined with them on young seals, the seals' bellies opened up and filled with roasted seabirds (to which we added from our near-depleted stores, as the most unfamiliar of condiments, bay leaves).

And we barely got home. The season was changing; the sun was setting earlier and earlier each afternoon; storms were blowing in from the north. The brave Icelandic pilot who came to collect us in a blizzard, in the near dark, was killed the very next day, flying his Cessna into a cliff in that mythically ghastly part of north Iceland called the Claw. When we heard the news, we had been celebrating, eating cream cakes in the Savoy in London: We slunk away to our various homes in an awful, somber silence.

But however memorable Greenland was as an experience, it was as science that our little expedition had its greatest value.

Ours was not the only expedition of its kind. At around the same time Oxford was sending people to Spitzbergen, Arctic Canada, Finnmark, and elsewhere in the Arctic, with much the same scientific aims; and other universities and institutions did likewise. There was a great curiosity, particularly among geologists, about what could be discovered in what they called "the Great White."

What we found out—or, more precisely, what was found out by others in distant laboratories once they had examined the rock samples we brought back—helped to prove a theory that, then still young enough to need the sturdiness of proof, turns out today in its maturity and certainty to have the most profound relevance to Krakatoa and to the key story of this book.

The frigid black-and-white coast of east Greenland may be a very long way from the lushly tropical green islands west of Java.

But geologically the two places in fact have a good deal in common. They are not formed from sandstone, or from shale, or from
soft layers of fossil-filled chalk. They owe their origins instead to
fire and brimstone. Both Greenland and Java are volcanic places,
seared and branded into place by the earth's most elemental
processes. And more than that: Each place was made, and stands
where it does on the planet's surface, thanks to the workings of a
once mysterious, much derided mechanism that was initially
uncovered (or, at least, more than partly confirmed) by those who
examined the collections, results, and observations that our trifling
little expedition and others like it brought back home.

Our collecting boxes (and many that we were obliged by the
foul weather to leave behind and collect the following summer season) were filled with scores upon scores of carefully drilled and
numbered samples of the country rock of east Greenland. And that
means volcanic rock. At the latitude where we had been on the
island's east coast, Greenland is more or less entirely underlain by
basalt—a dark gray, fine-grained variety of volcanic rock that was
laid down in layer upon layer during Tertiary times, thirty million
years ago.

Basalt is outwardly an unremarkable rock, generally scenically
unpromising (except in its columnar variety, when it forms beguil-
ing spectacles like the Giant's Causeway in County Antrim, or Fin-
gal's Cave on Staffa). On close examination it is not at all pretty: It
does not hold a candle to decorative igneous rocks like granite or
gabbro or varieties of the more exotic porphyries; nor does it often
compete as building stone with Jurassic limestones or Italian marble. But for the six of us trudging across the snows that summer,
charged with this very particular task, the basalts did have one
uniquely interesting feature, one that may not have been readily
apparent to the casual glance, but that was crucial for what was then
a fledgling program of geophysical detective work.

Cooling basalts, it turns out, contain small crystals of iron-oxide
compounds—principally of the cubic spinel mineral magnetite,
Fe_3O_4—that are highly magnetic. During the molten, plastic phase

of their cooling process these crystals tend to act as miniature compasses, swinging hither and yon in the still viscous mix, in an elegant harmony that is extremely susceptible to the lines of magnetic force that radiate between the earth's North and South Poles.

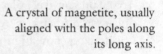

Once the cooling has concluded, and the magma has passed what is known as the Curie point* and become solid (technically *frozen*, though the word used here

A crystal of magnetite, usually aligned with the poles along its long axis.

has nothing to do with the freezing of water and snow, which takes place at a much lower temperature), the alignment of the swarms of magnetite compasses becomes fixed, and is set for all time. And each, crucially, is then aligned to where the North and South Poles were at the time the rocks froze—in this east Greenland case, thirty million years ago. The compasses are thus powerful forensic tools: They tell us where the poles were in relation to the rocks, or the rocks in relation to the poles, a long, long while ago.

Our task—which had been set by a small number of curious laboratories around the world, and whose professors some months before had persuaded a variety of grant-bestowing bodies to give us sufficient money to charter the icebreaker and buy our Nansen sleds and our pemmican and hardtack and then to set off—was to collect hundreds of samples of these basalts. If and when we got them home, they would be sent off to the researchers, who would measure the strength and the direction of the relict magnetism of the iron-oxide particles contained within them. In truth the strength of the age-old magnetism only claimed their most cursory interest. It was the *direction* of magnetism that beguiled them

---

*The Curie point, defined as the temperature at which the remanent magnetism in the cooling lava becomes permanently fixed, is 582°C for the mineral magnetite.

most—and in connection with that we were asked to note very carefully indeed, while taking the samples, the way they were aligned, relative to the North and South Poles of today.

So we spent hours—day and night, bathed in the Arctic summer's continual sunshine—painstakingly drilling basalt samples out of the sheer walls of the local *nunataks*. We used a portable power drill (with a diamond-studded bit that needed to be cooled by dousing it with bucketfuls of snow) and a very accurate sun compass* to make sure we always knew where each sample lay relative to the present-day poles. The selected cores, eight inches long, two inches in diameter, were then indelibly marked with their geological horizon (how high up each had been found among the other layers of rock) and their sun-compass orientation, wrapped in plastic, and set aside in specially made boxes of strong and waterproof fiberboard.

The science that required the presence of all those basalts cores was elegantly simple and, by the standards of the present day, somewhat mundane. The interested scientists all wanted to work on the samples to satisfy a suspicion that had been growing during the early sixties. This burgeoning belief was rooted in an unsayable, almost (in some quarters) still-heretical view of the making of the modern world, and held that the magnetic alignments of crystals in rocks laid down in the past might, just might, be substantially different from the alignments of the rocks themselves today. This was a something that, if it could be demonstrated, would spawn a profound revolution in contemporary geophysical knowledge and thinking. If it were to be proved, then the unsayable would suddenly become something to preach, and the heresy would, overnight, become dogma.

---

*This was necessary because of an added complication: The lines of magnetism in very high latitudes in northern North America (of which Greenland is, technically, a part) are so close, and vary so wildly each year, that a normal magnetic compass is worse than useless. All fishermen and navigators working these waters know this well; we suspected it might be the case, and decided in Oxford to obtain all our alignment readings via the one true invariable, the position of the sun.

And that is just what happened. The rocks we brought back did eventually (though many months later) prove essentially and exactly what everyone wanted them to. A comparison of the directions of the spinel compasses (the science of studying such phenomena had a century before been given the hybrid Latin-Greek name *paleomagnetism*) confirmed what a very large number of scientists suspected. There was a very considerable difference between the direction in which the Tertiary magnets were pointing and the position of the poles today. All of the millions of magnets were lined up not towards the present poles at all, but towards a point some fifteen degrees to the east of it. It was a simple realization, but it had a stunning effect.

This meant one of two things: that either the poles had moved in relation to the rocks or the rocks had shifted in relation to the poles. The first was initially the more tempting possibility, since it seemed quite simple for the North Pole, which was after all an invisible and rather mysterious entity, to have somehow shifted itself around by fifteen degrees. But the scientists who were doing the work on our rocks had an advantage, in that they knew this could not be the case. They were already performing palaeomagnetic studies on the other Arctic rocks of the same age, on samples from Spitsbergen and the Faeroes and Norway. What they found was that the computed polar positions suggested by the spinels in these rocks varied wildly—so much so that it looked not like the poles were wandering, but rather like there were scores of North Poles, all existing at the same time.

So if there was no polar wandering, then the alternative could be the only explanation. It was an explanation that proved a "Eureka moment," a truly life-changing epiphany for many of those researchers back in the sixties. It was the realization from the record of the rocks that *the basalts of east Greenland had moved*. Somehow the basalts of east Greenland had drifted westward, through fifteen degrees or so of longitude, in the thirty million years since they had been extruded from the earth.

In other words, the long-imagined (but until this moment generally discounted) phenomenon of continental drift had—and now, moreover, had provably and incontrovertibly—occurred. During Tertiary times the sea floor beneath the Atlantic Ocean had clearly and demonstrably spread open. Now, from the Greenland basalts, there was powerful evidence to suggest that the theories of continental movement—with the world beginning as one supercontinent, Pangaea, which had then broken up and spread itself over the global surface—which Alfred Wegener had advanced so obsessively and which had been so widely dismissed by the scientific establishment during most of the half century beforehand, had at last been more or less decisively confirmed.

Later in the 1960s, and with a growing amassment of evidence like that from east Greenland now safely in the bag, a great scientific sea change was in the offing. It first came about unexpectedly. It first came about by accident, and in a much more appropriate piece of geography, off the coast of Java, almost within sight of the island of Krakatoa. And it first came about, as sea changes are wont to do, at sea.

There were two initial discoveries. One advanced the evidence of sea-floor spreading and the consequent drift of continents from the merely circumstantial into the happily incontrovertible. The other provided skeptics with something that Wegener had never managed to produce: the model for a mechanism that explained how spreading and drift might work.

This first came to light after a series of unrelated experiments conducted off the south coast of Java—quite properly for the present story, and particularly so in that the scientist who performed the work was a Dutchman, from Delft Technical University, by the name of Felix Vening Meinesz.

His initial intellectual interest was quite unrelated to continental drift: Vening Meinesz was solely concerned with the very accurate measurement, at various places on the globe, of the earth's gravity. He was especially concerned with the measurement of gravity in the mysterious world below the very deep oceans. Since gravity is an acceleration, it is extremely difficult to measure from something that is itself moving—like, say, a ship. So Vening Meinesz had to try out a series of homemade and very specialized measuring devices.

Although the wholesale acceptance of continental drift was not to come about for fully forty years after Wegener's death, Vening Meinesz performed his early work while the unsung pioneer was still alive, between 1923 and 1927. He had taken a crude gravimeter, consisting of a pair of pendulums swinging in opposite directions, and mounted it on gimbals inside the most stable seaborne vessel he could imagine: a submarine. He then had the Dutch navy, using submarines with the somewhat unimaginative names of *Her Majesty K II* and *Her Majesty K XIII,** perform a series of shallow dives off Java—and found to his astonishment that about 190 miles off the southern coasts of both Java and Sumatra, there was a dramatic lowering in the strength of the local gravitational field.

This enormous gravity anomaly coincided precisely with the existence of a tremendously deep and long gash in the sea floor known as the Java Trench. To appreciate the depth of the trench, one can imagine someone moving eastward (driving some seabed-crawling machine, if that is not straining credulity too much) two miles beneath the surface of the sea, somewhere off Christmas Island. Without warning the seabed starts to slope downward at a gradient of 10 percent or more. It keeps on going down, down, down—until, at the dark and ice-cold bottom of the Trench toward which the slope is heading, the sea is nearly five miles, or 24,440 feet, deep. Then, even more abruptly, the bed begins to climb back up, dipping briefly down, once, and then rising one final time clear up to the continental shelf, the shallows, the fringing reefs and finally the beaches of south Java itself. Within two hundred miles of the shore are some of the deepest parts of the

---

*Vening Meinesz was a rather overlarge person, uncomfortable to have in a small submarine. The captain had to remind his crew to draw in their breath every time the good doctor sat down.

world ocean—and directly above them, Vening Meinesz discovered, some of the lowest and weakest gravity accelerations to be found anywhere.

The Dutchman was promptly invited to the United States by a young Princeton scientist named Harry Hess; and, together with two other young men who were to go on to becoming rising stars in the new field, Maurice Ewing and Teddy (later Sir Edward) Bullard, they took off in a boat called the *Barracuda* to see if the Java anomalies could be found above the submarine trenches known to exist in the Caribbean. They did, spectacularly so. The four excitedly discussed why this might be—with Harry Hess and Vening Meinesz openly speculating that they were caused by some mysterious force dragging the rocks of the seabed downward and (as it were) dragging the gravity down with them. Hess wrote a seminal paper in 1939:

> Recently an important new concept concerning the origins of the negative strip of gravity anomalies . . . has been set forward. . . . It is based on model experiments in which . . . by means of horizontal rotating cylinders, convection currents were set up in a fluid layer beneath the "crust" and a convection cell was formed. A down-buckle in the crust . . . was developed where two opposing currents meet and plunge downwards. So long as the currents are in operation, the down-buckle is maintained . . . the currents would have velocities in nature of one to ten centimetres a year.

Perhaps, at last, we had a theoretical mechanism. Perhaps there were currents below the solid surface of the planet, currents that dragged the continents along on top of them and that then plunged downwards, dragging the continents down with them. Continents might thus be moving toward or away from one another at between half an inch and four inches every year—a rate that, miraculously, was entirely consonant (once someone hurriedly did the arithmetic) with what Wegener had proposed for the breakup of Gondwanaland more than twenty years before. But this was 1939, and the world was being enveloped in a manmade turbulence of a very different kind: Harry Hess and his bold theory would have to wait until the war was over.

But when it was over, a wholly new and unexpected piece of evidence for crustal movement came to light—also at sea, but this time not off Java but off the northwestern United States. And while the first groundbreaking prewar study had to do with anomalies in the earth's gravitational field, this next series of experiments had to do with a study of the earth's magnetism. More especially, they were concerned with what was known as the remanent magnetism that

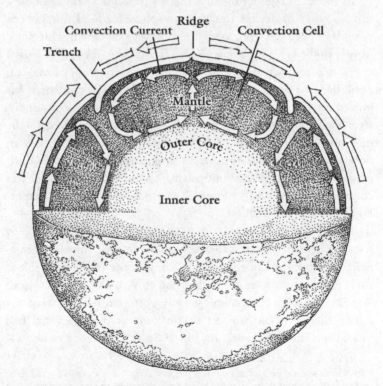

The process of convection inside the earth's mantle, the driving mechanism behind continental drift and plate movement.

might be held in the rocks of the sea floor, the old magnetic signatures that we later studied in east Greenland.

The pioneering work on remanent magnetism had been accomplished largely by an ebullient redheaded Mancunian named Keith Runcorn. He was a lateral-thinking geophysicist who wondered out loud if perhaps the magnetic field of the earth somehow varied over time—if it had perhaps varied in its strength or its direction. And if it had varied, then perhaps a record of those variations would be discoverable by examining the remanent magnetism of rocks laid down during the time that was being investigated.*

In the early 1950s Runcorn and his colleagues and associates, using a range of devices (including a sphere made of thirty-seven pounds of pure gold, borrowed from a very skeptical British Royal Mint), studied this fossil magnetism in a variety of rocks of a variety of ages across Britain. The conclusion to which they came was published in a paper in 1954. The evidence showed that there was indeed significant variations in the magnetism held in rocks of different geological eras, and this could really be accounted for by only one of two happenings: Either the magnetic poles were wandering about relative to the earth's landmasses, or the earth's landmasses were wandering about relative to the poles. The latter was continental drift—and to Keith Runcorn it looked a most temptingly plausible explanation.

Further compelling evidence to support this notion was soon to come in. If the rocks on other continents showed the same remanent magnetic evidence for a wandering of the poles, then it was likely that yes, the poles themselves had moved. On the other hand, if the results from continents far away from one another were different, then it would suggest that it was not the poles that had moved but the continents. And in 1956 that evidence came in—

---

*The idea of remanent magnetism was first put forward by Pierre Curie, who discovered that rocks that are cooled in a magnetic field assume the polarity and direction of that field as they do so. The crucial temperature at which the magnetism is locked in (for study by later geologists) is known as the Curie point, and varies from rock type to rock type.

showing, to the delight of people like Keith Runcorn and Harry Hess, widely different degrees of magnetic variation. The differences could be accounted for only by the continents having drifted apart from one another. The noose was closing. Wegener's theories were being resurrected, and fast.

The clincher came quite accidentally, in work that began in August 1955, in the cold seas between America's most westerly point, Cape Mendocino in California, and the southern tip of Canada's Queen Charlotte Islands. An English geophysicist named Ron Mason, on sabbatical at Caltech, was vaguely aware of highly classified U.S. government research into underwater magnetism—classified, it was said, because the U.S. Navy was looking at deep-sea hiding places for its long-range submarines.

Over coffee one morning he asked whether it might be possible for him to join Project Magnet, as it was called, and, without interfering with the government work, to tow a magnetometer behind the project's ship and make his own maps of any magnetic anomalies he might find down on the sea floor. The project director agreed, and that summer Mason arranged for a long, floating, fish-like object—known formally, in the kind of language heard in science fiction, as an ASQ-3A fluxgate magnetometer*   to be towed behind the U.S. Coast Guard's ship *Pioneer* as it searched for seemingly more important things on behalf of the Pentagon.

Ron Mason's instruments up in the *Pioneer*'s operations room, hooked by wire to the magnetometer that was trailing behind and below the ship, recorded variations in the strength and direction of the magnetism of the seabed rocks below. The recording paper unrolling from its drum (which in due course was to be superimposed on a topographical map of the sea floor and the Pacific's coastline) indicated the data by way of an intricate trace of lines—

---

*The instrument was originally designed to be mounted in a low-flying aircraft, with the aim of detecting enemy submarines in the waters below: Its acronym, MAD, indicated that it was a *m*agnetic *a*irborne *d*etector. It was easy enough to modify the instrument so that it could be towed behind a ship in the water itself, by placing it inside a fish-shaped, nonmagnetic container.

with some of the lines showing rocks that had certain properties, and others indicating rocks that had properties that were precisely the opposite.

As the hundreds of miles of ink traces began to plot out on the screens, Mason could hardly believe what he was seeing. At first the marks on the ever rolling drum seemed meaningless, no more than a random mess of indecipherable hieroglyphs. Within a few hours, however, the traces quite inexplicably began to display an absolutely regular, absolutely consistent pattern. As the ship cut steadily through the water, and as the floating ASQ-3A behind it recorded the magnetic fields in the rocks hundreds of fathoms below, so the traces began to arrange themselves in an unmistakable pattern of parallel and linear *stripes,* just like those found on the skin of a zebra or tiger.

The stripes became more and more obvious as, day by day, week by week, and eventually month by month, the little ship worked itself steadily along its recording route. The vessel would scuttle back and forth on its preordained track, the midshipman keeping it at steady five knots for hour after hour, as it traced and retraced its passage along the specific sector of sea floor that Ron Mason had decided to measure. And as it did so, so the zebra-stripe pattern from the rocks below steadily built itself up—with the recording paper eventually being covered with long black-and-white patches that alternated from black to white to black and back to white again in an even more uncannily regular manner.

All the stripes, moreover, were not simply regular. After just a few passes of the *Pioneer,* their arrangement could be seen: They all pointed not merely in parallel but essentially *only to the north and to the south.* Or, to be more accurate, they pointed along the long axis of the ocean in which they were found.

The ship might go east-west, it might go diagonally; it might reverse its voyage—but no matter. Pass after pass after pass, the magnetism in rocks below was recorded as a series of stripes arranged in long north-south trending patterns that made the plot of the seabed look like an acrylic bedsheet from Frederick's of

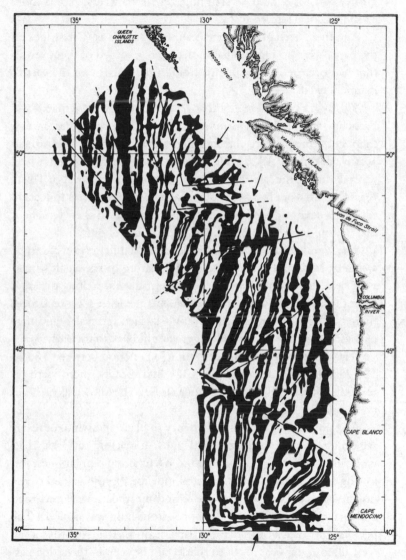

The magnetic "zebra stripes" discovered on the seabed of the
northwestern Pacific in 1955, which were finally to confirm the
idea of sea floor spreading.

Hollywood, or a herd of standing zebras or tigers, rather more exotic than it deserved to be.

And then in a flash everyone realized what it was. The stripes on the paper were noting anomalies that were a record of the reversals that had occurred from time to time in the polarity of the earth's magnetic field.

This was a curious reality that had been dimly recognized by a Frenchman named Jean Brunhes in the early part of the century, then confirmed (by a Japanese geophysicist named Motonari Matuyama) in the 1920s. He showed that there had been a field reversal during the late Pleistocene, ten thousand years ago. Thirty years later his work had been fully confirmed: basalt layers in Iceland showed a series of back-and-forth field reversals in their remanent magnetism. From then on there was no doubt about it.

The work in Iceland and elsewhere showed field reversals to be a standard (if stubbornly inexplicable) feature of the earth's magnetic field, just as can happen in any manmade self-exciting dynamo. Over the last 76 million years, the paleomagneticists soon reckoned, there had been no fewer than seventy-six such reversals—and they had continued to a measurable degree right back to the early Jurassic, 150 million years ago (except for a long period between 85 and 110 million years ago when, equally inexplicably, there were no reversals at all, during what has since come to be known as the Cretaceous Quiet Zone).

And now here, at last, was evidence in black and white of reversals in the rocks' magnetic field and, moreover, reversals that, when plotted onto a map, seemed to occur along regular lines, following a pattern that was imprinted indelibly on the seabed of the northeastern Pacific. And, as more and more data were recorded and analyzed, something even more astonishing was noticed: The peculiar intricacies of the stripe pattern on one side of the ocean were almost identical to those that could be seen in the stripe patterns on the other, and that there was a point, or axis, in mid-ocean, on which this symmetry seemed to hinge. In an instant the explanation for that became clear too.

The north-south trend of the magnetic stripes suggested that the rocks below were moving outward, were on each side flowing away from this central point, just like rainwater hitting the ridgepole of a roof, with some water then going down one side of the roof, some the other.

And the conclusion to be drawn from this turned out to be the single missing part of the mechanism for continental drift, the one that Alfred Wegener had never been able to imagine. The mid-oceanic north-south axis was a place where—presumably, logically, but suddenly somehow astonishingly—*entire tracts of brand-new seabed were somehow being created.* The midocean axis was a ridge of making, where the world was welling up from inside itself and spilling out onto the floor, to be carried out and away, thus making ever more room for the making of more.

And as this brand-spanking-new submarine real estate was being slowly and steadily made over millions of years, and as it spread off away from the deep-sea ridgepole both to the east and to the west, so the remanent magnetism of its rocks, with the record of the earth's polarity reversing itself every few tens of thousands of years, remained locked there on the sea floor for all to see. By seeing this, by understanding what it was, scientists were able, at long last, to claim that they had incontrovertible evidence of sea floor spreading. Once radiochemical dating techniques had been perfected, then the timing of the spreading, of the making of new seabed, as well as the dating of the continental drift could all be fully established too.

Now that there was a mechanism and a date, the rest suddenly became almost too obvious. The conclusions began rushing in, filling the intellectual void that had so plagued the science for the decades since Wegener had died forty years before.

In 1962, armed with the new evidence from the *Pioneer* and all the other armadas that dragged schools of fish-shaped magnetometers behind them,* Harry Hess decided he would revisit the

---

*And which showed, for every ocean, ridges where new sea floor was being made, and giant rafts of magnetized rock moving steadily away from them—the Atlantic Ocean being by far the most spectacularly regular of them all.

nagging question that the war had so rudely interrupted. There was now solid new evidence to underpin his notion that convection currents were at work under the earth's crust, and that the continents were indeed being moved about upon it, like gigantic rafts, colliding and bouncing and plunging down back into the earth's molten heart, in a ceaseless supra-terrestrial dance. "One may quibble over the details," he wrote in a paper that year, which is now a classic,* "but the general picture on palaeomagnetism is sufficiently compelling that it is more reasonable to accept than disregard it." Continental drift was most assuredly taking place. There could be no doubt—and, indeed, since the publication of Harry Hess's paper, there never has been.

By curious chance, I came to know, if only slightly, both Harry Hess and Keith Runcorn. The encounter with Hess was to provide the more embarrassing memory, at which I blush to this day.

It was the early spring of 1966, and I was approaching twenty-two. I had lately—more because it was my turn rather than because of any innate talent—been elected president of the Oxford University Geological Society. In that capacity, and because I had often visited the United States—and Princeton University as well, as it happened—I had managed to persuade Professor Hess to speak to the final, celebratory Society meeting of that Hilary Term. Hess was by then a famous figure, and his acceptance of an invitation written by a mere undergraduate was a signal honor for all. Senior academics from most of Oxford's science departments were eagerly gathering to hear the great man speak.

It is a tradition of the society that the event's protagonists— including the guest—wear black tie. Accompanied by my lieutenants, I met Hess (he was clothed as requested, though he called it his "tux" and it looked suitably moth-eaten and academic) at Oxford Railway Station. We had decided to give him (and our-

---

*"History of Ocean Basins," in *Petrologic Studies: A Volume in Honor of A. F. Buddington,* Geological Society of America, pp. 599–620.

selves—we were spending official OUGS funds) a good feed at a well-known riverside inn out in the country, perhaps ten miles from town. So the five of us—society secretary, treasurer, vice president, Hess, and me—drove off in my 1935 Morris 8, a venerable car much prized by my friends but, as it turned out, of lamentably timed unreliability.

We dined early and very well, beside an immense inglenook fire, in a building composed of ancient thatch and mid-Jurassic corbels, on soup and lamb and a '59 Aloxe-Corton. Our date with the assembled worthies of the university was at eight-thirty, in the museum. We left at ten minutes to eight—ample time to get to the venue. Harry Hess had enjoyed the Corton, and we had opened, as I recall, three bottles. He was a happy man.

The car, however, was less so. It broke down in some flat and nameless swamp, five miles equidistant from inn and Oxford. It was foggy, cold, and dark. We had no telephone and no way of finding one. Our only option was to walk, taking a path that our treasurer claimed to recognize and that he laughably described as a shortcut. Well-shined black patent shoes are less than suitable for the Oxfordshire mud. We found a pub, made a telephone call— but the person who answered hadn't heard about any meeting or any museum. Hess had a couple of whiskeys.

We arrived back in Oxford at ten o'clock—muddy, wet, cold, and, in the case of Harry Hess, agreeably and pleasurably drunk. Our audience had stayed, not knowing what else to do. The speech was a disaster. Someone tripped over, and much of what was said was incoherent. Maps fell down. The projector fused. The front row glared at us, deeming us responsible for the supposed slur on this monarch of high academia.

I was not asked to convene any further meetings of the Geological Society; but Harry Hess wrote later to say he could not recall a more amusing or satisfying evening in recent years, and hoped that we would remain in contact—which we did, until he died three years later.

I had known Keith Runcorn well during my time as a reporter in

Newcastle-upon-Tyne, where he was professor of geophysics. He had taken me under his wing, seeing me as a local reporter interested in science who might help him publicize his research on deep-ocean tides. He had an arrangement with Cable & Wireless to make use of one of their disused trans-Pacific telegraph cables, and had set up monitoring equipment on Fanning Island, in what is now the Republic of Kiribati, to measure the tiny electrical impulses generated in the cable by the deepwater tidal movements in midocean. For readers in the grim and grisly winters of Tyneside, newspaper stories about the blue waters and eternal sunny skies of a South Pacific atoll made a welcome change; I wrote about him often, liked him enormously, and only wished that my paper had the budget to let me go out, as he had often suggested, to spend a season on Fanning, measuring the currents, soaking up the Polynesian way.

Once I left the northeast of England, Keith Runcorn and I lost touch—one might rightly say that we drifted apart. From time to time I would see his name on various celebratory papers, or on programs for conferences called to note this anniversary or that, and all to do with the theories of continental drift of which, along with Hess and a small army of others, he was now world-renowned as an architect.

And then, in December 1995, I read that he had been found, brutally murdered, in a hotel room in San Diego. He had been on his way to the annual meeting, in San Francisco, of the American Geophysical Union. He was to give a paper about his new interest, the magnetic fields that were then being measured on the surface of the moon.

DEFINITION

It turned out to be but a short step from a wholesale acceptance of the mechanics of continental drift to the creation of the new theory that would be christened "plate tectonics"—that set of abiding planetary principles that today is generally acknowledged to be the first properly global theory to have been adduced and accepted in the history of earth science.

The crucial first mention of the theory came on July 24, 1965, in the British journal *Nature*, under the authorship of the genial Canadian who is now most generally accepted as the "father" of plate tectonics. Many hundreds of scientists had already spent thousands of man-years working away with their gravimeters and magnetometers, their polarizing microscopes and their hammers, to create a vague and pointillist image of the emerging theory. But it was not fully recognized until that summer morning in 1965, when the thick, red-covered, and always in those days infuriatingly tightly rolled magazine thudded onto desks in geological laboratories across the world. And then, though scores had been involved, no scores seemed to need settling: There was no contest for this particular moment of glory. It was the bluff, genial, approachable, and down-to-earth University of Toronto professor J. Tuzo Wilson who, by opening his four-page essay with the following declarative paragraph, in essence created the new science:

> Many geologists have maintained that movements of the Earth's crust are concentrated in mobile belts, which may take the form of mountains, mid-ocean ridges or major faults. . . . this article suggests that these features are not isolated, that few come to dead ends, but that they are connected into a continuous network of mobile belts about the Earth which divide the surface into several large rigid plates.

His realization was essentially twofold: the one discovery announced in this seminal 1965 *Nature* paper; the other, inextricably linked to the first and made two years beforehand, when the reacceptance of continental drift was well under way, but in need of as much hard evidence in support as could be found.

The first discovery came when J. Tuzo Wilson looked at the Hawaiian Islands. After thinking about what he saw, he came up with a piece of evidence that added mightily to all the news about tiger-striped remanent magnetics and basalt conveyor belts on the seabed. In the tradition of good science—close observation followed by prescient deduction, with in this case the addition of the basilisk glare of his own geophysical insight—he saw something, something that the ancient Hawaiians themselves had suspected for centuries.

The Hawaiian Island chain stretches for more than 2,500 miles, across thirty degrees of Pacific longitude. Most of the islands one would not think of as Hawaiian. Nihoa, Necker, Tern, Disappearing, Laysan, Lisianski, Kittery, and Seal Islands, for example, are almost unknown: Only Midway and Ocean Islands, at the northwestern end of the chain, are household names, and those mainly because of wars and naval battles. What is currently (but technically wrongly) thought of as the Hawaiian Islands is the four-hundred-mile line of just nine bodies of rock and palm reaching from the outermost pinnacle of Kaula, via the northwesternmost (and still privately owned) island of Niihau, to that great chunk of basalt at the southeast called Hawaii, which is known by most non-Polynesian visitors as the Big Island.

Hawaiian legend has long acknowledged what casual visitors may notice too: that the dusty, half-dead island of Niihau looks much wearier and older than the feisty, bubbling and fiery island of Hawaii. The dank black Waialeale swamp at the summit of Kauai—the wettest place on earth, the locals say—looks prehistoric; the fresh crags of Diamond Head on Oahu (the "diamonds" being glittery and new-looking olivine crystals) look young.

To the ancient Polynesians, who closely examined the islands' soil erosion and vegetation, the difference in age was also self-evident. They incorporated the difference in age into their stories. Pele, the goddess of volcanoes, once lived in Kauai. But then she was attacked by her older sister Namakaokahai, the goddess of the sea. And so she fled, southeast, to Oahu. Namakaokahai attacked again, and Pele moved southeast once more, to Maui. And then a third time, whereupon Pele moved yet again, this time to the Halemaumau Crater of Kilauea, on the summit of Hawaii itself. She had moved three hundred miles southeastward, hopping from island to island, as one volcano after another exploded and died behind her.

Like many legends, this old yarn has its basis in fact. The sea attacks volcanoes—the waters and the waves erode the fresh-laid rocks. And this is why Pele herself moved, shifting always to the younger and newer volcanoes, and relentlessly away from the older and worn-out islands of the northwest, and down toward their more recently created and unspoiled cousins in the southeast.

It was in 1963, in the cold of faraway Toronto, that J. Tuzo Wilson first looked at these stories, examined the geological maps, peered at the records of the various eruptions and lava flows on the Hawaiian main chain, and came up with an idea. Volcanoes had existed in this area of the world for what seemed a very long period of time. So it seemed likely that, for some reason, there was at this particular point in the earth's mantle a deep and stationary hot spot.

The heat from this fiercely hot zone partly melted the rocks of the upper mantle lying above it. The magma created by this melting was lighter than the surrounding solid rock; it burst through the remainder of the mantle into and through the crust; and then it erupted through the ocean floor to produce either a seamount that would remain always below the surface, or an undersea volcano that would grow and grow until it breached the surface of the sea and emerged as an island.

That part seemed simple and logical. But what Wilson then deduced, both from the Polynesian legends and from his own observations, was that while the hot spot remained static, the upper mantle and crust above had moved, carrying one island at a time away from the hot spot and abruptly stopping the volcanism there. After a while, magma erupted again at a new site that lay in the wake, as it were, of the first moving island, and another fresh island was made altogether. It was rather like moving a baking pan full of batter slowly away from a single intense gas burner, creating a line of ripples and partially cooked batter—the batter cooked at one end of the pan, uncooked at the other, and all the stages in between.

Here in the Pacific, on a grand scale, was a four-hundred-mile-long trail of volcanoes—in various stages of cooking. The older rocks were in this case at the northwestern end of the trail, the newer rocks at the southeastern. If the radiometric dating confirmed this, then it could be incontrovertibly proved that the mantle and crust along the Hawaiian chain was itself moving northwestward over a hot spot that was now and long had been sited directly under where the Big Island is today.

And the dating of the rocks proved it, precisely. The Kauai basalts were shown to be 5.5 million years old; those on Hawaii itself were on average less than 0.7 million years old; and in the craters where Pele is said now to be lurking, they are still being created, millions of tons of hot-spot-induced basaltic melt baked afresh every day.

The dating clinched the idea. Wilson was shown to be exactly right. Science could now add to the evidence of the tiger-stripe magnetic anomalies off Seattle this second indication of the undersea geological conveyor belt. The magnetic evidence had been found below the surface; this from Hawaii was by contrast all on the surface. Together they presented science with a sort of gigantic rock-hewn tape-recorder that showed not only the mid-Pacific movement that had been taking place over the last five million years but also the expansion of the sea floor, with the continents on each side of it, undeniably and indisputably, drifting.

The second realization—a true *Eureka!* moment of epiphany—came after Wilson starting experimenting not with rocks and magnetic traces or radioactive rubidium, but with paper and a set of scissors. The title of his famous *Nature* paper, "A New Class of Faults and Their Bearing on Continental Drift," suggests why he was indulging himself so.

John Dewey, who went on from Cambridge to become professor of geology at Oxford, remembers the moment. It was

> . . . one early autumn morning in 1964, I was sitting in my room in the Sedgwick Museum in Cambridge . . . when

Toronto's Tuzo Wilson, on sabbatical leave, sauntered in clearly bursting to tell anyone who would listen about his new ideas. He had discovered that I was the new lecturer in structural geology and said: "Dewey, I have discovered a new class of fault." "Rubbish," I said, "we know about the geology and kinematics of every kind of fault known to mankind." Tuzo grinned, and produced a simple colored folded paper version of his now famous ridge/transform/ridge model, and proceeded to open and close, open and close it with that wonderful smile on his face. I was transfixed both by the realization that I was seeing something profoundly new and important, and by the fact that I was talking to a very clever and original man.

Wilson's achievement was to show what would happen if a geological process caused a fault—in which one body of rock breaks and slips alongside another—across a ridge that, for some reason, happened to be spreading open. He was interested because new bathymetric survey data—the soundings of the sea—was showing, to everyone's astonishment, that the great ridges that ran down the centers of both the Atlantic and the Pacific Oceans were incised by dozens of deep gashes. Geologists initially thought these gashes were spectacular versions of nothing more than classic faults, tearing the ridge asunder as they sheared to the left or to the right as random happenstance directed.

But Tuzo Wilson looked at the bathymetric pictures of these gashes again and said to himself *no!*—if the ridge in the center happened itself to be opening up, for some wholly unconnected reason, then the directions of the fault slippage would be exactly the opposite of those that might happen across a line that was totally stable. He gave a name to the new phenomenon: A normal left-to-right or right-to-left fault was known as a transcurrent fault, but he named the fault he was now predicting, which had never been seen or described before (because no one had ever imagined a ridge spreading itself open), a *transform fault*.

He drew diagrams of how this should appear—how its oppo-

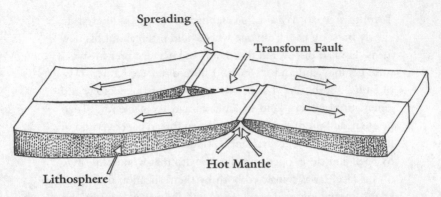

Spreading

Transform Fault

Hot Mantle

Lithosphere

J. Tuzo Wilson's famous transform fault structure, which finally and definitively demonstrated exactly what happens when sea-floor spreading occurs at a mid-ocean ridge.

siteness-from-normal should manifest itself. He created paper-and-crayon diagrams of the model, cut them out, and kept them in his wallet, showing them to anyone who would give him the time of day. This friendly Canadian gentleman—scrupulously dressed, nicely mannered—with his cut-out papers in his pockets soon became a familiar figure on more than a few university campuses in the early 1960s.

And, just like John Dewey, everyone who saw Wilson's demonstration knew in an instant that he had it right there, *the proof.* All that was needed was the test. And for this, the Cold War—of blessed memory; at least blessed to many scientists—came promptly to the rescue. Just as the U.S. Navy had been happy to tow magnetometers behind the *Pioneer* in the fifties, so in the sixties it was content to let scientists make use of a worldwide seismographic network it had deployed to listen for Soviet and Chinese nuclear weapons tests. It was a system with scores of highly sensitive pieces of seismic equipment scattered around the planet. It was ideally positioned to find out, in addition to the sites of bomb detonations, something that it had never been designed to discern: the direction of slippage along some of the great fault-line gashes in the mid-Pacific and the mid-Atlantic.

The system worked its magic. The Pentagon agreed the result should be free from military classification, which had so slowed earlier research. And so in 1967, from a laboratory at Columbia University that had examined the numbers, the results came tumbling out: The great faults in the middle of both oceans were *indeed* transform faults, their slip directions were the *exact* opposite of what would be expected had they been standard transcurrent faults. Tuzo Wilson's ideas had been vindicated yet again: first with the evidence from the conveyor-belt island chain of Hawaii, and now with his supposition about the world in general, and what was going on in the unseen middle of its oceans.

And of that, there was now no doubt at all. The rocks underpinning the oceans were most assuredly splitting themselves apart. The academic battle that had been fought for the previous fifty years was now over, for good. The stabilists*—as were called those who believed, as most once had, that the world and its continents had always been in approximately the same place—had finally to yield. The day belonged to the mobilists, who had since Wegener's time argued that the continents wandered, with what are now known to be dramatic and highly visible effects—such as the creation of the modern map of the world.

It was, for science, an incredibly exciting time. An odd time too—for geologists soon began to realize they had been studying their earth backwards. It was as though an entomologist had begun studying a bee by looking not at the bee as a whole but at the microscopic structure of the yellow hairs on its abdomen; or as though a botanist had tried to get to know an oak tree by first using an electron microscope to peer at a cross-section of an acorn. It is all a consequence of man's own insignificant size compared to

---

*As in an essay by William Dickinson, emeritus professor of Geosciences at the University of Arizona: "In my youth as a geoscientist I was a casual stabilist, assuming that the continents had maintained their relative positions on the globe throughout geologic time. That outmoded stance stemmed less from informed conviction than from sheer ignorance." Professor Dickinson is now a leading evangelist for the theory, and has as one of his proudest possessions a china plate, given to him and adorned with the mottoes HERO OF PLATE TECTONICS and IN SUBDUCTION WE TRUST.

what he is studying, of course. But the fact is, when plate tectonics came along, it was realized that geology had been spending its previous two millennia as a major science looking in great detail at sandstones, gneisses, rift valleys, and ammonites—but had never been able to stand back and look at the planet as a whole and then to work out the details, as happens with most sciences where man is generally bigger than whatever it is he is studying.

Plate tectonics offered for the first time an intellectual mechanism for taking the earth and looking at it as an entity—and the fact that its emergence as a brand-new science coincided so nicely with the development of satellites that could look at the planet as a whole was fortuitous, to say the least. One might say that all this meant that, for the first time, geologists were able to begin looking at things *right side up*.

And what a vision was now laid out before them! The oceans were coming apart at the seams. The crust and the upper mantle beneath them were then spreading out across the deep-ocean floors, moving in opposite directions on each flank of the midocean ridges. Then, when the mobile material reached the edge of the ocean— the edge of the *plate,* to use Tuzo Wilson's now universally accepted term—it plunged down below whatever it met, readying itself to be recycled in the deepest recesses of the planet, to return to the half-molten underneath of the world and in due course to thrust itself up once more through the oceanic centers and begin the long circular process all over again.

Plate tectonics is, in essence, the way by which the world deals with its steady loss of heat. A vast amount of heat accumulated during the formation of the planet a little over 4,500 million years ago—and natural radioactivity, particularly the decay of isotopes of potassium, uranium, and thorium, served only to add to the ferocity of the internal fire. But that heat is now ebbing away, and the means by which it is transferred from the deep interior to the surface tends to be by way of convection currents, just like those one sees working in a vat of vegetable soup simmering on the stovetop.

The rate of movement of these convection currents is very, very small, generally measured in only a certain number of millimeters each year. Currents of material of the inner earth rise up from the red-hot region, maybe a thousand miles down, pass right through a swimmy, hot, weakly plastic region called the asthenosphere, and then, when they reach a region above this, anywhere from four to twenty miles of the surface, slow, stop and eventually—in classic convection-current, vegetable-soup fashion—turn back downwards once again.

The currents, with the soft, pliable and plastic rocks of which they are composed, turn tail at this point on their upward journey, because they encounter an inconveniently brittle and rigid layer of the earth: the upper part of the mantle and the entirety of the crust, which is today called the lithosphere. And it is in the lithosphere that the tectonic plates themselves exist.

In the oceans the lithosphere is thin—maybe four miles thick— and it is young, no more than two hundred million years old. The lithosphere at the continents, with all their long-ago-made limestones and granites and shales and gabbros and schists and gneisses piled on top, can on the other hand be as much as twenty miles thick and much, much older—typically, in fact, as much as two *billion* years old. It is from these two kinds of lithosphere that the tectonic plates are made; the convection process that is going on below them, which is slowly cooling the earth to its ultimate frigid darkness, is what drives them to move and shuffle and shunt and bang their way around the surface of the earth. It is a deeply complex subject, the stuff of mathematical modeling and the employment of banks of supercomputers. It is still far from being fully understood.

But out of the early unraveling and understanding of the process was born in those early years yet another new term: *a subduction zone*. This, it seemed, was where the real business of the world was done. This was the comparatively narrow strip of territory below which, at the edges of the plates, the moving material collided, and one plate slipped beneath the other and began to head back downward, to balance the creation of new lithospheric material at the oceanic seams. Most of the plates that converge do in fact display subduction. And so by finding out just where the world's subduction zones were, science could find the boundaries of most of the planet's tectonic plates and by identifying them, work out what was happening to them, where, how fast and why.

Ever since the beginning of the 1970s—the phrase *subduction zone* appears to have been first used in *Nature* (once again) in the November 14, 1970, issue—a huge proportion of the world's geophysicists have been engaged in studying their bewildering complexities and their seminal influence on the arrangements of the earth. What happens deep within their hot and roiling mysteries is crucial to the understanding of the processes of the earth's making. All the greatest dramas of the solid world's evolution happen within them.

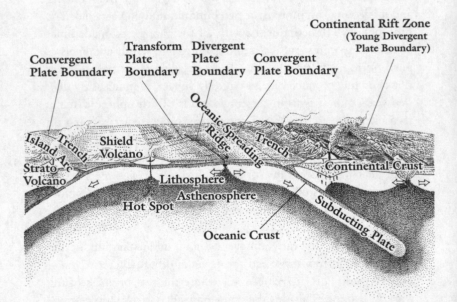

A subduction zone, like that responsible for the Sunda Strait eruption—and, indeed, for 95 percent of the world's most violent volcanoes.

And dramas that also include the realization, which was published in an otherwise obscure book in 1980, that it is "subduction along the Java Trench, where the Indo-Australian Plate is moving under the Indonesian island chain . . . that fuelled the 1883 eruption of Krakatoa."

The earth's surface appears to be armored with between six and thirty-six of these rigid plates, depending on how they are defined and counted. Many of the boundaries between plates have been drawn on the basis only of educated guesswork; what happens deep in the interior of China, for instance, where one plate is thought to meet another, has not yet been fully explained. Nor is it known what goes on around what is called the Scotia Arc in the chilly South Atlantic, east of the Falklands. But, generally speaking, modern tectonics accepts the undisputed existence of about a dozen major plates. The boundaries between these, where the mechanics of the planet's architecture are best exhibited, are by now very well drawn.

Exactly what happens at the boundaries depends on a number of factors. If the plate is entirely of *oceanic material*—that is, basically basalt—and it collides more or less head-on with another plate *of the same type,* then there is an almost randomly decided subduction, with one plate indeed sliding beneath the other. An arc-shaped arrangement of small volcanic islands is usually created, the arc being shaped according to the way the plates move relative to one another. The chain of islands stretching along the International Date Line south of Tonga, for example, is a classic illustration of what happens when two oceanic plates bump into one another. For the purposes of this account, they are best overlooked.

If both plates are, on the other hand, composed mainly of *continental material*—which can best be thought of as a light scum of ancient granites, other sediments, and metamorphic rocks lying on top of the oceanic basalts—then their collision may not result in a subduction at all. As the continental material is made up of lighter rocks, it generally resists being thrust down into the earth's subsurface; both plates may instead stay on top in line with one another, and buckle and crumple, forming a chain of mountains that is in most places not volcanic at all. India and Asia are a pair of continental plates that are colliding, for example—with the effect that the earth's crust where they smash into one another has crumpled itself up and more than doubled in thickness, to form the world's highest mountains, the Himalayas.

The Africa Plate is similarly moving slowly northward and colliding with the Eurasian Plate. The highly unstable mountains of the Caucasus, the earthquake-prone hills of Turkey, the ever restless fracture zone of the Balkans, the ski slopes of Iran's Mt. Damavand are all a consequence of this entirely continental plate collision. But where there is some small amount of subduction, as near Vesuvius and Etna, there are volcanoes. Shallow-focus, very dangerous earthquakes are to be found everywhere along a continental collision zone: Volcanoes are more likely where there is subduction.

It is also possible, given the complicated geometry of the earth, that two plates may well not hit each other head-on but *slide alongside one another,* like a tire sliding past the curb during a lame attempt at parking. The world's best-known example of this kind of meeting is the San Andreas Fault in California, notorious for what humankind regards as its capricious behavior (that a geologist would not think capricious at all). It causes earthquakes in abundance and has enabled Hollywood to come up with some heroically awful films about volcanoes in Pasadena, tsunamis—lethally massive sea-waves—scooping surfers off Pebble Beach, and faraway politicians uttering such droll slogans as The Coast Is Toast.

The fault is a classic example of what is known as a *conservative plate boundary,* where there is no collision, no crumpling, no subduction. The immense Pacific Plate is here simply moving northward relative to the North American Plate, sliding along it at about half an inch a year. Extraordinary stresses can build up if this movement is somehow hindered—by friction, for example. And when this stress becomes overwhelming and is suddenly released, terrible earthquakes can and do occur. During the infamous San Francisco Earthquake of 1906 parts of the fault, which in recent years has accelerated its rate of sliding to almost four inches a year, shot past each other a total of twenty-one feet in a matter of twenty seconds!

Finally, and most important for the understanding of what happened at Krakatoa, there are those events that occur when two plates of *different types* collide—when one plate that is all oceanic basalt smashes itself into another plate that is loaded up with continental crust. These, so far as this story is concerned, are the important ones.

What takes place when plates of different composition run into one another, as might be expected (but was actually not, until that sudden realization made in 1967), is that the heavier of the two, the oceanic basalt plate, dips itself underneath the edge of the lighter continent-laden plate. Immediately it has started to do so, matters directly beneath the point of collision—matters within the classically formed subduction zone, that is—become complicated beyond belief and, to the geophysicist, endlessly fascinating. What takes place within the zone also has a profound effect on the making of the region above and on the lives of the human and animal inhabitants there—not least because, intermittently, the results of the plates' meeting and the making of a subduction zone can be very dangerous indeed.

In essence the process at a subduction zone begins when the colliding oceanic basalt plate begins to head off down back into the earth. It drags some of the continental plate down with it and, in doing so, causes a pinch—a trench—in the bed of the sea. In the case of Java, the five-mile-deep Java Trench, two hundred miles off the coast, provides vivid evidence of the oceanic basalt plate as it begins barreling its way downward. Behind the plate—the coral reefs of Cocos (Keeling) Island, Christmas Island, the flat blue Indian Ocean and, in the far distance, the coast of Australia—all is placid, paradisal, serene. In front of the down-racing plate is the trench, a line of offshore islands, and then Java and Sumatra—one of the most volcanically unstable pairs of islands ever known.

They are unstable because of a well-understood, if wonderfully complicated, mechanism. The down-racing plate heads into the heat, dragged with it billions upon billions of tons of additional material—most crucially of all, water. This it gets from the huge thickness of the waterlogged sediments that are being dragged down from the seabed. Once this waterlogged material is stirred into the mix and reaches a critical depth, quite unexpectedly to those who first noticed the phenomenon, *it begins to melt*—doing so because the addition of this water has lowered the melting temperature of the mix.

And so what starts off beneath the seas as cold and solid now moves down toward the hot mantle; and turns viscous and runny; its fluid components begin to "sweat out"—suddenly bubbling and frothing and coursing and, because they are light and volatile, so rising back up again, passing into the solid mantle through which all the downsliding ingredients had passed. To make matters more complicated still, as they course upwards through this material, they begin to melt that too.

Suddenly a Hadean nightmare is created miles beneath the subducted continental crust: Immense volumes of boiling, gaseous, white-hot magma, alive with bubbles, energy, and restless muscle, seethe in vaults and chambers of unimaginable size and temperature. The Promethean material searches ceaselessly for some weakened spot in the crust above it. Every so often it finds one, a crack, crevice, or fault, and then forces its way up into a holding chamber. Before long the accumulating pressure of the uprushing material becomes too great, and the temperature too high, and the proportion of dissolved gas becomes too large, and it explodes out into the open air in a vicious cannonade of destruction. A type of volcano that, because of its position at the edge of a subduction zone, is far more explosive and dangerous than any other of the world's many different kinds of volcano, suddenly—and if there are people around, invariably terrifyingly—erupts.

Two points of simple geography remain. The first is a matter of tectonic trivia. The Indo-Australian Plate, the culprit that creates all the volcanic *tamasha* so notoriously present on Java, Sumatra, and the tiny island that once lay between them in the Sunda Strait, is moving because the sea floor south of Australia is spreading open. It is possible to calculate both the speed of that opening and the pole of rotation around which this spreading is taking place.

The speed of the sea floor spreading, the rate at which the Indo-Australian Plate is moving northward and colliding with the Eurasian Plate, seems to be about four inches a year, a hundred yards every millennium. Put another way: When Java Man—the *Homo erectus* who first came to these parts from Africa—was living in central Java about 1.7 million years ago, Australia and Asia were rather more than a hundred miles farther apart than they are today. They have been moving toward each other ever since, and doing so in a direction that, unusual though it may seem, turns around a pole of rotation that is located a few miles to the south and east of Cairo.

And second: One delightful symmetry is noticeable, though imperfect at best and complex in the extreme, as more and more details become known. The tracks of the subduction zones that enfold the islands of Indonesia, the zones that create all the trenches, island arcs, volcanoes, and mayhem that goes on among the islands of the archipelago, more or less follow the invisible line of biology and botany that was first hinted at by Philip Sclater in 1857 and drawn a year later by Alfred Russel Wallace.

On one side: Australia, cassowaries, emus, and kangaroos. On the other: cows, monkeys, thrushes, and elephants. On one side: the Indo-Australian Plate; on the other: the Eurasian Plate. The middle, where the two plates meet, and where they come together very slowly but with immense and unthinkable raw power, is in consequence a serried line of the world's greatest, most dangerous, and most predictably unpredictable volcanoes—including, lurking just on the Asian side of an imagined extrapolation of the Wallace Line, the most demonstrably dangerous of them all, the once-great island of Krakatoa. It is an island that has exploded many more times than on the one occasion for which it is now so notorious.

4

The Moments When
the Mountain Moved

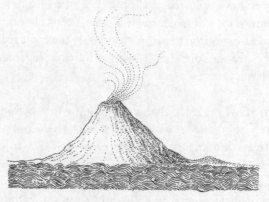

There was a sign from the sun, the like of which had never been seen or reported before. The sun became dark and its darkness lasted for eighteen months. Each day, it shone for about four hours, and still this light was only a feeble shadow. Everyone declared that the sun would never recover its full light again. The fruits did not ripen, and the wine tasted like sour grapes.

—Eleventh-century plagiarism by an Antioch patriarch called
Michael the Syrian, of a document supposedly written by the
sixth-century historian John of Ephesus, describing the
punishing climatic effects of an event that some believe to
have been an early eruption of Krakatoa

Humans have been recording their memories for about thirty thousand years, in cave paintings or songs, in carvings or writings, and during this time, the small cluster of volcanoes and off-islands that for the last three hundred years we have come to call collectively Krakatoa has exploded once, twice, four, or even eleven times, depending on just how the runes of geology, myth, and circumstance are read and interpreted.

Four of these eruptions are generally accepted to have emerged from the mists of uncertain history into the realms of possible reality. And yet, of these four, one is widely thought of now as most unlikely to have occurred at all; the date of a second is very reluctantly agreed to; a third is known to have been very poorly reported and subject to wanton hyperbole; and only the most recent survives as the one of the four that is incontrovertibly regarded as having taken place.

There is some evidence that a very long while before that—perhaps sixty thousand years ago or more—there once was a very much larger mountain that some geologists like to call Ancient Krakatoa, which they believe was something like six thousand feet high and centered on an almost perfectly circular island about nine miles in diameter. But then a gigantic eruption, witnessed only by gibbering hominids and Neanderthals, if indeed by anyone, may have devastated the island and its peak, blowing almost all to smithereens.

Once the dust had settled, what remained of Ancient Krakatoa was a group of four quite small and apparently stable-looking islands. At the northern end of the group were two low and crescent-shaped skerries—one to the east called Panjang, about three miles long, and to the west its larger colleague, four miles long, called Sertung.* Embraced within the parenthesis created by these twins was

---

*Geographical place names present a problem that is peculiarly endemic to those scores of places that have suffered under history's various colonial yokes, and often result in places having been given three names—the early indigenous name, the colonially applied name and then its postcolonial successor. The islands that make up the Krakatoa group display this complexity: Panjang—the ur-name—first became Lang Island, and is now Rakata Kecil; Sertung became Verlaten ("lonely, forsaken island") during Dutch times, and is currently back to being Sertung. Mercifully for this footnote, which would be further awash in explication, the island in the group that had the English name "Polish Hat" no longer exists: It disappeared, a victim of the eruption.

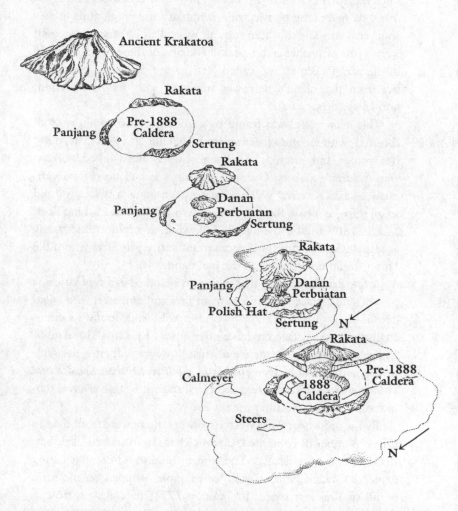

The presumed geological evolution of the Krakatoa islands, viewed from the north. The original great island—Ancient Krakatoa—is thought to have exploded some sixty thousand years ago.

both the Polish Hat, a tiny chunk of the fine-grained volcanic rock known as andesite, and the northern end of the island that we truly once regarded as Krakatoa proper. This was then a lozenge-shaped mass, six miles long by two wide, and half a mile high at its higher southern end. The southern summit was called Rakata; to its north were a pair of smaller crater-peaks. The first of these peaks, roughly in the island's center, was called Danan; and nudging up into the sky from the island's narrower northern spur was the second, known as Perboewatan.

This main island was found by early visitors to be always well forested, with an abundance of fresh-looking lava flows, steaming hot springs, and outcrops of sulfur that were once worked by Batavian dynamite makers. Over the centuries it had been used variously as a VOC naval reconnaissance station, as a place to build small ships, as a base for a small northern Sunda Strait fishing fleet, and, in 1809 and for the decade following, as a remote and barely accessible Alcatraz for those recalcitrant native prisoners whom the Dutch could not control on the mainland.

A few have suggested that it was an island of evil reputation, a pirates' lair, a place that visitors from Java and Sumatra, who sailed over in their *prahus* to collect wood or wild fruit, found so disenchanting that they failed to leave offerings for Krakatoa's local gods. In fact, there is little evidence of this, however seductive the symbolism and the symmetry—*the island of vile repute that exploded and killed thousands*. From most credible accounts, Krakatoa's real reputation was actually rather the reverse.

By the reports of most Western visitors, there were from time to time a number of contented, if somewhat impoverished, little settlements on the island. The great English circumnavigating explorer Captain James Cook, for example, stopped on the main island of Krakatoa twice. In January 1771 his colleague Joseph Banks, the renowned botanist and polymath, noted: "At night Anchor'd under a high Island call[ed] in the draughts *Cracatoa* and by the Indian *Pulo Racatta* . . . this morn when we rose we saw that there were many houses and much Cultivation upon Cracatoa, so that probably a ship might meet with refreshments who chose to touch here." Six years later, when Cook called in again, a village and fields of cultivation were still to be found—pepper was being grown and harvested, as well as other cash crops.

Three years later still, the *Resolution* and the *Discovery* stopped at Krakatoa one further time—but on this occasion, in February 1780, they were without James Cook himself, since he had been bludgeoned to death in Hawaii the previous November. They stopped in the crescent-shaped roads between Krakatoa and Panjang islands for five days, giving the fleet artist, John Webber, time to make elaborate drawings of the village houses and the luxuriant vegetation—palms, tall grasses, ferns—in the valley on Krakatoa between the two more southerly peaks of Danan and Rakata. The expedition's daybook recorded the details of the party's sojourn on the island:

> The *Resolution* refilled its barrels at a stream located at the southern extremity of the little island, a short distance from the shoreline. A little to the south, one finds a heat-source, where the islanders bathe. While we were on a level with the southern extremity of this island the Master went to find an *aiguade* [a spring], but he disembarked with difficulty, and returned without finding any sweet water.
>
> The island of Cracatoa is considered very healthy in comparison to those thereabouts. She offers elevated ground which rises little-by-little from the shoreline on all sides; she is covered with trees, except in several places where the islanders have cleared them for growing rice. The population is not very considerable. The chief is subject to the King of Bantam, just as those of the other islands in the Strait. One finds on the coral reef a large quantity of little turtles, refreshments which elsewhere are very rare, and have an enormous price.

The pepper groves had all but disappeared ten years later, when the Dutch administrator visited; such people as lived on Krakatoa

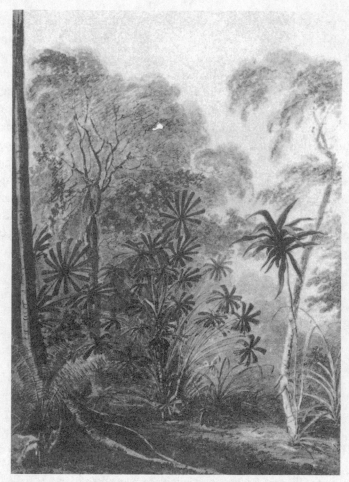

The lush coastal jungle of Krakatoa island, drawn by John
Webber, expedition artist for Captain James Cook, during the
fleet's visit in 1780.

were then raising chickens and goats, and made small sums selling
firewood, water, and food to stopping ships. There may have been
people living on the island at the time of each of the early eruptions; at the time of the cataclysm Krakatoa was, quite wisely,
vacant. But there is no suggestion that it had an evil name.

The three early occasions that, according to most modern history books, possibly saw eruptions were (according to the Western calendar, only lately adopted in Java) the *anno Domini* years 416, 535, and 1680. There is the vaguest of suggestions that between the ninth and the sixteenth centuries there were also no fewer than seven additional eruptive episodes, and that during the single century when the Buddhist kings of the Cailendra Dynasty were on the throne in central Java,* Krakatoa became so notoriously active that it was called "the fire-mountain." This figure of seven, when added to the three possible eruptions and the single certain catastrophe, gives the tally of eleven—a number that is rather doubtful, it must be said.

The one subsequent occasion when Krakatoa went wild is the only year that is fixed with total certainty, and that is 1883. Of the three earlier occasions (I am going to dismiss the Cailendra Seven—for want of any evidence at all, other than the somewhat less than illuminating "fire-mountain" remark in a Buddhist manuscript) we are rather less sure just when or what took place, if indeed anything did at all.

For the sake of completeness, though, it is probably worth first exploring something of the shadowy world of old Java, to see if it is possible to establish with any certainty the early volcanic history of the island—if for no other reason than that the establishment of Krakatoa's eruptive past may offer us some useful clue as to its probable future.

THE POSSIBLE ERUPTION OF A.D. 416

The most frequently quoted source for the first (and, perhaps, also the second) of these three rather questionable event-dates is a monumental modern history written in the nineteenth century,

---

*Building among other monuments the immense and ornate temple of Borobudur, the world's largest Buddhist structure.

known as the Javanese *Book of Kings*. Its author, a Javanese court poet called Raden Ngabahi Ranggawarsita, was well connected to the Dutch colonial establishment of the day, as well as being a key figure in the most distinguished of traditional circles. He worked in the *kraton,* or sultanly estate, of Solo, which is unarguably the most refined of all Javanese courts, a place where the gamelan playing, the *wayang kulit* puppetry, and the poetry still have few equals for their elegance, style, and cultural purity.

But for Ranggawarsita, who was intellectually as much scholar as poet, and who knew the European historical record well, the courtly life was evidently not enough. He had as his principal ambition something more than the fashioning of lyrical poems to mark significant moments in the life of the sultan of Solo. He wanted instead to create something of rather greater and enduring value: a truly comprehensive history of the entire island of Java, a document that could hold up its head alongside the great European tracts about their own countries and peoples. He spent most of his adult years working on the project—and in the 1860s, after decades of what must have been the most thankless toil, he finally achieved his goal, completing and uttering for publication a series of fascinating but undisciplined ramblings that make up what is possibly the world's longest book.* Unfortunately, according to his many critics, most of what he wrote he made up.

That being so, there has been an understandably cautious reaction among scholars who have tried to use Ranggawarsita's tome as the basis for serious historical research. Scientists who have read it have been among the most skeptical of all—and most particularly vulcanologists, who have been both drawn to, and confused by, the following highly alluring passage:

---

*By dint of writing three pages every day for thirty years, he wrote six million words, all in the elaborate language known as Court Javanese. The supposed reference to a possible fifth-century eruption of Krakatoa appears in an early part of Ranggawarsita's history, called "The Book of Ancient Kings," the *Pustaka Raja Purwah.*

the whole world was greatly shaken, and violent thundering, accompanied by heavy rain and storms took place, but not only did not this heavy rain extinguish the eruption of the fire of the mountain Kapi, but augmented the fire; the noise was fearful, at last the mountain Kapi with a tremendous roar burst into pieces and sank into the deepest of the earth. The water of the sea rose and inundated the land, the country to the east of the mountain Batuwara, to the mountain Raja Basa, was inundated by the sea; the inhabitants of the northern part of the Sunda country to the mountain Raja Basa were drowned and swept away with all their property . . .

What does it mean? To which mountain—since *Kapi* is not a name known today—does the passage refer?* And when did whatever happened take place? Geologists, more familiar with gazing at fossils or down microscopes, have pored over this one paragraph of elegant Javanese prose and gone over it with a fine-tooth comb.

It all would have been a good deal more helpful and a sight less confusing if Ranggawarsita had only written once about the supposed great eruption. In fact he went back to it twice. The paragraph quoted above comes from the 1869 version of his book. By the time of his second edition of 1885, he had decided to take another look at it (whatever, wherever, and whenever it was), and he wrote about it as follows:[†]

. . . in the year Saka 338 [that is, A.D. 416] a thundering noise was heard from the mountain Batuwara, which was answered by a similar noise coming from the mountain Kapi, lying west-

---

*By the time Ranggawarsita was writing, the name *Krakatoa*, or at the very least *Krakatau*, was in common use. So it is puzzling that he chose to use the name *Kapi*, unless of course it was a direct transliteration of the original documents he was using for his research.

†According to an English translation published in *Nature*—then awash in narratives of the 1883 eruption of Krakatoa—in August 1889.

ward of the modern Bantam. A great glaring fire which reached
to the sky came out of the last-named mountain. The whole
world was greatly shaken and violent thundering accompanied
by heavy rains and storms took place.

But not only did this heavy rain not extinguish the eruption
of fire of the mountain Kapi, but it augmented the fire. The
noise was fearful. At last the mountain Kapi burst into two
pieces with a tremendous roar and sank into the deepest of the
earth. The water of the sea rose and inundated the land.

The country to the east of the mountain called Batuwara, to
the mountain Kamula, and westward to the mountain Raja
Basa, was inundated by the sea.

The inhabitants of the northern part of the Sunda country
to the mountain Raja Basa were drowned and swept away with
all their property.

After the water subsided the mountain Kapi which had
burst into pieces and the surrounding land became sea and the
single island [of Java-Sumatra] divided into two parts. The city
of Samaskuta, which was situated in the interior of Sumatra,
became sea, the water of which was very clear, and which was
afterwards called the lake Sinkara. This event was the origin of
the separation of Sumatra and Java.

The second of the two passages is more amply filled with geographical detail: the references to Sumatra and Java splitting apart,
the identification of "the Sunda country" and of volcanoes like
Kamula—which is now known from other sources to be the western Javan volcano called Gede, south of today's Jakarta—Batuwara,
in western Banten, and Rajabasa, which is a four-thousand-foot
peak at the southern tip of Sumatra, that still bears that name. Given
all these circumstantial pieces of cartography, it looks more than
likely that "the mountain Kapi" is in fact Krakatoa.

But there is a small difficulty. The second description, filled as it
is with journalistic color—glaring fires, fearful noises, marine inundations, sea-destroyed villages—was written some two years *after*

the 1883 eruption of Krakatoa. And since it is a much richer account than that written in 1869, to the skeptical eye it looks as though Ranggawarsita—having read all the newspapers and perhaps interviewed some residents of Banten (where most of the 1883 flood destruction took place) and spoken to friendly Dutch officials and perhaps to a few scientists too—took his first account of 1869 and fifteen years later, to put it bluntly, embellished it. He seems to have scattered some 1883 details, like glitter, on top of his rather bland descriptions of that earlier event.

And, by doing so, if that is what indeed he did, the poet suffers the inevitable fate of any writer of nonfiction who makes *anything* up: He finds that readers call into question *everything* he writes. His entire history is then suspected of being flawed—and readers look even at his first description of the explosion of "Kapi" and the inundations and deaths that followed, and wonder not just when it happened, but if it happened at all.

Except, of course (and here I spring to Ranggawarsita's defense), at the time he wrote his first edition in 1869 he did not have any means of embellishing what he wrote. Yes, he might have seen accounts in the history books of the supposed eruption that took place in May 1680. But could he have made *all* of it up? Probably not. Most probably his writings for the first edition of the *Book of Kings* do come from his research into original sources. The trouble is—which sources? We know little of his historiography: we can only surmise that his description of an eruption in the Sunda Strait probably came from those ancient, peculiarly beautiful, and very fragile Javanese manuscripts written on palm leaves, and of which the *kraton* library at Solo had a very great number.

Matters here start to become somewhat vague. Of the ten thousand or so surviving palm-leaf texts—which are written in a variety of manuscript styles, including Balinese, Court Javanese, and a curiously complex calligraphy called Mountain Script, with which Ranggawarsita (and a mere handful of others) was familiar—only a very small number actually relate the ancient history of Java. Most of them appear to stop (or start) in about the ninth century. Before that, very little of which we can be certain appears to be known about Java at all.

To give some of the flavor of this uncertainty: there are references in Ranggawarsita's work to a mysterious monarch called Jayabaya, who supposedly oversaw the writing of the accounts of the eruption that are to be found in some of the older palm-leaf texts. Jayabaya in turn is supposed to have been told about the eruption by a Hindu god named Naraddha, who descended from heaven and related fabulous stories that, he insisted to the king, should be inserted into all and any accounts of the history of Java.

So here we have it: an account that arrives from an ancient god, via an enigmatic tenth-century king who dictated material to scribes who wrote on leaves, and then through the good offices of a nineteenth-century court poet who, yes, could read those leaves but who was clearly also a man much taken to flights of fancy. Small wonder that the scientific community has received this basic story of Krakatoa's first recorded eruption with a healthy degree of skepticism.

Except for two things. First, the details are very good (even the relatively unadorned first account of 1869) and fairly convincing geologically. And second, there is a firm date given for the event: A.D. 416. Of that Ranggawarsita has no doubt: The Hindu god seemingly told Jayabaya that the event took place in the 338th year of the Shaka Calendar. This is a widely recognized Hindu dating system used only on these islands, which began in Western terms in the year A.D. 78, when Hinduism was supposedly brought south from the Champa Kingdom in what is today's Vietnam. To get the modern date of 416, one simply does the arithmetic.

But there is not one other shred of independent contemporary evidence—not a line of dust in any of the ice cores taken from either polar icecap, not a millimeter of shrinkage in the size of any tree ring* taken from any of the world's ancient forests—to sug-

---

*Volcanoes that eject immense amounts of dust into the atmosphere leave behind two kinds of easily recognizable signature. Slender bands of the deposited dust can be found trapped in ice cores. And—because the dust causes a filtering of the sun and a lowering of worldwide

gest that there was a volcanic eruption anywhere in the world in the early part of the fifth century. The error rates from these techniques are in the range of plus or minus twenty-five years. So it seems clear from scientific evidence that no volcano of consequence erupted in the world during the first quarter of the fifth century; that much appears certain, the Shaka year, those glaring fires, the inundations of the sea and the explosion of the mountain Kapi notwithstanding. Without firm scientific evidence, and with the historical evidence little more than a vividly told tale from a Victorian poet-fantasist, one has to wonder seriously whether there was a fifth-century eruption at all.

THE CONFUSIONS OF A.D. 416 OR A.D. 535

There is, however, quite abundant evidence that some kind of titanic event occurred about a century later. It can be seen from a very considerable body of tree-ring, ice-core, and worldwide anecdotal evidence that a massive volcanic eruption most likely occurred somewhere in the region of Java or Sumatra around A.D. 535. Could this perhaps also be Krakatoa? Could perhaps the Shaka date in Ranggawarsita's history be an error—but all the other supposed observations, at least for his first account, be substantially correct?

A slender and somewhat tenuous body of evidence suggests error to be the explanation. This evidence derives from the observation that volcanic eruptions, especially truly large ones that occur close to large places of habitation, trigger widespread social dislocation. People die in their hundreds, communications are severed, there is disease, ruin, a collapse of social order. And a barely

---

ambient temperatures—there is also evidence that the growth of trees is at the same time briefly slowed. Tree rings appear much closer together in the cooler years, when there is slower growth. So the ring record provides a neat and easy way (always provided that a cut tree of suitable antiquity is at hand) of establishing the climatic record for hundreds of past years.

recognized consequence of all this mayhem is that historical record keeping suffers. It becomes patchy and incomplete. Historians, like everyone else, have little other than personal survival on their minds.

And the Javanese *Book of Kings* exhibits this patchiness, at just the right time. For all of the fifth century, the book contains its fairly routine and regular selection of entries—there appears to have been no interruption, no trauma, that ever caused a historian to stop writing about Javanese trivia for a protracted period during the hundred years between A.D. 400 and A.D. 500. That century on Java seems to have been blessed with an absence of awfulness. The year A.D. 416 appears with the account of the explosion itself, but, very pointedly, seems not to have been followed by years of disruption, with historians taking time off from their duties to attend to their own needs. There is just as much inconsequential detail in the Javanese palm-leaf history books for the ten-year period between 420 and 430 as there is for the period between 400 and 410.

But in the next century, the sixth, this is not the case at all. For the first thirty years of the sixth century, fully three-quarters of the years have journal entries—the normal strike rate, as it were. But for the eighteen years that follow the crucial year of A.D. 535, however, fewer than one-fifth display journal entries. And then for the thirty years following that, the rate of mention of ordinary matters rises back to a little under three-quarters the normal rate. So it looks very much as though *something* occurred in or around A.D. 535, something that sent the palm-leaf scribes of Java into a tailspin, into a state of historical catalepsy, for almost the following two decades.

And what is more—there is confirming ice-core and tree-ring evidence aplenty from the world outside as well. Dating of both cores and trees is these days highly accurate, and it does now seem that between the years A.D. 510 and 560 some major event that sifted dust around the world and caused the sun to dim and arboreal growing conditions to slacken surely did happen (even if the error rate of twenty-five years is taken into consideration). There is also a Chinese record attesting to a huge detonation heard at around that time—an account that intriguingly adds, moreover, that the noise came *from the south*—to the south of China, in other words. And Krakatoa is of course due south of China.

It would be stretching matters to say that any of this evidence is wholly convincing: But it seems fair to say that of those two supposed early eruptions, only the later—that occurring in 535—seems likely to have involved Krakatoa. The other record, if it is not simply a mistaken date, may have related to an eruption of any one of Java's other twenty active volcanoes—or it may simply be a mélange of myth, confusion, legend, and embellishment, and may never have happened at all.

THE LIKELY ERUPTION OF A.D. 535

Assuming, then, that the first Krakatoa eruption did occur in 535, and not a century beforehand, it is fair to say from what little we know of the social conditions in the western Java of the time, that there were probably precious few people around to be aware of it. In particular there seems to have been no city worthy of the name within a hundred miles of the eruption. Chinese traders, ships' captains who made voyages along the northern and eastern coasts of Java and Sumatra, appear to have left the most comprehensive written records of the region: A number of these speak of a community of the time known as Si-tiao, which was most probably on Java and which was possessed of "fertile lands and communities with streets."

Another state, which in the annals of the Peking mandarinate was called P'u-tei—probably in southern Sumatra—was populated by a race said to be "as black as lacquer"; its members liked to row out from their settlements to sell chickens and fruit to the emperor's passing sailing junks. There are also records of a rather dyspeptic-sounding community called Holotan, whose people grumbled to the Chinese that they were unceasingly subject to attacks by their neighbors. And cannibals "with tails"—probably early versions of the ceremonial long headdresses still to be found in parts of southern Borneo—also caught the attention of the Chinese court's punctilious scribes.

But neither the Indian nor the Chinese traders who passed by with their gold, jade, sandalwood, and cloves noticed a large city or any kind of sophisticated urban population in Java or Sumatra. Fires might have lit the sky, and torrents of ash and pumice rained down from the clouds, but the people of *Nusantara* (the old Malay word meaning "the islands between") who saw and heard and were duly astonished, horrified, or hurt, were villagers only, artless country folk whose descriptions of those early and startling events were inevitably vague and imaginative.

In 1999 a British television documentary, based on a remarkable book, *Catastrophe,* by a London-based author named David Keys, suggested powerfully that an A.D. 535 eruption of Krakatoa not only happened but was the primary cause of an extraordinary number of seemingly unrelated yet world-changing events.

The climate changes triggered by the eruption—if that is what it was—helped to bring about an astonishing series of utterly apocalyptic events: among these, the television program suggested, were occurrences of no less magnitude than the fall of the Roman Empire, the outbreak of the rat-borne Great Plague, the historyless miseries of Dark Ages, the birth of Islam, the invasion of Europe by the barbarians, the collapse of Central America's Mayan civilization and the birth of at least four new Mediterranean states—the list goes on and on. And though the arguments promoting such ideas appear at times more than a little speculative, everything eventually distills into a single fact: Something enormous did take place somewhere in the world in the first half of the sixth century A.D., and it had a staggering effect on the world's climate. But what exactly *was* that something?

A volcano, by all accounts. Evidence used to suggest that nothing at all took place in A.D. 416 is the same evidence that indicates something very major indeed occurred 119 years later. Dust in the ice-cores, acid snows in Greenland and Antarctica, compellingly seductive data from thousands of tree-ring samples—all point to an event, somewhere, in the first half of the sixth century. And thanks initially to Ranggawarsita's writings, however unreliable, the finger points alluringly to Krakatoa as the site for whatever that event might have been.

Surprisingly, very few halfway reliable scientific tests have ever been undertaken to indicate the date of Krakatoa's previous eruptions. It somewhat defies belief that science has been generally content to allow historians to determine the story of Krakatoa's past, when rather more accurate methods—radiometric dating principally among them—could give the answers to a degree of accuracy that poets to the court of Solo unsurprisingly could never really match. The television program–makers saw to it that this sorry state of affairs was remedied.

In 1999, as a response to the wide interest over the remarkable suggestions made in *Catastrophe,* the resident specialist on Krakatoa at the University of Rhode Island, Haraldur Sigurdsson, went on an expedition to Krakatoa* to use the magic of modern chemistry to try to find a definitive answer to the puzzle that Ranggawarsita had set. He attempted this by taking samples of charcoal from a number of various and evidently very ancient lava flows that he found on Rakata, the surviving relic-mountain of the 1883 eruption. He performed dating tests on the samples using the well-known half-life of the carbon-14 isotope.

The results were, however, only moderately conclusive. The event that had burned the charcoal had occurred, Professor Sigurdsson was able to say, between A.D. 1 and A.D. 1200. There had, in other words, been a very large volcanic eruption on Krakatoa during the first twelve hundred years of the Christian era—and it may well have been a large enough event to trigger the climate changes that would in turn cause the economic and social dislocations (and the migration of plague-carrying rats) that would cause the profound events that are the central thesis of *Catastrophe.*

---

*Financed by the British television company Channel Four.

But as to whether that single event can be pinned down to any one year—and whether that year is likely to be A.D. 416 or A.D. 535, there is no ready answer still. Perhaps only Ranggawarsita really had any idea.

THE NEAR CERTAIN ERUPTION OF A.D. 1680

Pure fancy was not so much of a factor eleven hundred years later, however, when Krakatoa may have lifted its skirts once again. *May have,* though, has to be the operative phrase. We have already seen how possibly vague and imaginatively drawn were the descriptions of the supposed events of 416 and 535. There is no evidence at all for any eruptions during the Buddhist Cailendra Dynasty. The descriptions of whatever took place in 1680 do not exactly amount to a paragon of scrupulous exactitude either.

As we have seen, there were just three European witnesses to what might have occurred: the silver assayer named Johan Vilhelm Vogel, who first saw evidence of ruin on the island; the writer named Elias Hesse, who produced his all too vivid account of an event that, according to him, was still playing itself out more than a year later; and the unnamed captain of a Bengal trading vessel *Aardenburgh.*

The two first of these witnesses wrote accounts of a volcanic event that seem highly colored, to say the least. And while the *Aardenburgh*'s master may well have spoken robustly of an eruption to attentive and well-lubricated audiences in Batavia's dockside bars, he seems to have written nothing in his log that was of sufficient interest to alert his masters in the castle: The official records, comprehensive in all other respects, are silent. Something did definitely occur on Krakatoa, of that there is little doubt; but whatever it was, it was probably much less significant a happening than the eruption that seems to have taken place in the sixth century; not to mention what took place two centuries later, in the nineteenth century.

It did nonetheless take place within a respectable distance of a newly settled urban population. For Batavia, the primary company town of the VOC, was in 1680 a full eight decades old, and it had attained some kind of settled maturity. It had a walled quarter and turreted administrative offices, a Chinatown and a collection of godowns, and any number of small terraced houses with streets, canals, and taverns, forming a dreamy, steamy simulacrum of the VOC employees' much missed homes back in faraway Amsterdam or Leiden, Delft or Utrecht.

The Dutch who settled in seventeenth-century Batavia were not an especially content people, by all accounts. Their lives were a succession of uncomfortable trials. They were seldom fully well: They tended to succumb to a variety of tropical ills—malaria, cholera, dengue fever—and they were morbidly afraid of the air on which they suspected the responsible germs were borne. The foul smells from the coast,* brought in by the sea breezes that usually began to blow at breakfast time, prompted everyone in the city to keep their doors and windows firmly shut until dusk; and then to shut them once more soon after, to keep out the evening mosquitoes. Everyone bathed in the city canals. "The ladies unblushingly dived into these general public bathtubs," wrote one Bernard Vlekke, "a custom which was vainly forbidden . . . because the canals were used as sewers and were therefore, *rather* filthy."

And then there was the drinking and partying and the smoking, and the slightly desperate merriment of colonial life. No less an authority than Jan Pieterszoon Coen himself, the founder of it all, had decreed the beneficial effects of spirits: "Our nation must drink or die," he is quoted as having once remarked—an epithet the Dutch distilling industry still likes to remember today. The average seventeenth-century Hollander in Batavia would take a glass of neat genever before his breakfast, and would then while away the day consuming glass after glass of arrack as he sat swel-

---

*The locals liked to use the beaches for their ablutions.

tering in his dark, airless house. And with small Dutch cigars cost-
ing no more than a few guilders for a box of a thousand, with big
Havana cigars only a few guilders more, and a nut of dark shag big
enough to fill a meerschaum bowl going for just a few cents, the
Batavian air was always blue with tobacco smoke.

Down by the Jakarta waterfront at Sunda Kelapa there is still the
slenderest of reminders of those times, barely preserved in a city that
is eternally busy, endlessly and boisterously changing, and (given its
bewildering jumble of races and religions) all too often bickering
with itself, sometimes very violently. Down where the oily waters of
the dock slap against the splintered timber piers, the old Dutch
spice godowns remain, huddled among the tenements of the fish
market. There is a massively built ship-repair dock, with flagstone
floors and teak beams, now turned into a restaurant that serves
rijsttafel and ice-cold Bintang beer. There is a portion of the old city
wall, with a round-topped lozenge of a sentry box, badly in need of
care and repair. And there is the Culemborg bastion, with a nineteenth-century watchtower and customs office, Chinese characters
carved into the floor, and, it is said, a marker showing the hour
meridian, the center point of Java Time (Bali being so far east that
it is an hour later, Aceh in Sumatra, to the west, an hour earlier).

But the loveliest sight in Sunda Kelapa has nothing to do with
the Dutch, the VOC or, seemingly, with Krakatoa at all. It is a far
more elemental, far more timeless vision: the sight of the serried
ranks of enormous and gaily painted wooden sailing ships—maybe
fifty of them, on a busy day—that are to be found tied up alongside
the long quay. They are a type of schooner known as *pinisi,* sailed
by rough-and-ready Bugis from Sulawesi, or matelots from Kali-
mantan and the outer islands of the archipelago. They bring in
wood—most often illegally cut from the great forests of Borneo and
Sulawesi—and sell it in the Jakarta market before sailing home with
televisions, washing machines, and other necessities less easy to
come by on the distant islands.

In one of the local waterfront bars I had fallen in with a forester
who had been kind enough to explain to me, over a number of

beers, the nature of the distribution of harvestable and protected hardwood trees throughout the archipelago.* Later in the afternoon, when it was still insufferably hot, he took me down to the cool of the docks, and (to illustrate a point he was making about the ruin of Indonesia's rain forests) we walked past the ranks of traders' boats—all moored on a slant, their enormous prows arching over the quayside, young men and pye-dogs sleeping in the welcome shadows they cast.

The forester had mixed views about the traders who owned and sailed the ships. He admired them for their courage, their seamanship, their derring-do. He knew they traveled vast distances without proper navigational equipment; he liked their songs and poetry and wild romancing; he knew they would rarely permit a woman to sail on a vessel with them, that their seafaring tradition was everything. He marveled at the physical strength of the sailors—was amazed still (though he had lived on Java for twenty years) at the way the minute, wiry, and barefoot Bugis sailors would unload huge balks of mahogany, pieces weighing twice as much as a man, and carry them down narrow and slippery gangplanks. But it was the timber they carried that vexed him: These seamen were unwittingly a part of the distribution chain that turned rare Javanese teak into baubles for Western living rooms, and they should, in his view, be stopped.

We went back to his apartment, in a miniature skyscraper on the edge of the fish market. He lived on the twentieth floor, with his Chinese wife and three children. He had recently installed a steel front door, four inches thick, with bolts that made it look like the entrance to a bank vault. It was because of the anti-Chinese riots of 1998, he said—the scars and scorchmarks of which were still everywhere to be seen in Glodok, the Chinese area south of the docks. Mobs of Javanese had raided his apartment that terrible day, driven

---

*The western trees—mainly dipterocarps, with winged sepals—are very different from the eucalypts and gum trees that are found in the forests of the eastern islands. The former are distinctly Asian species, the latter Australian, separated by a Wallace Line of their own, echoing as one would expect the distribution of the islands' other living beings, such as the cassowaries, cockatoos, thrushes, kangaroos, and apes.

his wife and family into the street, stolen everything they owned. They had left his wife alone only because he was a Westerner, a foreigner.

But though this was all fascinating, and though his views on the rain forests and the Bugis traders were admirable, I found myself rather more enthralled by a picture that hung on his living-room wall. It was a fine eighteenth-century etching by the well-known Dutch cartographer Jan van Schley. It showed two well-laden company ships, both heeling slightly before a good breeze, passing in front of an island with a pointed mountain. There was a row of trees a few feet up from the island's shore, but otherwise the island was made up of naked rock. Flaming from its summit were great tongues of fire and a roiling, boiling mass of dark smoke that towered above the pair of galleons, and almost mingled with the clouds above. The picture was called simply *Het Brandende Eiland*—"The Burning Island"—and it was a depiction, without a doubt, of the otherwise little-chronicled eruption that supposedly took place in 1680.

He had picked up the etching in an antique shop in Jogjakarta some years before. He knew that it looked somewhat fantastical and was more likely to be a figment of van Schley's normally restrained imagination than an accurate picture of a real happening. But it was a powerful image nonetheless—a reminder for his living-room wall of the awful power of the volcano near which they all then lived. And a more general and emblematic reminder, given his own domestic circumstances, of the highly precarious, often highly tenuous nature of life in the East Indies.

Before the Certain Eruption of a.d. 1883

Two centuries after the 1680 eruption that was pictured or imagined by van Schley, when Krakatoa did finally make up its mind to explode, and did so cataclysmically, the city of Batavia had become in outward appearance an entirely different place.

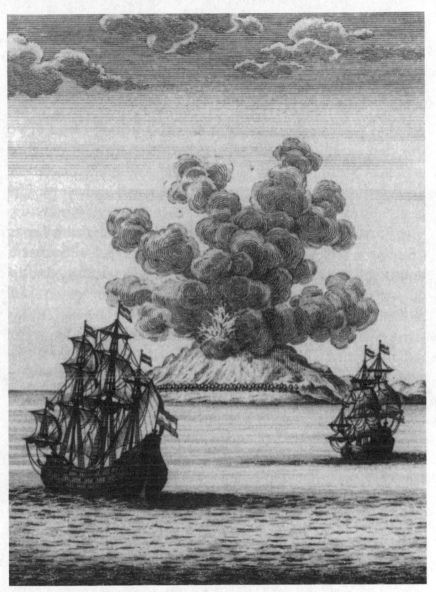

Jan van Schley's early etching *Het Brandende Eiland*, showing two caravels passing in front of what is presumed to be Krakatoa in full eruption, in what is further presumed to be 1680.

In the 1880s it was no longer simply a company town, for a start. War with Britain a century beforehand had essentially brought that to an end. Naval blockades during the war had starved the VOC of cash and had kept thousands of tons of Java's choicest exports marooned in the Batavia warehouses; and then the Treaty of Paris of 1782, which ended the fighting in 1784* broke the Dutch trading monopoly in the region and placed huge financial burdens on the once grand, formerly secure company. The Dutch share in the world spice trade promptly dwindled, the revenues from the newly imported plantation crops of coffee, tea, and quinine were never sufficient to balance the VOC's enormous expenses, and after two centuries of operation the weary old company had been declared bankrupt. Its commission to govern formally expired in 1799, and the Dutch government took charge of the East Indies as a purely colonial possession.

Six years later Holland was overrun by forces of the globally ambitious French emperor Napoleon I, who promptly installed his brother Louis on the Dutch throne. War with England then became, and remained for the next ten years, the dominant reality of all European life. It was a reality that spread east to the Indies too, and specifically to Batavia in its new role as the capital city of what was now called the Dutch East Indies. The British were on the prowl in the Eastern seas; the Napoleonic Wars that raged in Europe had a considerable effect, much of it still very apparent today, on the way that the city of Batavia and the government that was run from it were allowed to develop.

It was indeed a Napoleonic *maréchal*, one Herman Willem Daendels, who was to make the most evident changes, and who would determine the shape of the city that existed at the time of Krakatoa's grand ultimate explosion. His orders, which he issued

---

*France and the newborn United States of America concluded their own peace with Britain; Holland, however, kept fighting for some while more, placing still heavier burdens on its faraway colonials.

promptly on his appointment to the job of governor-general in 1808, were simple: by any means possible and with no expense spared he was to render Batavia proof from any possible British military attack.

He did this by formally abandoning all the castles, warehouses, and fortifications at the seaside, effectively closing down what was dismissively called *benedenstad,* the lower town, the old Batavia, and building an entirely new capital five miles inland, made secure by its geography from any possible sea-borne ambuscades.

Old Batavia, the now squalid, cramped and run-down creation of Jan Pieterszoon Coen, had by the eighteenth century become notorious across the Orient as a pox-ridden graveyard for its resident Europeans. The new Batavia to the south would truly be worthy, he decided, of its long-unrealized sobriquet: *The Queen of the East.* With this in mind he decided to call his new uptown suburb *Weltevreden,* or Well Contented, since that is what he fully expected its grateful Dutch residents to be.

He tore down Batavia Castle and built for himself a palace on a vast plain of neatly clipped lawns, where the governors-general of the East Indies would maintain their city headquarters for decades to come. And as he moved the epicenter of administration, so genteel society moved with him, to the extent that by the middle of the nineteenth century a vacuum had been left in the old town, as the famous British colonizer Thomas Stamford Raffles* was to note in 1817:

---

*Although Raffles is best known for having essentially founded Singapore, he was also lieutenant-governor of the East Indies from 1811 to 1816, Java's brief British-ruled interregnum during the Napoleonic Wars. The Dutch have long since accepted with equanimity the five years of their territory's rule from British Calcutta: A portrait of Lord Minto—Raffles's superior—hangs today without remark among other Dutch governors-general in the Regents' Room of the old Dutch Colonial Office in The Hague. Lady Raffles is buried in a pretty Grecian tomb in the Botanical Gardens at Bogor, south of Jakarta. Her husband is memorialized in many ways: by having discovered and restored the marvelous Buddhist temple at Borobudur, in central Java; by having written a near-definitive *History of Java*; and by having named after him the planet's largest flower, *Rafflesia arnoldii,* which has blooms a yard wide, weighs as much as twenty-four pounds, and has a memorably horrible smell.

Streets have been pulled down, canals half filled up, forts
demolished and palaces levelled with the dust. The Stadhuis,
where the supreme court of justice and magistracy still assemble, remains; merchants transact their business in town during
the day, and its warehouses still contain the richest productions
of the island, but few Europeans of respectability sleep within
its limits.

By 1858 a passing Dutchman, A. W. P. Weitzel, was to amplify the
theme:

> After sundown Batavia is silent and empty; not only the offices
> and large warehouses but even the shops are closed; no carriage
> is heard any more and the few indigenous people who move
> along the streets make no noise at all on their bare feet; and if
> the police had not ordered them to carry torches, they would
> wander there as dark shadows.

And though the first *Thomas Cook* guide was written in 1903,
twenty years after the eruption, it did manage to catch something
of the atmosphere of the new city that Daendels had created out of
the wreck of the old company town, and that was well into its full
flowering at the century's end:

> Whilst Batavia proper—the lower town with its counting
> houses and shops, its native and Chinese populations, its canals
> and moats, its dust and dirt, and old-fashioned mansions—
> makes anything but a charming impression, the upper town, to
> which all Europeans return in the evening, reminds one of a
> gigantic park, in which villas are built in rows, and great trees
> shade the broad and gravelly paths, and spacious squares bring
> air and wind.

The years immediately before the eruption were stable, prosperous, expansive. Whatever setbacks the city had endured in the

aftermath of the VOC's collapse at the end of the eighteenth century had been more than amply reversed in the cosmopolitan, buccaneering, free-marketeering years of the latter half of the nineteenth. The Suez Canal opened in 1869: an appetite for goods from the East—now far closer to European markets, with the shortcut through the desert and ever faster ships on routes— meant that new trading houses were springing up, and new markets for new tropical commodities and wares were emerging on all sides. The population of Batavia, like that of other favored Eastern cities, began to grow, and fast: it jumped from half a million in 1866 to well over a million at the time of the eruption (merely 12,000 of those were Europeans, 80,000 Chinese, the rest all native East Indians understandably eager to take advantage of the growing wealth around them). As early as 1832, according to an account written by the renowned Dutch scholar of Javanese, Professor P. P. Roorda van Eysinga, there were widespread signs of an accelerating prosperity:

> . . . hundreds of carriages of European officials and shop-people raising clouds of dust on the streets, while the Chinese, with their recognizable and unpleasant features, with their long pigtails and silk caps, are everywhere busy hammering, sawing, painting, sewing, building and so on. Rich Arabs and Chinese rode through the streets, half-clad Javanese carried heavy loads, shabby Eurasian clerks walked beneath their sunshades to their offices, old women sold cakes, an Indian sat calmly eating his rice on a banana leaf, and vegetable, milk and fruit sellers, butchers and hill-dwellers offering monkeys and birds all mingled together in the crowd.

This gathering crush of people and the ever accelerating bustle and pace of Batavian life had a further effect: Just as in British India, so the wealthier and more influential of the European townspeople moved from the sweltering streets of the capital city up into the hills, to a cool and green town that the government promptly

christened *Buitenzorg*, or "Without a Care," the equivalent of the
French *sans souci* (it has now reverted to the native name it had
before, the decidedly less carefree-sounding Bogor).

Buitenzorg, just like its corresponding hill stations at Simla and
Murree and Ootacamund in India, had already become a fashion-
able place at the beginning of the century, its reputation firmly
anchored once Daendels had decided, in 1808, to shift his personal headquarters to a great country house that had been built
there by an eighteenth-century VOC commander. His officials
warned him against trying to make the journey: In the torrential
winter rains, they said, it would take thirty teams of horses to get
him there. "I shall use thirty-one," he snorted—and promptly
moved into what would be the principal palace for all the governors-
general that followed him, Raffles included.

The house survives today, essentially unchanged: a vast, one-
storied gleaming white palace, its immense airy rooms currently
filled with pictures of well-endowed and slightly clad young
women, the pictures and their subjects both equally assiduously
collected, it is said, by Sukarno,* who brought modern Indonesia
to independence. The parklands around the *istana* are alive with
thousands of roe deer, imported by the Dutch a century ago to
provide affordable and acceptable meat for the official banquets
they so delighted in giving.

The modern world then came with an unexpected suddenness to
the Dutch-created world of colonial Java.

On May 24, 1844, Samuel Morse transmitted the famous bib-
lical message "What hath God wrought!" from the Supreme
Court in Washington to his colleague, Alfred Vail, forty miles away
in Baltimore. Twelve years later, over almost exactly the same dis-

---

*Sukarno's private study is dominated by an immense oil painting of happy-looking workers
toiling in a Russian soviet. It was the gift of a visiting Politburo chief, and an indication of
the clear left-wing leanings of the man who was eventually to be replaced by General
Suharto in an American-backed coup d'état, which led to the corruption and civil strife that
disfigure Indonesia still.

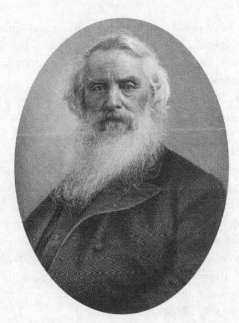

Samuel Morse.

tance, the electric telegraph—which was to play such a crucial role in the spreading of the story of Krakatoa—was formally introduced to the Indies, with the connection of the great palace at Buiten-zorg in the summer of 1856 to the colonial offices down in Batavia. From then on the pace of technological innovation in Java quickened. The island was connected internationally in 1859, acquiring an undersea line to Singapore (though this failed after a few days) and then, in 1870, links to both the Malay States and Australia. These proved to be totally stable: and so by the time of the Krakatoa eruption, the places where the explosion was seen and felt and heard and suffered were all connected, by the dots and dashes of Mr. Morse's code, to the world beyond.

The eruption of Krakatoa was, indeed, the first true catastrophe in the world to take place after the establishment of a worldwide network of telegraph cables—a network that allowed news of disaster to be flashed around the planet in double-quick time. The implications of this rapid and near-ubiquitous spread of information were profound—and to such an extent that they deserve a separate chapter, which follows later.

Houses in Batavia were connected to one another by telephone from 1882; gossip could spread, as well as news, warnings, and invitations. There had been a gasworks in the city, and the luxury of piped gas for cooking and street lighting, since 1862 (the present gasworks still has one ancient gaslit street lamp outside the office, which sputters to life from time to time). There was a tramway—drawn by horses from 1869 (with ten of the tiny Sumbawan tram ponies dying every week) and then by noisy, dirty, ever chuffing steam engines that ran from 1882 onward (the poor—meaning the Indonesians—were forced to ride in separate carriages from the Europeans). Railway trains were first brought to Batavia by the Dutch government in 1869, with the first line—up to Buitenzorg, naturally—completed in 1873.

And in 1870 that most potent symbol of European civility brought across to the Orient finally made it to Batavia: An iceworks was built. No longer did the genever-drinking Hollanders have to drum their fingers on the teakwood bars at the Harmonie or the Concordia Military Club, as they waited for the ice ships from Boston, the vessels of what the Americans called "the frozen-water trade," to appear on the horizon off Sunda Kelapa. Now they could have locally made ice delivered to their fretworked porte cochère doorsteps in Weltevreden every day, or sent to the *Indische* who worked behind the club bar. And they could be content in the knowledge that if the dinner party proved too boisterous or the numbers of guests in the club altogether too many, the host could simply send out for more ice and it would be there, available for delivery, fresh and dripping cold, every day of every following year. "Everything in Batavia," wrote a well-satisfied traveler named W. A. van Rees in 1881, "is spacious, airy and elegant."

Thus was Batavia quite comfortably arranged at the approach of the critical year of 1883. There was an air of calm contentment

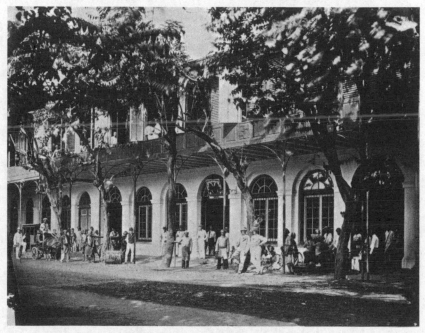

A Batavia city scene, around 1865.

in the land (a long-drawn-out guerrilla war between the Dutch and Islamic-led forces in Aceh in northern Sumatra was proving costly for the Batavian government, but was otherwise of no great moment to the citizenry). The traders in Batavia were doing well. Just as in Kansas City, everything here was up-to-date: Modern conveniences were being installed for the Europeans, and the huge numbers of *indigènes* flooding into the city each day were, if nothing else, testament to capital's ever growing wealth.

The planters out in the countryside were rich as well, but their contentment was tinged by the perennial caution of the farmers, dependent on so many more factors than their own efforts. The newspapers of the day display an increasing concern in the 1880s, for example, that disease and blight might well ruin the crops; and so the Java planters began a program of experiment and diversification—bringing in Robusta coffee plants from the Congo, for example, to ease the times when blight affected their Arabica or Liberia bushes.

Prospectors found tin, and began working it on the islands of Bangka and Billiton (with the brother of the then Dutch king, Willem III, the main promoter of the Billiton Tin Company). Wildcatters drilling shallow wells in Borneo found oil, though it would be some years before they thought it worth extracting, and many more years before the establishment of the huge combine that would one day become Royal Dutch Oil.

Plant importers, already delighted by the success of the cinchona trees that yielded as much as a seventh of their weight in pure quinine, brought in new rubber trees from Brazil (one of the ships carrying the precious seeds, bound for the Botanical Gardens in Buitenzorg, was passing through the Sunda Strait on the very day of the eruption).

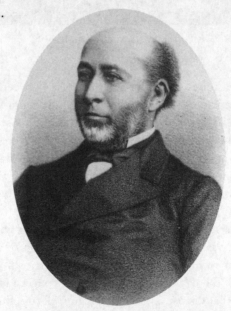

Frederik 's Jacob, governor-general of the East Indies, in 1880.

Back in Holland, the moody and clumsily undiplomatic King Willem III was on the throne. In November 1880 he made a surprise appointment: Frederik 's Jacob, a fifty-eight-year-old former sailor, mapmaker, and sugar manufacturer who for the previous ten years had been director-general of the Dutch Railways, was to go to Batavia as governor-general. He took up his appointment the following April and conducted himself during his first years in office with an unassuming efficiency.

There was a great deal of ceremonial about the tasks of a Dutch colonial governor-general—and it was not at all out of the ordinary to see Mr. 's Jacob in an official uniform that was adorned (the better both to impress the natives and to maintain the morale of the settlers) with yards of ornate gold brocade trim, silver and enamel stars, ribbons and garters of red, white, and blue, knee breeches and a tall felt hat with a feathered brim and a wild cockade. He wore it for the first official function of 1883, on February 19, the formal celebration of the official birthday of his faraway Dutch king.

Although the birthday party in Batavia was a fine and formal affair, marred by nothing more untoward than a general eagerness for strong drink, nearly ninety miles away in a straight line to the west of the palace something was stirring, deep within one of the menacingly dark and forest-clad pinnacles at the northern end of the Sunda Strait. The island with its pointed mountain, first noticed and named almost exactly three hundred years before, and which had been placid and beautiful for so very long, was now becoming restless again.

5

The Unchaining of the Gates of Hell

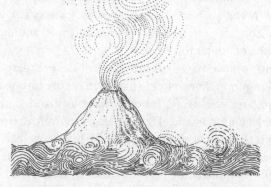

And I thought: The world is our relentless adversary, rarely
outwitted, never tiring.

—a Dutch pilot caught near Anjer, quoted in "Krakatau,"
a short story by the novelist Jim Shepard, in his collection
Batting Against Castro, 1996

The rains are usually terrible in Java in the early part of the
year, and the February of 1883 was no exception. There
were floods in the lower-lying parts of old Batavia, and
dozens of country roads were impassable. When the adjutants of
the garrison artillery regiments inspected the city parade grounds
on Waterlooplein on the afternoon of Sunday the eighteenth, they
found the glutinous red mud too deep for their cannon wheels. So
they sent a message, via the aide-de-camp, to Governor-General
s' Jacob up at his great white palace at Buitenzorg: The traditional
military parade that had been planned for the following morning
to offer loyal birthday greetings to his faraway imperial majesty
King Willem III could not, it was greatly regretted, go ahead.

Some might have seen the cancellation as an omen. It was in

any case not a happy time for the Dutch monarchy. The then head
of the House of Orange—whose official titles included King of the
Netherlands and Grand Duke of Luxembourg, and whose full and
magisterial-sounding name was Willem Alexander Paul Frederik
Lodewijk—belonged firmly to the glittering European aristocracy
of a *belle époque,* an aristocracy that was only a very few years away
from the corrosive effects of revolution and war, and neither wise
nor prescient enough to realize it.

Willem, whose sixty-sixth birthday fell on February 19, was a
less than inspiring monarch. He had been on the throne for thirty-
four years and was weary, moody, rigorously anti-Catholic, and
notoriously undiplomatic. He was also keenly frustrated by the
limitations that had been imposed on the powers he had known as
crown prince, which had been savagely cut back by the constitu-
tional reforms initiated by his own father, Willem II.

But out in the distant colonies of the East, such regal maunder-
ings were either unheard or ignored. Monday, February 19, King's
Day, was the culminating moment of a whole weekend of midwin-
ter festivities. The party started on the Saturday, in the Willem III
Grammar School, with a dance that the élite of the capital were
expected to attend. The governor-general came down by special
train from his mansion, made a speech welcoming everyone to the
three-day saturnalia, inaugurated the dancing with a waltz with his
wife, Leonie, admired the young girls' extravagantly full ball gowns
(waspishly noted by the columnist of the *Locomotive* as being fully
a year behind the fashions of Paris), took part in the traditional
eleven o'clock conga, and then took himself off home—leaving the
partygoers to dance themselves into the kind of frenzy that was only
permissible in a servant-rich colony with a long weekend of indo-
lence and celebration stretching ahead.

It was a busy time for the governor-general. On Sunday he had
church service to attend, alms to dispense, military parades to
bless—and then came the unwelcome news about the weather and
the consequent cancellation of the following day's cavalry tattoo.
And on the next day, Monday, s' Jacob had to get himself gussied

up in his finest raiment once again and travel down from Buitenzorg to the sweltering and now rain-soaked capital for the much
more formal (but paradeless) day of festivities. Every Dutch building was festooned with flags of red, white, and blue, and the ships
in the harbor flew pennants and signal bunting. In Waterlooplein,
with its palaces and military barracks,* thousands of soldiers—both
the regular officers from Holland and levies from the "loyal races"
throughout the islands—were arrayed in tidy ranks for inspection.

The king's representative in the Indies, who, when he deigned
to come down to hot and steamy Batavia, held court at his immense
and newly completed white marble Doric-columned palace on the
Koningsplein—the King's Square—staged an official morning audience that day. He ordered to be arrayed before him his entire council for the Indies, his senior civil servants, the generals, the bishops,
the foreign diplomats (including the British consul-general, a Mr.
Cameron, and his American counterpart, Oscar Hatfield) and the
favored elite of Batavian high society. He announced, as was customary, an amnesty for a slew of prisoners. He also told his sweltering audience that a dozen or so banned persons—the concept of
"banning," so much a feature of later apartheid rules in that other
former Dutch possession in southern Africa, was very much in force
in the East Indies at the time—were to be relieved of their ordeal,
as a birthday bounty from the representative of his majesty.

The rest of the happy day was then devoted to sports—including cricket, which had infected the East Indies after a visit from the
Singapore Cricket Club ten years before—or to parties at the two
principal clubs, the Harmonie and the Concordia Military, or to a
massive public display of fireworks. "The Governor-General gave a
gala dinner at the palace," reported the next day's *Javasche
Courant,* in the breathlessly sycophantic tones of the times. "The
palace was magnificently decorated and illuminated. All of the high

---

*There was a tall monument to the victory at Waterloo (Willem's father had been Dutch
commander in the field) surmounted by a very small lion—so small that it was regarded with
amused contempt by most Batavians, who thought it looked like a poodle.

authorities of the colony were invited, and they cheered with delight when the Governor-General raised his glass to the health of the King. Then there were wonderful fireworks on the Koningsplein, and thousands of people came out in the evening to enjoy the sight."

Whether the worse than usual rains and the attendant cancellations were seen as any kind of augury or not, all seemed otherwise perfectly normal in a colony that this day of celebrations had already demonstrated was run with strict and meticulous order. The colonized people were at this time in their history by and large content, the merchants were prosperous, the big Javanese cities—Batavia, Surabaya, Bandung, Jogjakarta, Solo, and Anjer among them—were generally peaceful and well mannered.

And yet, unknown to all but a few academics and seers, and to a few in those deeper recesses of Javanese and Sumatran *kratons* where the sultans lived and ruled, a multitude of strange and unrecognized forces were busying themselves—social, political, religious, and economic forces that would erupt across the colony within just a very short while. The seeds of impending troubles—of uprisings, insurgencies, and militancies that would in time mature to become generally anti-imperialist and specifically anti-Dutch—were just then starting to take root, were growing unseen, like mushrooms in the dark. We shall return to them later; for now they provide the faintest, barely audible *basso continuo* sounding beneath the more dramatic events that were to dominate the year.

Whatever might have then been going on beneath the outwardly content surface of East Indies society, it happened that at this very same time quite another array of unseen forces, which some historians would later link with the coming societal changes, were also gathering. These, though, were physical forces, forces related to the then unrecognized phenomena of tectonics, subductions, fault zones, and sea-floor spreading, forces that were regrouping after years of quiescence and that—just as invisible as the coming changes in politics and social life—had begun to prepare for their reawakening.

They were at work particularly beneath the surface of western Java, readying themselves for a six-month period of violent and nightmarish activity. They would first become apparent, and violently so, just a scant ninety days after the smoke, flames, and thunder of the last fireworks of his majesty's birthday party had died away.

It began with a sudden trembling. At first it was slight, more of a quivering of the air, a series of windy rumblings, of vague, barely noticeable atmospheric flutterings. In normal times these would probably have passed without comment, other than by some Dutch planter in a bar who might point with amusement to the surface of his genever and get others to peer at how shaken and rippled it was. Seismic events, where the rocks themselves moved, were a commonplace throughout the islands. Earthquakes and eruptions seemed as everyday as thunderstorms and plagues of mosquitoes.

But this one was a little different. For a start, the year had already been curiously quiet across the entire colony. Between January and May the observatory at Batavia had logged only fourteen earthquakes, and, of these, four were in eastern Java and seven in Sumatra. The year was calm; people were lulled into a kind of easy complacency.

And western Java was in any case a quietish corner of the archipelago, seismically speaking. Everyone had heard the stories about ancient eruptions, true; and there were those who looked at maps and thought they had heard tell of when Java and Sumatra were one island that had broken apart during some terrific volcanic event ages ago. But most believed that Krakatoa was long since extinct, inactive, peaceful, and, most likely, dead. The Batavians who came out from town to swim on the clean white western beaches; the local bird sellers who hawked pretty Anjer swallows to passengers on ships setting off for the long sea passage back to Europe;* the bumboat skippers who sold fruit, coral, and curios to seaborne passersby: All thought of this corner of their islands as endlessly tranquil, a place far removed from the more violent activity that was so common near mountains like Bromo, Merapi, or Merbapu.

But then the vibrations began. It was just after midnight, early on the morning of Thursday, May 10, when the lighthouse keeper at what was then called First Point—the more southerly of a pair of lights on the enormous rocky headland at the southeastern entrance to the Sunda Strait, known to approaching mariners as Java Head—felt what he knew only too well was a tremor in the air. The lighthouse suddenly seemed to shift on its foundations. The sea outside whitened, appeared to freeze briefly (as we now know it does above a depth charge), became uncannily smooth like a mirror, shivered slightly, and then returned to its usual sickly swell.

It was really nothing much. There was no suggestion of where the vibration might have come from. The keeper checked his records: The last volcano to erupt anywhere nearby was Lamongan, and that lay six hundred miles to the east. Whatever had prompted this particular rumbling was probably closer than that. Though it caused no obvious damage, it was strong enough, unusual enough in pattern, and curious enough in location, for the lighthouse keeper to log it, and to note in a report that he wrote during the following weekend and sent off to Batavia with his other weekly summaries.

Five days after the initial vibration the same happened again, except that this time it was stronger and more sustained, more widely felt. What had been felt in west Java was now felt on the other side of the Sunda Strait, in Sumatra. The Dutch *contrôleur* in

---

*The birds, reported the *Illustrated London News* of the day, invariably died as soon as the ship encountered chilly weather: Very few made it all the way to Europe.

the south Sumatran town of Ketimbang, Willem Beyerinck, was sufficiently roused by the thudding, rumbling bangs occurring beneath his feet on the night of May 15 to send a telegram, an official confidential message presenting all the facts to his superior, the resident of Lampong. It took him five further days to pluck up the courage to send it: He had to make sure he was right. In the end he made the call: Powerful tremors, he reported, were now being felt continuously up and down the Sumatran coast on the northern and western side of the strait.

This was the first official word that something untoward was in the offing. And coming from a civil servant with the rank of controller, the news had considerable weight. Controllers in the colonial service of Holland were not a breed of men given to panic.*

The ships were the next to take note. The Sunda Strait—seventeen miles across at its narrowest—was then, as today, a frantically busy waterway. The number of vessels passing between Java and Sumatra on their way to and from Europe and the Americas on the one hand, and China and the more distant reaches of the East on the other, was prodigious in the 1880s. There were at least ten ships in the vicinity when Krakatoa's first eruption began: Those definitely known to be nearby included the American brig *A. R. Thomas;* the British barque *Actaea,* commanded by a Captain Walker; the Dutch mail packet *Zeeland,* commanded by a Captain MacKenzie, on her way from Batavia to the Indian Ocean and then by long sea to the Netherlands; the *Sunda,* a steam ferryboat skipping her way from Batavia to a series of local ports; the *Archer,*

---

*A *contrôleur* was one of the more junior grades in the Dutch colonial service, presiding over a subdivision of a residency known as an *afdeling*, or department; but junior or no, a candidate had to spend four years at the College of Delft and pass with honors a rigorous examination that included the Javanese and Malay languages (both very similar, to be sure); French, German, and English language and literature; Islamic law, algebra, geometry, trigonometry, geology (no doubt helpful for his present task), drawing, land surveying and leveling, as well as a host of other easier disciplines including, for some less explicable reason, the subtle mysteries of "Italian book-keeping." An official warning of troublesome seismic activity sent by a *contrôleur* would certainly cause senior members of the colonial administration to sit up and take notice.

an Australian passenger steamship of the Queensland Royal Mail Line; the *Conrad,* a Dutch mailboat heading northwards from Europe to Batavia; a Dutch barque, the *Haag,* under the command of a Captain Ross; the German warship *Elisabeth,* heading south from Singapore; and, more mundane than romantic, the hoppers *Samarang* and *Bintaing,* both shuttling between Java and Sumatra, performing harbor drudgery on behalf of the Batavia Port Authority.

Each had a story to tell. The first to log something unusual was the *Elisabeth,* a German corvette returning home after two years in China and Japan, time spent doing picket duty on the Imperial Navy's Far Eastern Station.* The ship had called briefly at Singapore and then, though it bypassed Batavia, had stopped instead at Anjer to coal, take on water, and drop off a single passenger. The *Elisabeth* left the tiny port's quayside on the morning of Sunday, May 20. She turned toward the south, trimmed her sails, and set a course that would take her down the Sunda Strait and out into the open ocean.

Her master, a Captain Hollmann, then became the first European to see the very beginning of the eruption of the mountain. He was the first to write a report about what now seemed to be behind all the rumbling and trembling that had been noticed around the region since the lighthouse keeper's summary of the curious vibrations that he had experienced earlier in the month. It was 10:30 on the hot, cloudless summer morning. Captain Hollmann was looking from his bridge directly across to starboard toward the 2,625-foot southern summit of Krakatoa,* when sud-

---

*German commercial and military interests in the Far East were growing fast in the late 1880s, not least because the Manchus in Peking's Forbidden City, vastly impressed by the Prussian-led strengthening of the Reich after 1871, had asked the Germans to help modernize their own armed forces. Matters were soon to turn sour, however; and in 1898 Germany annexed for its own naval use—for vessels like the *Elisabeth*—the port of Tsingtao on the Shandong Peninsula. The influence of the next fifteen years of German rule lingers still: German architecture remains highly visible in the town today, and Tsingtao beer, one of China's better-known exports, was for years prepared under the supervision of a Bavarian *Braumeister.*

denly something took place that he never imagined possible. Without warning:

> . . . we saw from the island a white cumulus cloud, rising fast. It rose almost vertically until, after about half an hour, it had reached a height of about 11,000 meters. Here it started to spread like an umbrella, probably because it had reached the height of the anti-trade winds, so that soon only a small part of blue sky was seen on the horizon. When at about 4.00 in the afternoon a light SSE breeze started, it brought a fine ash dust which increased strongly . . . until the entire ship was covered in all parts with a uniform fine grey dust layer.

Faced with the astonishing sight of a white cloud that was streaking up into the heavens, reaching to what his navigating officer calculated was fully seven miles into the clear blue sky, the ship's marine chaplain, Father Heims, allowed himself a little more latitude than his superior in the imperial navy, writing:

> . . . the crew had assembled on the upper deck in clean Sunday clothes, to be mustered in divisions. The commander had just looked at the parading crew and started to inspect his pretty clean ship, when a certain motion was noticed among the officers which were assembled on the upper deck and bridge in their Sunday clothes. Glasses and heads all turned towards the lonely countryside in which the shores of Sumatra and Java coincided with the small island of Krakatau: there, at least 17

---

*This summit, Rakata, lay at the southern end of Krakatoa and, since it was by far the highest point, was the most visible feature of the thickly forested, lozenge-shaped island, which measured about six miles along its north-south axis, and was, at its widest point in the south, some two miles across. The other high points, apparent only on a close inspection, were Danan, a cratered peak in the middle of the island rising to 1,496 feet, and Perboewatan, on the narrow northern end of the island, which was 399 feet high. Two other islands are associated—Lang Island, lying two miles to the northeast, Verlaten a similar distance off to the west. Both, as suggested before, are probably relics of an even greater super-Krakatoa of earlier times.

nautical miles distant, an enormous shining wide vapour column rose extremely rapidly to half the horizon, and reached
within a short time the colossal height not below 11,000
meters, contrasting in its light-coloured snow-like appearance
with the clear blue sky. It was convoluted like a giant wide coral
stock.

At this point the right reverend's prose turns rather more purple
than he probably would on reflection have wished. He compares
Krakatoa's rising plume of steam and smoke first to a giant cauli-
flower, then to a billy club, next to "the convoluting steam column
from the smoke stack of a gigantic standing steam locomotive"
and finally to an odd confection that he christens "three-
dimensional steam balls."

Mercifully, after only a few lines of such description, he aban-
dons his quest for literary permanence and returns to writing for his
parishioners back home, who were no doubt eagerly awaiting his
reappearance in the pulpit. In doing so, he provides some highly
useful evidence for those who would later study this first phase of
the eruption:

> We did not hear any detonations. The veil over the sky was so
> dense and uniform that the almost full moon was only barely
> visible during the night . . . on the next morning . . . the ship,
> which was so clean 24 hours ago, looked very strange: it looked
> like a mill ship or, more precisely, like a floating cement factory.
> On the outside everything—ship's wall, torpedo pipes, the
> entire masts, etc.—was covered uniformly with a grey sticky
> dust . . . it had accumulated thick and heavy on the sails; the
> steps of the crew sounded muffled. . . . The people enjoyed
> collecting the lava dust as polishing material, and it was not
> very heavy work to collect the stuff in sacks and boxes.
>
> The sky above this ash rain disaster appeared like a large bell
> made of rather dull milky glass in which the sun hung like a
> light blue lamp . . . for another 75 German miles we had to sit

in the evening with our faces looking backwards as we sat together trying to get some air. The distribution of the ash-fall would be over an area at least as large as Germany.

One by one the other ships in the neighborhood reported their news. Some of the reports were made public within days or weeks of the event; in later years the logs were found and published, or private letters surfaced from their amazed commanders or crew, as well as messages from passengers who had been aboard and knew that they had seen something strange and wanted, earnestly, to tell of it.

The British ship *Actaea*, for instance, which was sailing eighty miles west of Krakatoa, noticed a "peculiar green colour" in the morning skies to the east-southeast; by midafternoon her sails and rigging were covered in fine ash and dust; and when the sun set it did so as "a silver ball."

The hopper *Samarang*, en route to the port of Merak, felt a sudden swell, massive enough to lift her and her screw clear out of the water.

The *Zeeland*, sailing with her full complement of passengers and mail back toward Holland, passed within five miles of Krakatoa.* Her compass needle suddenly began to spin around and around, uselessly. When it settled it showed a deviation from normal of twelve degrees. Steam and debris then roared up from the most northerly of the three cones that could be seen on the island, and to the crew a deafening noise began to sound, seeming to combine the thunder of heavy artillery and rattle of continuous machine-gun fire.

---

*To get into the Sunda Strait the *Zeeland*, like all other vessels passing along this great waterway connecting the Indian Ocean and the South China Sea, had perforce to skirt around another—noneruptive—island that seemed almost to block the northern entrance to the channel. The island is known today as Pulau Sangiang. In the nineteenth century, however, at a time when so many of the coastal features sported English names—Pepper Bay, Welcome Bay, First Point, Java Head, and Polish Hat are all along the Sunda Strait—this navigational nuisance of rock and forest was officially called, on the Admiralty charts, Thwart-the-Way Island.

Captain MacKenzie next saw to his astonishment a huge column of black cloud. Not white—despite what others had said he was sure it was, at this moment, black. It rose swiftly above the mountain, with lightning flashes deep within the clouds, and a continuous crackling sound. The sea all around the island was punctured by immense gray waterspouts rising into the air; it became so dark he had to reduce speed to five knots. There being no radio at the time, he frantically raised a string of colored signal flags, alerting all who might be watching him. The Lloyd's agent at Anjer certainly was, for he noted in his log that MacKenzie was fast *standing into danger*, and others who came near him might suffer the same fate.

And then came the moment when a Delft dinner plate fell off a dining-room table in the old part of Batavia and broke into a thousand pieces.

The plate had belonged to Mrs. van der Stok, a middle-aged Dutch lady who at the time of its breakage—shortly after ten minutes to eleven on the Sunday morning—was quite probably setting her table for family luncheon. It had been a part of her trousseau on the day she had married Dr. J. P. van der Stok—the distinguished scientist from Utrecht who had brought her out to Batavia some years before on his appointment as director of the colony's Magnetic and Meteorological Observatory. The couple lived in a single-story house attached to the observatory, and on that hot and cloudless Sunday morning both could not help but notice that something, somewhere, had gone badly awry.

First there was the plate, lying smashed on the marble floor of the dining room. Then in the living room Dr. van der Stok himself, reading the Sunday edition of the Java *Bode*—the word means "messenger"—heard all the windows and doors rattle and bang. From somewhere to the west came a low, rumbling sound, like that of distant artillery. He got out his pocket watch and noted the time: 10:55 A.M.

He walked across to the observatory, noticing immediately that the needles and pens suspended by their cocoon-threads on his magnetic declinometer were ticking and trembling violently—not in the usual side-to-side sweeps that one might expect from an earthquake, but in a series of buzzing up-and-down motions that did not register properly on the paper rolling from the drum. The more he thought about it, the more he realized something odd: The vibrations were not so much being felt through his feet, as if they had emanated from somewhere deep in the earth; they were in fact being felt *in the air*. True, there were ground tremors, and buildings were shaking—this was self-evident. But most of the shaking was coming through the very atmosphere itself. And vibration of this kind was the very particular hallmark of an erupting volcano, not of the subterranean shaking of an earthquake.

Later he would see more commonplace left-to-right activity on his declinometer—but only when the ash had begun to drift down over the city that evening. Then he surmised quite rightly that the magnets were going wild simply because the falling ash was rich in iron, like a blizzard of tiny compasses.

For now, though, it was all vibration and rumbling and the occasional period of low and menacing thuds. Van der Stok knelt down and put his ear to the ground. Nothing. Nothing deep-seated. And what was more, these vibrations were continuing for long, long periods—an hour already, with no signs of abating. Earthquake vibrations last for matters of seconds only, a few minutes at the most—followed by periods of quiet, and then aftershocks, and then more movement and mayhem. This was very different.

And all of this made him now, at noon on that Sunday, quite certain that he knew what was happening. These particular kinds of vibrations were absolutely typical of those caused by a volcano. And so each time a fresh explosion of trembling started again, he noted the time in his official logbook. Had it been an earthquake, there would be less need to catalog the aftershocks, the timings of which were in any case mathematically predictable. But this was evidence of a volcano, somewhere, and the measuring of its palpitations might give some clue as to its future behavior.

All the while van der Stok was kept busy fending off inquiries from worried Batavians—men and women who flooded to his observatory, even on this sunny Sunday, wanting to know *just what was happening*. This, most of them said, was like nothing they had ever known before. There had been a curious trajectory about each person's morning on that day. They had awakened to the unusual sounds, and they had been then merely puzzled. By the time they breakfasted, they had become concerned. The Christians among them had gone off to their churches, feeling moderately alarmed. After matins they had ventured back out onto the streets, by now in their droves, and they were, at least privately, at first quite agitated and, as the thunder wore on, very apprehensive indeed.

The director noted with the dispassion of a trained observer a distinct racial difference in public attitudes to the events. The Dutch who called in to see him appeared to be outwardly calm, the men displaying either the equanimity of their long experience in the Orient, or the public stoicism, the stiff upper lip, the code of *pas devant,* that they felt appropriate to their standing as proudly indifferent colonials. The native Javanese, on the other hand, seemed to be much more deeply concerned, with many frightened and, in more than a few cases, quite terror-struck.

To the Javanese and Sumatrans, and especially to the coastal people called the Sundanese—who were and still are a very distinct group well known for their mystical beliefs and deep piety—an event like this out in the Sunda Strait was heavy with suggestive power. In their view the eruptions showed that spirit of the mountain, the widely feared Orang Alijeh, had for some reason been made angry, was now roaming abroad, breathing fire and smoke and displaying his displeasure. Such a happening could only bode ill for all. Most of the colonials officially and instinctively disapproved of the natives' superstitions, and van der Stok himself was icily dismissive. And yet some of the wiser old colonials did wonder about such things—at least, they would wonder after the terror was fully over, many months hence: *What had it really meant?*

For the moment, though, what was of more immediate concern

was the need for a widespread battening down of the hatches, with the inhabitants of western Java and southern Sumatra waiting to see just what nature—or the gods—had in mind to hurl down at them. For the next two days they had a great deal to keep them anxious.

The little south Sumatran town of Ketimbang is at the best of times a dreadfully vulnerable place: not only is it on a funnel-shaped bay, where the spring tides can rush in most dangerously; not only is it sited on mangrove swamps and mud-flats, ready to be inundated by every rising of the waters; but it also lies directly beneath a small volcano called Raja Basa—the mountain's flanks rise steeply to more than four thousand feet immediately behind the cluster of coastal houses and their tiny fishing port. Willem Beyerinck, the colonial controller who had been one of the first to note the ominous initial rumblings five days before, was in time to be tested harshly by what happened. His wife maintained a close watch on the entire affair, from beginning to end, and kept a detailed journal. In May, or so it seemed from her detached and insouciant tone, the early stages of the eruption appeared to her more inconvenient than truly alarming.

When it all began on the Sunday morning she had been taking the air on her veranda, idly watching the ships that passed by up and down the ever crowded strait. She would derive hours of pleasure from watching the long-distance boats fresh out from Batavia, their sails all slipped and bellying in the breeze as the craft began scudding their way toward Europe. There was always a great deal of shipping to see: The view from the controller's elegant little house was magnificent.

But then, without warning, she was jolted from her reverie: Another hammer blow, another set of violent tremors started up once more. "We were much bothered," she wrote. The vibrations were best seen in the water barrels kept stored in the bathroom, she said, since their surfaces rippled prettily with every detonation. She took up her diary and began to make notes of yet another subterranean interruption.

She was writing all this in her journal when, quite unannounced, a *prahu* arrived and was hurriedly slithered up on to the mud, its long bamboo outrigger propping it to one side. Eight frightened-looking fishermen leaped from it, ran up the beach, and made straight for Beyerinck's office. They jabbered excitedly, in a mixture of Sundanese and pidgin Dutch. They were from the island of Sebesi, they said, and they had all been together that morning—on Krakatoa.

They had gone there to gather wood for boat building. They had been felling trees and stacking cords, chatting contentedly among themselves, when suddenly they heard what they assumed was a burst of cannon fire from a warship. It was probably, they thought, a Dutch man-o'-war exercising out in the Strait. At first they paid little attention, and carried on felling trees—until there was a second, terrifyingly loud report, which made them dash down to the beach to see what was going on.

As soon as they got there, they said, they saw the very beach itself split wide open, and jets of black ash and red-hot stones roar out into the air. They fled in terror, running for safety and then diving into the sea to swim out to where they had left the boat. The tide had risen dramatically, they said: Just an hour beforehand they had been able to wade to the mooring place.

The controller's wife, by all accounts a skeptical and somewhat hard-bitten lady, was in no mood to listen to these excitable natives. She told her husband, acidly, that it was simply impossible for *a beach to erupt*. He was ready to agree, to shoo the fishermen away. But then Beyerinck's superior suddenly arrived from his own headquarters farther up Lampong Bay, in the little port city of Telok Betong.

He was a Mr. Altheer, he was then just a month away from the end of his five-year posting as resident of Lampong, and he was keenly eager to do the right thing and leave with his reputation in good standing. He had just been telegraphed by the governor-general, he told his junior; he had been ordered to investigate a situation that could be heard and was now fast alarming the entire

Batavian citizenry. Whatever the risks, he and Beyerinck must leave for Krakatoa, immediately.

And so the two men jumped into the official government launch that had brought the resident down from Telok Betong, and they bumped out into the rough waters of the bay, turned due south, and sped through what turned out to be, quite bizarrely, wave upon wave of floating pumice stone. They headed past the two islands of Sebuku and Sebesi, which hid Krakatoa from their view, dodged masses of pumice and charred and floating trees, were drenched by massive sudden waves, enveloped in clouds of choking gas and a miasma of falling ash. It is twenty-four miles from Ketimbang to Krakatoa: It took the men four hours to bring the coastline of the island into some kind of view.

And though we do not know from their records whether the pair actually landed, we do know that they spotted exactly what it was that had so alarmed the fishermen before them: The north-ernmost beach of Krakatoa was indeed belching fire and smoke, and the smallest and most northerly of the three cones of the island, Perboewatan, was in the process of erupting, the roaring and belching noise of its concussions getting stronger with every minute.

The two officials turned tail, their worries for their own safety finally taking precedence over Mr. Altheer's concern for his colonial career. They sped back through an ever more dense coagulation of hot gray pumice, making their way through the fast-falling tropical night to the coast and the Ketimbang telegraph station. Shortly before midnight they sent a hurriedly composed message in Morse code, marked for the eyes of the governor-general only.

Within minutes came a reply: The resident of Bantam, on the western side of Java across the strait, had reported seeing essentially the same thing: flames, belches of fire, rafts of floating rock, ash falls. And he had heard it too: the noise of what was undoubtedly an eruption, memorably terrible, and sounding to him like the crashing, screeching roar of a great ship's anchor chain that was being endlessly raised, its shackles banging hard and rustily against the vessel's sides.

More evidence had meanwhile been streaming in from all quarters. The northbound sailing vessel *Conrad,* enveloped in an inferno of choking ash and dust in which the ambient temperature was at least ten degrees higher than it had been elsewhere, had battled for well over five hours to pass through a near-impenetrable field of floating pumice, its gray masses packed tightly, like ice floes. Forests on the flanks of Krakatoa were seen to be ablaze. The doctor on board the southbound *Sunda* saw bright-red clusters of fire, "like sheaves of wheat," bursting from the column of smoke that poured from Perboewatan. He saw a new crater on the island's western side spurt a torrent of dark-red fire. Later that evening, when the *Sunda* was thirty miles away from the eruption and almost in the open ocean, the doctor asked a sailor to drop a bucket into the sea, and the man pulled back up only pumice, with barely any water at all.

Peculiar, ominous sounds had been heard up in Singapore, more than five hundred miles to the north. An English plant-collector named Forbes, working more than thirteen hundred miles away in Timor, reported a sprinkling of ash around his grass hut. There were some fanciful reports as well: Someone reported that all the chronometers in the government observatory at Surabaya in east Java, five hundred miles away, had mysteriously jumped forward, and that the time ball by which ships in harbor might learn the hour had somehow stuck on its shaft. These later turned out to be nonsense.

On the other hand the usually circumspect Lloyd's agent—a man who in Anjer owned a small *pension,* the Anjer Hotel, down by the docks*—turned in what sounded like an entirely responsi-

---

*The agent was named Mr. Schuit. But there is ample opportunity for confusion ahead, for in Anjer at the same time there happened to be a lighthouse keeper also called Mr. Schuit, an unrelated widow named Mrs. Schuit, and a newly appointed telegraph master who was called Mr. Schruit. Since all played major roles in the August cataclysm, it is as well to be forewarned.

ble report. He scribbled a hasty telegram to his head office in Batavia, for eventual transmission to the insurance exchange in London, with the first impression of what he was seeing: "Krakatan [*sic*] casting forth fire, smoke and ash, accompanied by explosions and distant rumblings." Only later in the day was he able to send a rather more discursive account.

So now, more than a week after the initial feeble quiverings had first been felt at the lighthouse, the cause and the source of it all had been seen, and in action. It was now well beyond any doubt that the long-dormant (but now apparently not extinct) island of Krakatoa was in an entirely new phase of its geological development, active once again and fast starting to erupt. What would happen next, the scale of the catastrophe that would envelop the region in a little more than ten weeks' time, would positively beggar belief.

And yet two days later, after its alarming opening salvo, the island quieted down again. A thin plume of white smoke and steam still rose above the Perboewatan crater, hinting that something was continuing to roil beneath the surface. But outwardly, all looked tranquil once more. The low, triple-cratered island and its neighbors slumbered hotly, surrounded by a calm and deep-blue sea; and when viewed from the ports of west Java, they again became near-invisible, compared to the hazily violet and distantly looming silhouettes of the truly enormous volcanoes of Sumatra.

The two peaceful days went past, then three. After a fourth the governor-general decided that if all was really quiet, it might now be prudent to go and take a closer look at Krakatoa, both to see what had already transpired and, more importantly, to see if such an event looked likely to take place again.

The first-ever government inspector to visit the island had done so just three years before. He was a mining engineer from Doorn named Rogier Diederik Marius Verbeek, and he would make his name in later years with his monumental, 546-page study of the great eruption of '83. But in July 1880, when he first set foot on the "geologically completely unknown terrain" of the island, he was himself also quite unknown, a specialist only in the coal mines of east Borneo. He found himself near Krakatoa while on temporary attachment to Holland's splendidly named Imperial Beacons & Coastal Lighting Service, which had a small ship named the *Egeron* bound for an inspection of a lighthouse on a clifftop with the rather less regal name of Flat Corner. "On my return trip to Batavia I was able to pay a short visit* to the islands in the Sunda Strait," of which Krakatoa, he said, was by far the most interesting.

He sketched the four islands of the group; he took a small boat to the northern end, close to the soon-to-be-notorious four-hundred-foot peak of Perboewatan; he chipped away with his hammer at what was evidently a recently made lava flow; he took samples of what he later decided was a rather unusual dark andesitic obsidian—a glassy,† evidently very rapidly melted-and-cooled rock that in this case had, most interestingly, a highly acidic character.

Its composition was in fact much more than simply interesting: It was highly suggestive of its having come from a melt of the half-oceanic, half-continental mix of materials that had been made deep inside what geology now knows to be a typical subduction zone. But Verbeek could not possibly have known this, nor could he have made any but the most primitive surmises about the rock's observed and rather curious acidity. The very concept of a subduction zone was wholly unknown back in the nineteenth century; along with the mysteries of sea-floor spreading and continental drift, these zones were only to be understood almost a century later.

And in any event Verbeek was not going to be allowed to make any further observations, since the crew of the *Egeron*'s pinnace were champing at the bit, eager to steam on home, and they kept blowing their siren to bid him stop hammering. He was frustrated

---

*Which he later said was for no more than "a couple of hours."

†The Ancient Romans used shards of Vesuvius obsidian as razors, since the rock had the property of fracturing with extremely sharp edges.

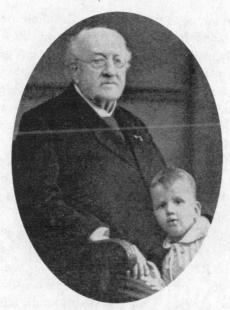

Rogier Verbeek.

too: it turned out to be far too difficult for him to pass along the Krakatoa shore through the dense jungle that stretched clear down to the sea. "There was no time," he admitted lamely, "to collect any rock samples from the southern part . . . that is, of the peak." But, he added drily, "little did I think that the places where I hammered rocks would disappear altogether three years later."

Three years later Dr. Verbeek very nearly missed what would be for him the culminating event of his geological career. He had left for Utrecht, to supervise the making of a geological map of south-western Sumatra, and only by the greatest good fortune did he come back to Java on leave in the summer of 1883. He actually completely missed the first part of the eruption, even so—his ship steamed past the fire-torn and half-ruined island in July, six weeks after these first eruptions had begun. His absence forced Governor-General s' Jacob to choose in his place one of his deputies, an obscure mining engineer named A. L. Schuurman, to make the hazardous first journey across to Krakatoa. His mission was simple: to see what might be seen, and to make an official report on whether anything devastating was likely to happen again.

There was no problem in finding a suitable boat for Mr. Schuurman. A combination of a widespread popular fascination with what had taken place and the eagerness of local shipowners to satisfy that fascination—the force of the market, in other words—were to supply him with exactly what he needed.

The British chaplain in Batavia, the Reverend Philip Neale, was to write later that since "the spectacle was regarded by the Dutch as a curiosity . . . an agreeable excursion was made to the island by one of the mail steamers trading in the Java Sea." The Netherlands Indies Steamship Company* was the first to recognize the tourist potential of the event, and came up in short order with an excursion vessel, the 1,239-ton *Gouverneur-Generaal Loudon*. On Saturday, May 26, representatives from the company tacked up notices in the Harmonie and Concordia Clubs advertising the delights of such an "agreeable excursion" and announcing a competitive price of only twenty-five guilders. By Sunday morning they had closed their lists, such was the surge of interest—and on Sunday evening the party set off. It was seventeen days after the first vibration, only a week after the first eruption. The *Loudon* was filled to its capacity of eighty-six passengers, and the government's Mr. Schuurman was very much among them.

After steaming through the night toward a "purple, fiery glow" that could be seen in the middle of Sunda Strait, the passengers watched as dawn finally broke. As Schuurman wrote in his report:

---

*The timetables of the various shipping lines that served the region make for delightful reading. The Netherlands Royal Mail Line, founded in 1870, "maintains a regular fortnightly service between Europe and Java, leaving Amsterdam every alternate Saturday, calling at Southampton, Lisbon, Genoa, Port-Saïd, Suez, Colombo (occasionally), Sabang (Sumatra), Singapore and the beautiful island of Java as terminus. (Batavia, Samarang, Sourabaya, etc.) The best-equipped and most comfortable liners of today. Excellent cuisine. They carry the Royal Netherlands and Royal Italian mails to the Far East." The single fare from Southampton's Extension Pier (leaving every other Tuesday at a time that connected with the London boat train from Waterloo) was sixty-five pounds.

The view of the island was fantastic: it was bare and dry, instead of rich with tropical forests, and smoke rose from it like smoke coming from ovens. Only the high peak [of Rakata] had some green left, but the flat northern slope [of Perboewatan] was covered with a dark gray ash layer, here and there showing a few bare tree stumps as meager relics of the impenetrable forests which not too long ago covered the island. Horrible was the view of that somber and empty landscape, which portrayed itself as a picture of total destruction rising from the sea, and from which, with incredible beauty and thundering power, rose a column of smoke. The cloud was only several dozens of meters wide at its foot, wheeling to a height of 1,000 to 1,200 meters while widening, then rising from there to 2,000 to 3,000 meters in height and in the meantime fading in color, delivering its ash to the eastern wind which, falling as a dark fog, formed the background of the tableau.

The *Loudon*'s captain, T. H. Lindeman, kept well away from the island. But he loaned Schuurman a small boat, in which the engineer and a small party of curious daredevils approached the northern end of Krakatoa. The beach was covered with pumice; they struggled onshore through ash, into which they sank up to their calves.

Following the tracks of the most courageous, or perhaps the most stupid, we climbed inland with no further obstacle than the ashes which gave under our feet, the route being over a hill from where we could see, emerging from the ash, some broken tree trunks showing signs that their branches had been violently stripped off. The wood was dried, but nothing indicated that it had been alight, or smouldering. No leaf or branch could be found in the ash, and it is therefore likely that the deforestation must be attributed to a whirlwind.

The foolhardiness of the explorers knew no bounds. They climbed the crater, knowing how dangerous and unpredictable it had to be, and stared down in amazement into the deep, dish-shaped basin. Its bottom, Schuurman noted, was covered with a "dull, shiny crust," which occasionally emitted a rosy glow, through which a powerful column of smoke escaped with what he then admitted was a truly frightening noise.

> The clouds of smoke appeared to break through as with diffi-culty but with unmatched force, and they seemed to flee in numerous but closely linked, tremendous bubbles whose inter-nal friction caused the turning and convoluting movement of the clouds in the lower part of the column of smoke . . . only at the edges of the point of eruption could the exhalation of steam from a number of cracks and gaps be observed.

The men stayed for most of the day, burning the soles of their shoes, coughing and spluttering in the clouds of ash, occasionally darting for shelter when the crater burped out a greater than usual bubble of smoke and sulfurous gas. And then, just after six o'clock, the tropical darkness began to fall (as Krakatoa is only six degrees south of the equator), and Captain Lindeman sounded the *Loudon*'s steam horn to urge everyone to get off the island. One passenger, a Mr. Hamburg, stayed a few moments longer to take photographs. Then everyone pulled out. "We started our return trip to Batavia at 8 o'clock in the evening," Schuurman noted at the end of his official report, "thankful for the beauty and for a specta-cle which made a deep impression on all, and an unforgettable one on most."

For the next eight weeks, all seemed quiet—so quiet that, even though technically speaking the eruption was still going on, with smoke screeching from the Perboewatan crater and ash blowing high into the skies, "visitors to Batavia, unless they had made inquiries, might have failed to hear of its existence at all." The geologist H. O. Forbes, pleading for information about this unknown period of Krakata's life, added that many of the ships' reports that had come in from this time seemed to have been written "either with the mind bewildered and confused by the terrifying incidents amid which the officers found themselves, or from the after-recollection of the events, of which under such circumstances the important dry facts of time, place and succession are liable to be unconsciously misstated."

One seemingly reliable report that did come to light—though half a century later—was from a young Liverpool seaman named R. J. Dalby, who in June was aboard the barque *Hope,* six months out from South Wales bound for Saigon. While his vessel called in at Anjer for telegraphed orders—this was in the day before ships' radios, of course—Dalby was given shore leave, and he took a canoe across the strait. The view on all sides, he remembered for a radio audience in 1937, was

> . . . a real paradise, a profusion of vegetation rising from the seashore to the summit of hills several thousand feet high. I well remember one particular evening, just at the time when the land and sea breezes were at rest, the very atmosphere impressed one with a mystical awe. It was enhanced by the subtle scent of the spice trees, so plentiful on the island and, to crown it all, the sweet yet weird and melancholy chant of some natives, paddling their canoe close in to the dark shore. There were three of us in the boat, and we rested a long time trying to take in the strange grandeur of our surroundings; it was at this time that we noticed a long straight column of black smoke, going up from the peak of Krakatoa Island.

Was Dalby's recollection new evidence, perhaps, that by June the high peak of Rakata had now joined Perboewatan in erupting? Certainly a second crater had opened up later on that month—after a stiff wind had died down on June 24, people on the Javan coast could see quite clearly that two separate columns of smoke were rising, and that the most northerly of the two was rising majestically. The controller of Ketimbang, the doughty Mr. Beyerinck who had first paddled out to the island in May, went back again in July and found two craters—but the more northerly was not on Rakata, but at the foot of the insignificant peak in the island's center, Danan.

Dr Verbeek himself then saw Krakatoa, on July 3, as he passed by on his way back to Batavia from Europe. One must suppose he knew nothing about what had just taken place: while Krakatoa had been busily erupting, he had been sunning himself, radioless, somewhere between Gibraltar and Suez, aboard the eastbound packet vessel *Prinses Marie*. The last time he had seen the island was in 1880: Now, in the dark of the small hours,* he could see very little more than a vague red glimmer on the vessel's port side.

And finally, on August 11, a Dutch army captain named H. J. G. Ferzenaar, who had been ordered to prepare a survey of the island for the military topographic service, landed and spent two days there. He went alone—the local governor ("unable to keep his promise") refused to go, and all the other officials he approached were too timid.

Ferzenaar found a bewildering variety of signs that the island might be readying itself anew for something quite spectacular. There were now at least three craters erupting—one, which he thought looked especially potent, was on the southern side of the midisland peak, Danan. All told, he counted *fourteen* vents in the rocky surface—fumaroles, one would call them now—from which grayish or pink smoke was rising. Most of these vents were also on this highly unstable-looking southern flank of Danan.

He paddled his *prahu* around the eastern coast of Krakatoa, turned around the northern headland, passed on the outer side of the small sliver of an island on the northwest side—and then called it a day. Heavy smoke made visibility difficult; navigation, especially in a vessel without power, was exceptionally trying. He drew a map

---

*That he was up and about at 3 A.M. suggests either that he had learned enough about the May eruptions, presumably from his stops in various ports like Port Said and Singapore, to be wanting to catch a glimpse, or that he was a chronic insomniac.

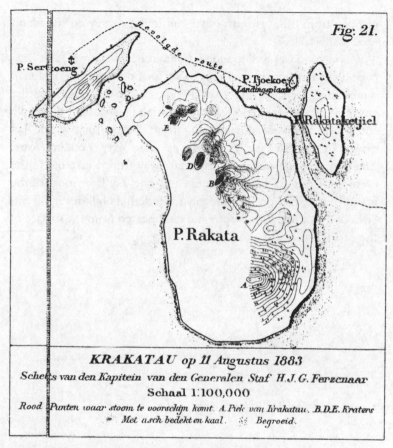

The last map ever made of Krakatoa, sketched sixteen days before the eruption by Captain H. J. G. Ferzenaar. All but the southernmost peak of Rakata vanished in the cataclysm.

that showed as much detail as possible, including the tiny spots and streaks of red from which the new eruptions were beginning.

This small and handsome map would have to do: Any proper survey of the island had "to wait until later, because measuring there is still too dangerous; at least, I would not like to accept the responsibility of sending a surveyor. A large portion could be mapped from other islands, but I consider a survey on the island itself inadvisable."

His caution was well founded. Captain Ferzenaar was, as it happened, the last human soul ever to set foot on Krakatoa. His map represented the final time that anyone would be able to see the entire fifteen square miles of tropical island—an island of people and forests and wild animals and visitors and history, which had existed in this place for at least the previous sixty thousand years. The good captain sailed his tiny craft away from Krakatoa on the evening of August 12. Two weeks and one day later most of the island that he had drawn suddenly exploded; its billions upon billions of tons turned into vapor and disappeared from the surface of the earth, forever.

6

A League from the Last of the Sun

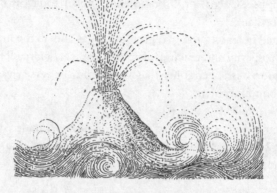

. . . it could not tell why the telegraph company caused it to be sent a full account of a flood in Shanghai, a massacre in Calcutta, a sailor fight in Bombay, hard frosts in Siberia, a missionary banquet in Madagascar, the price of kangaroo leather from Borneo and a lot of nice cheerful news from the Archipelagos—and not a line about the Muskegon fire.

—a contemporary account of the vexed mood of Michigan's *Alpena Evening Echo*, quoted as an example of information overload in *The Victorian Internet* by Tom Standage, 1998

The first that the outside, Western world knew of the extraordinary events that were starting to unfold in the distant and exotic East was a nineteen-word entry close to the bottom of the second column of page twelve of *The Times,* in London, on the morning of Thursday, May 24, 1883.

It appeared just below a story about a police raid on a suspected betting ring in a pub in Newcastle-upon-Tyne, and just above an announcement by the London police that the number of paupers known to be in town ("exclusive of lunatics in asylums")

was 52,032 indoors and 37,898 on the streets. There were adver-
tisements in columns beside the entry: Readers were invited to buy
Negro Head Gin for thirteen shillings and sixpence a gallon, John
Brinsmead's Pianos for thirty-five guineas, Moir's Mulligatawny
Soup, Epps's Cocoa, or, a brand still familiar today, Rose's Lime
Juice Cordial.

Slipped in among the two gripping sagas and all the intimations
of the prosperity and epicureanism of the Victorian readership, but
with a journalistic economy that verged on the terse, was the following announcement:

> **Volcanic Eruption.** Lloyd's Agent in Batavia, under date of
> May 23rd, telegraphs: "Strong Volcanic Eruption, Krakatowa
> Island, Sunda Straits."

It was perhaps appropriate that the first news of the explosion of
an island in the middle of the sea came through the agency, in
both senses, of the Society of Lloyd's. This was an organization
that by now was quite venerable, it having been more than two
centuries since London merchants met in Lloyd's coffeehouse to
discuss the coverage of risks to their far-flung fleet of cargo ships
and set up a mutual-aid arrangement to cover themselves in the
event of any losses. Lloyd's had become formally incorporated by
Parliament in 1871, and was by the latter half of the century
respected as the world's oldest and premier society of insurance
underwriters for ships. In that capacity the body employed or
retained scores of agents or subagents, as they were formally
known, at almost every port and capital city in the world.

The Lloyd's agency system, which still exists, had been set up in
1811. In port cities around the world, most often in the little streets
close to the docks, there will still be an office with a brass plaque
outside, or perhaps the enameled crest with its cross-and-anchor
badge and the words *Lloyd's Agent* picked out in scarlet. Filling the
posts of agents of the Committee of Lloyd's has long been, from
the Lloyd's point of view, quite simple: Men have invariably been

selected for being no more than "resident and well-established at the place concerned, and of high commercial status and integrity." From the applicants' point of view, it was less easy. The privilege of appointment was considerable: Many applied, and only a few could be chosen.

The task of an agent was in good times one of great simplicity, in bad times one of formidable complexity. Agents were initially bound by their contract to do no more than "to collect and to transmit to the Corporation information of likely interest to the Lloyd's Market, and insurers worldwide." But in those times, more frequent then than now, when ships went down or when there were collisions, strandings, piracy, or arguments over cargo, it turned out they were also there to settle suits, to adjudicate disputes, and to pay just claims on the policies of insurance that were underwritten by the syndicates of Lloyd's.

Although that first message about "Krakatowa" reportedly came from the Lloyd's agent in Batavia, the colonial capital, it did so purely because of reasons of protocol. It was thought more appropriate to have the formal report of a major event like this come from the agent—in this case a Scotsman, a Mr. McColl—who was based in the nation's heart, even if he was only transmitting secondhand information about an event he himself did not see. He may not have done; one of his deputies, however, most certainly did.

For Lloyd's, which had (and still has) a truly worldwide presence, also had as agent a man on the spot, someone with a bird's-eye view of Krakatoa and all that was going on there—too much of a bird's-eye view, as it would later turn out. He was the already encountered Mr. Schuit, the Dutch owner of the seafront Anjer Hotel that he had conveniently set down close to the docks in the small Javanese port of the same name.

The nature of Lloyd's business demanded that they have a presence in Anjer. Not only was it a bustling little coastal port in its own right, but it was the place where the northbound vessels, on passage for Batavia, would take on their pilots, and where southbound vessels would drop them. Anjer, principal pilot station of west Java, was the first port that newcomers would see in the island, their first landmark after passing the light on Java Head. It was a natural and necessary place for Lloyd's to have a man.

And Mr. Schuit had been chosen to fulfill the task—for which Lloyd's paid him a modest retainer—because of his view. His inn possessed a large wooden veranda overlooking the sea, and he and his guests would come of an evening to sit in lounge chairs there. A remarkably lovely spectacle was spread out before them: The island-filled, mountain-ringed, sunset-spectacular Sunda Strait, with the seemingly endless passage of ships sailing (or steaming: This was 1883, after all) along it, on their various ways between the Indian Ocean and the South China Sea. This last was what convinced the Lloyd's committee that Schuit was the ideal man for them in the town where they particularly needed to employ one of the best.

He was fascinated by the passing trade, and so were his guests. He bought and mounted a large brass telescope under his porch, so that he could identify the more distant vessels. With this he could see well enough to read the signal flags (there would be no marine radios for almost another thirty years) and could pass on messages to owners and agents as asked. He would look out especially for the distinctive arrangement of three flags that read "ZD2"—this short-hand was known by all agents to mean "Please Report My Passing to Lloyd's, London."

He knew the Anjer harbormaster, who kept him fully up-to-date about all the vessels that docked to load or unload cargoes, or that simply stopped by to bunker, victual, or derat. Schuit would thus send to London almost every day the names of ships that stopped, copies of the signals he had received, and statistics about the number of *quintals* of pepper, or *piculs* of coffee, *maunds* of this, and *catties* of that, which had passed across the Anjer quays. And as well as all the shipping and commercial information, he would also from time to time send to London entirely unrelated "information of likely interest"—of which the entirely unexpected burst of activity on Krakatoa was certainly a prime example.

It was of interest to the underwriters back home not merely because of the fascination provided by the explosion of any big volcano—but because this particular one lay almost directly athwart the main navigational passage of the strait, and so would be bound to interest any master whose vessel might soon be making passage in those waters.

In many ways the mechanics of Mr. Schuit's job were, in the closing years of the nineteenth century, changing fast. The ships themselves were altering their appearance, drastically. Sail as a means of moving them across the oceans was steadily giving way to steam. Wooden hulls were being replaced by steel, copper nails by iron rivets. The Suez Canal had opened for business, making passages to and from Europe more rapid and less risky. There was a steady growth in traffic, with more cargoes as world trade increased, and with ships from more nations. Congestion in the shipping lanes reflected the ebb and flow of global business and global politics; and it also reflected (as with the warship *Elisabeth*, which figured prominently in Schuit's reports, on her way home from a posting off China) the rise and fall of distant empires.

And the ways in which Mr. Schuit and his like were transmitting their reports was changing too. For most of the previous seventy years he and his predecessors would have sent their signals to Lloyd's by hand of messenger, the packages of bundled slips going back to London on the very homebound ships about which men like Schuit were reporting. But now, ever since midcentury, technology was beginning to make the lives of all kinds of far-flung intelligence gatherers—Lloyd's agents, diplomats, traders, and foreign correspondents among them—a good deal easier and much more efficient.

For the previous ten years Schuit had been sending all his messages from Anjer—including the information about the first explosion of Krakatoa, which the still extant records show was transmitted at precisely 3:47 A.M. on May 23—by way of two related and newly invented devices. The first was the electric telegraph, which, as we have already seen, came to the East Indies in

1856; the second was the submarine telegraph cable, which was to play a highly significant role in the unfolding of the Krakatoa story. This underwater cable arrived in Java after many fits and starts—the first one failed after sitting on the ocean bottom for only a month. But by 1870, thirteen years before Mr. Schuit needed to get his story out to London as quickly as possible, the international cable connecting Batavia was working well, and his message was received only a short while after he sent it.

It took an event like Krakatoa's eruption—which astonished and mystified an entire educated world—to underline the real revolution that this new technology was visiting upon the planet. True, other events had already been recounted by means of the new machinery; and its utility—to commerce, diplomacy, and news gathering in particular—was in no doubt. But with the explosion of Krakatoa came a phenomenon that in time would be seen as more profound. This eruption was so enormous an event, and had so many worldwide implications and effects, that for humankind to be able to learn and know about it, in detail, within days or even hours of its very happening entirely changed the world's view of itself. It would not be stretching a point to suggest that *the Global Village*—the phrase is modern, coined by Marshall McLuhan in 1960, referring to the world-shrinking effects of television, even presatellite*—was essentially born with the worldwide apprehension of, and fascination with, the events in Java that began in the summer of 1883. And Agent Schuit's first telegram to London was one small indication of that revolution's beginnings.

Although what *The Times* published was brief in the extreme,

---

*The phrase was born in the foreword to a book called *Explorations in Communication*, published in 1960:

> Post-literate man's electronic media contract the world to a village or tribe where every-thing happens to everyone at the same time: everyone knows about, and therefore par-ticipates in, everything that is happening the minute it happens. Television gives this quality of simultaneity to events in the global village.

The telegraphic transmission of news about Krakatoa, disseminated simultaneously through-out the entire newspaper-reading world, had much the same effect.

what Agent Schuit actually wrote by hand was considerably longer and more discursive. It started:

> On Sunday morning last, from six to ten o'clock, there was a tremendous eruption, with continuous earthquakes and heavy rain of ashes. On Sunday evening and Monday morning it was continued. The eruption was distinctly seen here until nine o'clock this morning, and smoke was seen until twelve o'clock; afterwards it cleared up a little, and at this moment the air is clouded again. Capt. Ross reports from Anjer that on May 22 he was sailing near Java's First Point and tried to get Prinsen Island in sight, but found that it was surrounded by clouds. Then he steered from Krakatan, but found it to be the same there. The captain observed that the lower island or mountain situated on the north side of Krakatan was totally surrounded by smoke, and from time to time flames arose with loud reports. Fire had broken out in several places, and it is very likely that the trees in the neighbourhood have caught fire. The mountain of Krakatan* has been covered all over on the north side with ashes.

Schuit wrote this dispatch in English, since that was the language in which Lloyd's conducted its business, and he marked it to be sent to London and copied to the Lloyd's agent in Batavia. He was

---

*The vexatious question of the proper spelling of the island arises here, not surprisingly, given that within the space of a single day, because of the telegraphic equivalent of Chinese whispers, the spelling changed from *Krakatau* to *Krakatan* to the rather puzzlingly elaborate *Krakatowa*. By the next day, however, May 25, 1883, *The Times* had settled for *Krakatoa*, which is how the name of the mountain remained, in most of the English-speaking world, for most of the following century, and since. An anniversary book published in 1983 tried gamely to turn its back on all but the supposedly proper spelling, *Krakatau*; but the old wrongheaded one persists, despite all best efforts to do away with it. Searching for the possible orthographic villains in the piece: The Lloyd's men seem innocent; the telegraph operators may have made an honest mistake; a *Times* subeditor working that night with the Lloyd's copy seems to have made one executive decision on the first night—*Krakatowa*—and then changed it the night after. All told, the newspapermen seem to have made the greater error—but in doing so created a name that has stuck.

evidently a night owl, because it appears that he wrote his message—longhand, and the prescribed telegraph form long after dinner, in fact very early on the morning of Wednesday, May 23. He then took the completed sheets down to the small white stucco building that housed the continually open office of the Dutch East Indies Post and Telegraph Office and handed them to the clerk.

Since the message was marked URGENT, the duty officer sat down at his wood-and-brass Morse tapper and register, and, with the lightning speed (and near-total accuracy, aside from reading Schuit's handwritten *Krakatau* as *Krakatan*) for which good operators were renowned, dispatched the signal to his opposite number in Batavia. From the Batavia telegraph office,* at the corner of Post Weg and Kerk Weg (or what is now Cathedral Street), the message then underwent a form of mitosis, with one copy of the signal going to the Lloyd's office and agent just down the street in central Batavia, the other going straight on to London. The version sent to London was dispatched from Batavia (or possibly originated at Anjer—we cannot be sure) at 3:47 A.M. on the Wednesday—in London, late on Tuesday evening.

The message took some long while to reach London—one reason why the first published report of the eruption appeared in *The Times* not on the Wednesday morning, as might be expected, but in the paper of Thursday, May 24. It took its time to make its way from the East Indies capital to its British equivalent because of the mechanics of early-nineteenth-century communication: it began its long journey by first going north of Java up to Singapore, and doing so by way of that newly made and wholly revolutionary invention, the submarine telegraph cable.

There is a nice coincidence of geography and botany in the story of the building of submarine telegraph cables—a business that had begun only in 1850, when the first cable was laid between Dover

---

*The telegraph office was a pair of single-story buildings that still stand, now forming the core of a convent for Jakarta's Ursuline nuns.

and Calais. What allowed the industry to flourish—something that really happened only when cables ceased their infuriating habit of breaking in midocean, and they could then be relied upon to work and carry their signals without interruption—was the discovery of a handsome evergreen tree called *Isonandra gutta,* from which oozed a rubbery, workable, and waterproof substance that was soon to be called gutta-percha.

A London firm called S. W. Silver & Company discovered that gutta-percha could be extruded like rubber, and could be made to cover copper wires that would then be totally waterproof. The firm's directors promptly changed the company name to the magnificently sonorous India Rubber, Gutta Percha & Telegraph Works Company—and under that rubric established factories that began to spin hundreds upon hundreds of miles of armored, waterproof telegraph cables for ships to lay deep down on the

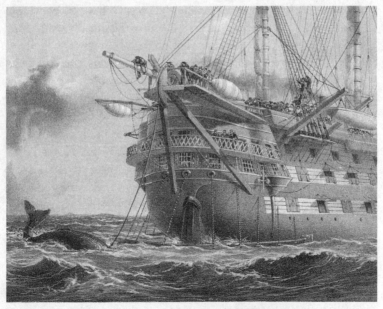

The converted warship *Agamemnon* laying the first transatlantic telegraph cable off the west coast of Ireland in 1857.

ocean floor. ("The moment the electric telegraph was invented," said a manual of the day, "so gutta-percha, the very substance it demanded, was discovered.")

By 1865 the India Rubber, Gutta Percha & Telegraph Works Company was producing submarine cables as fast as the world was able to connect itself, and they were happily transmitting electronic messages hither and yon at great speed and with near-total security. It was this guarantee of security and privacy that was their real value: The disruption of wars that kept breaking out like wildfires all over Europe was making the cables that passed overland risky means indeed of sending messages anywhere.

And the coincidence? It so happened that the handsome evergreen tree with its rubbery sap was found in abundance in just one corner of the planet: in Borneo, Sumatra, and Java. Gutta-percha, so critical for communication, became in the 1860s a major East Indian export, just like pepper, quinine, and coffee. Though unseen in day-to-day life, it was an export that was to have an enormous impact on the technological development of the outside world. And further: Though this substance that made cables work so well just happened to be found in Java, it was twenty years after the technology was first used that it was employed to connect Java with the rest of the world that was already using it. The former may have been pure coincidence; the latter was pure irony.

Mr. Schuit's signal thus passed along the 557-mile-long gutta-percha-covered cable that had been laid by the converted cargo ship *Hibernia* in 1870. A cable had first been laid along the route eleven years earlier, by the Dutch government—but the technology of 1859 was primitive, the demand slight, and when it broke, as it did after some four weeks, no one from the government bothered to order its repair.

On the second occasion the combination of commercial demand, market forces, and new technology succeeded where the pioneering had failed: A brand-new private cable was laid on the orders of the British-Australian Telegraph Company, designed not so much to connect Java with the outside world as to allow the vastly powerful Eastern Telegraph Company to connect its trans-India lines to the burgeoning populations in Australia and New Zealand.

And so the *Hibernia* laid the connecting cable to Batavia. Together with her sister-ship the *Edinburgh,* she then laid another from Banjoewangie in eastern Java to Port Darwin in northern Australia; engineers connected the two by a landline that ran the length of Java itself; and when finally in 1872 the Australian government completed its own landline across to the continent, then London and Sydney could exchange messages with each other with consummate ease and near-total commercial security.

Mr. Schuit's Morse-coded message reached Singapore by way of the British-Australian Telegraph Company's gutta-percha-and-jute-twine-covered four-strand copper cable, the fastest and most secure thus far designed. The signal was then amplified and sent on its way toward London. It had two choices: It could either go, slowly and insecurely, via the long chain of landlines that had been established midcentury, when telegraphy had first begun to expand; or it could go "Via Eastern." A customer could in those days specify on the telegram forms which cable should be used, and pay the costs that particular cable company charged. Specify *Via Eastern,* and it made most of its long journey by sea. Leave the cable-routing box blank, and the message went the long and slow way and, for most of its length, by land.

But the land journey had a fine romance about it, even so. From Singapore it went up the Malay coast to Penang and then by a short sea journey to Madras. Thence ever westward it progressed, by way of cities still known, or by towns obscure, long forgotten, or with names no longer used.

The lines of creosoted pitch-pine telegraph poles took the cable to Bombay, then to a switching center, from which it emerged as an armored cable and dipped beneath the Arabian Sea to reach Karachi, then rose back on pole-supported line to the village of

Hermak in Baluchistan, to Kerman, Teheran and Tabriz in Persia, to Tiflis in Georgia, on to Sukhumi on the Black Sea coast, along the corniche to Kertsch* in the Crimea, to Odessa, up across steppes and coalfields to the Polish city of Berdichev (with a population of 52,000, in those days before Hitler many of them Jews), through Warsaw, Berlin, and the North Sea port of Emden,† beneath the sea for one last time, before making landfall in East Anglia, and thence by telegraph pole for the final fifty miles to London. A telegram sent thus could take a week to reach its destination.

But if the message was marked *Via Eastern,* then it went swiftly and securely, beneath the sea. Thirty-five years earlier the first insulated cable had been dropped from a ship called the *Princess Clementine* moored in Folkestone Harbor and connected to a boat two miles away, with a message successfully sent between the two. Since that time the undersea cable had become fixed in the public consciousness. Tennyson had written a hymn to the romance of the idea of coded voices hurrying along the ocean floor; and so had Rudyard Kipling, whose brief poem "The Deep-Sea Cables" remains among his best loved:

The wrecks dissolve above us; their dust drops down from afar—
Down to the dark, to the utter dark, where the blind white sea-snakes are.
There is no sound, no echo of sound, in the deserts of the deep,
Or the great gray level plains of ooze where the shell-burred cables creep.

Here in the womb of the world—here on the tie-ribs of earth
Words, and the words of men, flicker and flutter and beat—
Warning, sorrow, and gain, salutation and mirth—
For a Power troubles the Still that has neither voice nor feet.

---

*Well fortified and "known for its mud volcanoes," according to the 1882 *Lippincott Gazetteer.*

†*Emden* was also, coincidentally, the name of the German surface raider that ransacked the cable station on Cocos Island, off southern Java, in 1916.

They have wakened the timeless Things; they have killed their father Time;
Joining hands in the gloom, a league from the last of the sun.
Hush! Men talk to-day o'er the waste of the ultimate slime,
And a new Word runs between: whispering, "Let us be one!"

The undersea cable connecting Singapore with London, after passing first by land in those days to Penang, crossed the Bay of Bengal to Madras, hopscotched across India and then plunged into the Arabian Sea on the long passage from Bombay to Aden. It was then put ashore briefly at Port Sudan, nosed up to its receiving stations at Suez and Alexandria, crossed the southern Mediterranean to the island of Malta, traveled through the northern part of the Mediterranean to Gibraltar, between the Pillars of Hercules to Carcavelos on an Atlantic headland ten miles west of Lisbon, up to windswept Vigo on Spain's western Galician coast, and finally it headed north, sheltered deep below the storms of the Bay of Biscay and the Western Approaches, arriving at its landing place at Porthcurno, on the southern tip of Cornwall, in England.

After this messages did their final two hundred miles by land, along conventional wires that were carefully tended, since they carried the international and imperial traffic of the hallowed and respected Eastern Telegraph Company. The signals would arrive at the receiving room of what was called London Station about three hours after they had been sent from the Morse tapper in the faraway East. In the case of Schuit's Anjer message reporting the eruption of "Krakatan," sent from Batavia at 3.47 A.M. local time on the Wednesday, the signals would have arrived at Lloyd's (subtracting the eight-hour time difference between London and Batavia, and adding the approximate three hours of transmission time) at about 10 P.M. on the night of Tuesday, May 22. Marked URGENT, they were decoded from their Morse and sent immediately to the delivery address—the Foreign Intelligence Office at Lloyd's.

And then, as it turned out, to *The Times.* The paper, then quite intimately involved with the British establishment, had a cosy arrangement with Lloyd's: Any signal deemed to be of likely interest would be passed promptly across to the paper's foreign news editor. The message was read in full by the duty editor; it was edited—heavily, since only seven words of it were actually published in the paper—and for some inexplicable reason the word *Krakatan*—itself an error, more probably the telegraph operator's than Schuit's—was changed to the similarly eccentric *Krakatowa*.

The cable was received early enough to be published in all editions of the paper of Thursday, May 24—including the Scottish edition, which was printed first, as well as the Final London Edition, which was printed last, on the best-quality paper at around 3:30 A.M. and delivered to all the embassies, palaces, and government offices of the capital.*

By midmorning all the loyal readers of what was then known as "the Thunderer"† came to learn of the event, and slowly but surely—since the volcano was to figure in the news dozens more times before the year was out—began to incorporate the name of this distant island, hitherto quite unknown, into their daily lexicon. And as it was incorporated into the lexicon, so it became, in short order, a prominent feature of received culture, all over the world.

Krakatoa achieved this happy status thanks in part to another of the great creations of the era: the news agency. Based in London, the first agency was founded by a German-Jewish businessman who, with great prescience, saw news and its fast delivery as a sal-

---

*In 1917 *The Times* started producing a limited-run late edition on very-high-quality paper that was christened the Library Edition. Five years later this was renamed the Royal Edition, and it was printed every day—except for a pause during World War II—until formally abandoned for budgetary reasons at the end of 1969.

†A *Times* columnist named Edward Sterling, known for his trenchant editorializing, wrote in 1829 that his paper had "thundered out" in support of social and political reform—the phrase was widely noticed, and for at least the next century and a half the epithet stuck.

able commodity. He gave it the name to which his immigrant parents had changed theirs: Reuter.*

In 1815 it had taken the news of Napoleon's defeat at Waterloo four days to reach readers in London; when Napoleon died on the Atlantic island of St. Helena six years later, most Britons remained in ignorance of the fact for the following two months. But twenty years later the telegraph had been invented, and by the 1840s its lines were spreading across Europe like the tentacles of kudzu; it was not long before entrepreneurs—like Julius Reuter, then a small-time newspaper publisher in Paris—decided that this new medium could allow news to be sped from place to place electrically and then sold to those in need of it, like any other commodity.

Speed—being first with the news, getting *the scoop,* beating the competition—was the most important thing to Julius Reuter. In those early days he was often too far ahead of the technology and had to come up with inventive solutions. To get news from Paris to Brussels before the telegraph link had been fully completed, for example, he used pigeons: The French news would be telegraphed to the border town of Aix-la-Chapelle; the messages would then be transcribed and taped to the legs of forty-five specially trained Reuter pigeons; two hours later the news would arrive in the center of Brussels. It was a series of stunts like this—as well as his agency's reputation for factual reliability—that enabled Reuter to win contract after contract to supply news, from all the world and to all the world. From October 8, 1858, when his service properly began, the list of successes and scoops under the Reuter byline is legendary indeed.

Reuter was, for example, four days ahead of the London papers with his account of—important then, at least—the king of Sardinia's address to the opening of his parliament. He beat everyone with the news of the Austrian defeat at the battle of Solferino in 1859. By 1861 he had a hundred correspondents, working world-

---

*Changed from the surname Josaphat.

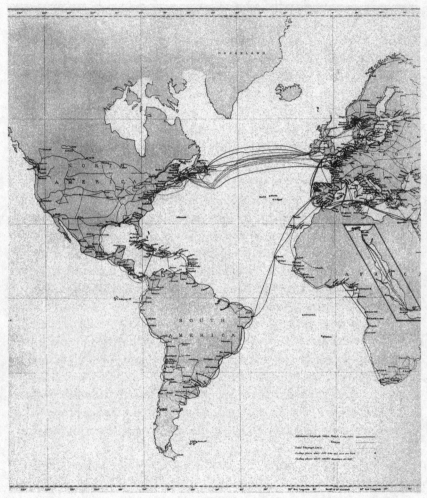

Part of the worldwide network of telegraph connections used by Reuter's agency for collecting and disseminating this most valuable commodity—news.

wide. During the American Civil War—with the Atlantic still not connected by cable—he arranged for telegraphed news to be rowed out to ships at remote ports on the Americas' eastern shores, places like Halifax and St. John's and Cape Race, and collected a week later by rowing boats sent out from Ireland. (When President Lincoln was shot on April 14, 1865, the news was carried aboard the SS *Nova Scotian,* taken off by whaling boat when the liner closed on Londonderry and telegraphed to London by Reuter in time to be published in the subscribing newspapers of April 26, twelve days after the event.)

By the time of Krakatoa's eruption, the entire East was fairly well connected with the newspaper-reading West, as Reuter had already so amply displayed by writing in timely fashion about the arrival of Admiral Perry and his American warships off Tokyo, and the subsequent restoration to the Chrysanthemum Throne of the Japanese emperor in 1868. Timely—but not instantly, since Japan was not to be connected fully with the worldwide telegraph system until the Danes laid a cable between Tokyo and Shanghai in 1872, and connected the reemergent state with the Great Northern Telegraph Company line that went between St. Petersburg, Copenhagen, and Paris, and on to London. From the time the Danes had completed this cable, all of the major Eastern cities—Shanghai and Peking, Manila and Tokyo, Saigon and Rangoon—were hooked into the system. They had become a part of the fast-growing network of international cables; and, because of this network, they had become beneficiaries of, as well as contributors to, the worldwide newsgathering operations of Reuter.

And Batavia was connected as well: It had a Reuter office and, in 1883, a stringer—a retained freelancer who fed such news as he found into the agency wires. His name was W. Brewer, and he would turn out to be the point man who would send most of the properly factual reports of the eruption for distribution around the world.

In May, though, the first that Reuter heard of the eruption came not from their Mr. Brewer but from London, via the Lloyd's telegram. The agency got it right, but it got it late—a full day late. When the rest of the world was given the story under the Reuter byline, it was already May 25. The largest Dutch newspaper of the day, the Rotterdam-based *Nieuwe Rotterdamsche Courant,* published it that Friday morning—having to be content with the undignified reality that a British newspaper had managed on the Thursday to scoop a Dutch paper with a story about an event in a Dutch colony.

Reuter, however, peddled its story hard. The paragraphs about the first eruptive event that had appeared in *The Times* of May 24 were fleshed out by Mr. Brewer in Batavia and could be found, translated into the vernacular where necessary, in that Friday's major newspapers in the United States, Southern Africa, India, France, and Germany.*

It was thanks to the combined agencies of all those involved—to Samuel Morse, to the directors of the India Rubber, Gutta Percha & Telegraph Works Company, to the Eastern Telegraph Company, to the Committee of Lloyd's, to Reuter, and to the small network of eager correspondents in Anjer, Batavia, and London—that this first remarkable story started to be told.

It was the first-ever story about a truly enormous natural event that was both *about* the world and was told *to* the world. Part of the planet's fabric had been ripped asunder: And part of that same planet, the part connected by cables and telegraphs and with access to newspapers, was now being informed of the event. And the very process of relating the dramatic happenings, especially in the weeks and months that followed, would enable all who heard, read, and understood it to share in the cruel intimacy of the moment. Millions of people hitherto unknown to one another began to involve

---

*These subsequent items settled on *Krakatoa* as the volcano's name.

themselves, for the first time ever, in looking beyond their hitherto limited horizons of self; they started to inhabit a new and outward-gazing world that these storytelling agencies, and this event they were relating, were unwittingly helping to create.

The story of Krakatoa had a small beginning—seven newsworthy words, buried well down in the pages of a single London newspaper. As the summer of 1883 wore on, it was to become a very much greater story indeed. And when it was over, three months later, it was to have implications for society—for the laying of the foundations of McLuhan's "global village"—that have reverberated in a far more important way, and for far, far longer, than anyone at the time could ever have supposed.

7

The Curious Case of
the Terrified Elephant

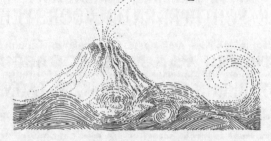

Eau de Cologne: A perfumed spirit invented by an Italian
chemist, Johann Maria Farina [1685–1766], who settled in
Cologne in 1709. The usual recipe prescribes twelve drops of
each of the essential oils bergamot, citron, neroli, orange and
rosemary, with one dram of Malabar cardamoms and a gallon
of rectified spirits, which are distilled together.

—definition from
Brewer's Dictionary of Phrase and Fable, 1959

The final Monday of July 1883 was one of the last quiet days
ever to be experienced on Krakatoa, an island that now had
precisely four weeks remaining before it was blasted out of
existence. It was also on that day, July 30, that John and Anna Wilson's Great World Circus, long expected and universally wel-
comed, finally arrived in Batavia.

They had been to Batavia many times before, the canny
Scotswoman Mrs. Wilson well knowing they could be certain of a
grand colonial crowd. On this particular occasion the performers
and their animals had sailed in on the liner from Singapore, and
they promised their audiences the staging of amazements and

This advertisement for Anna Wilson's Circus promises "a great and brilliant performance on behalf of the victims of the recent Banten catastrophe."

delights in an atmosphere of more sumptuous comfort than could
possibly be imagined.

Two years before, members of the audience had complained
they were too crowded and too hot. So for this visit Mrs. Wilson
had brought with her from New York a brand-new tent, a giant
she had christened the Mammoth. As the riggers hauled up the
immense acreage of bleached canvas on the west side of the Koningsplein, so passersby gasped at what they saw and heard. "Room
for 5,000 seats!" the posters exclaimed. "Real gas-lamp illumina-
tions!" blared the barker's horn. "Unbelievable new attractions!"
proclaimed the newspaper advertisements including, the first time
ever seen in the Orient, *Cannonball Holtum from Denmark, and*

His Incredible, Death-Defying Feats of Courage and Fearlessness.

This time, the Batavians said to themselves as they handed over their one-guilder coins for tickets and watched the final preparations being made, Wilson's Great Circus truly was going to be the greatest that Java had ever seen.

It was also to be—or so everyone thought at the time—the crowning glory of what had thus far turned out to be a delightful summer season. The volcanic grumblings and growlings over to the west in the Sunda Strait continued, true; but now that the initial surprise of the May eruption was over, the fact of all the fire and smoke and occasional tremblings of the earth had become part of the reality, an established routine. No one really gave it a thought—it meant no more than the occasional tremor might mean today to someone in Tokyo, Los Angeles, or along the San Andreas Fault. Batavians made jokes about it all. It was not going to be allowed to cause any disruption of the gay social round—though one of the Batavia papers, the East Indies edition of the *Algemeen Dagblad*, did sound rather exasperated in June when it remarked sourly, "What musical entertainment there is can hardly be enjoyed because of all the shaking noises of the doors and windows caused by Krakatau."

Nonetheless the social year that had begun so splendidly in February with the King's Day dance at the grammar school—about which the papers still talked in the summer, so successful was it said to have been—rolled on, unstoppably. In mid May, for example, horse racing was held up in the mountain cool of Buitenzorg, and a costume ball in the stewards' hall on the Sunday of the two-day meeting—Governor-General s' Jacob naturally lending his august presence.

Then a *soirée dansante* was held in the Batavia Plant and Animal Garden on May 27. Usually this would have been an uneventfully happy evening—but as it happened a small shadow was cast over it too. Not, however, by Krakatoa. Most of the guests had been hoping to see the royal Bengal tiger that, famously, had been presented to the garden earlier in the month by a German philanthropist named Schröder. But the Garden announced it had neither the space nor the wherewithal for housing this very large and fierce animal, and the curator ordered it put on the next boat to Melbourne—disappointing the dancers, who had hoped it might enliven their evening.

Any sense of chagrin that might have been felt by the dancers at the loss of their tiger was quite probably offset by another, much more practically consequential announcement that was also made in May. The 985-ton British steamship *Fiado,* under contract to the Australian Frozen Meat Company, had, its owners announced to the papers, been fitted with up-to-date refrigeration equipment and would begin a regular service to Batavia and Singapore, bringing cargoes of frozen beef, lamb, pork, and poultry.

The first such cargo arrived on July 20. To judge from the newspapers of the day, which greeted the docking of the *Fiado* as they might the unexpected arrival of a monarch or a star, a wave of gastronomic ecstasy promptly swept the colony. The Australian meat, never seen before, turned out to be of extraordinary quality. "Not even a large and well-bred Balinese ox could compete with this," a local gourmand was quoted as saying. Meat was no longer a luxury; henceforward every European in the colony could eat as well in Java as they once had in Amsterdam. Perhaps, now, even better.

The epicenter of Batavian gaiety in that summer of 1883 was the newly renovated and extended Concordia Military Club, the indisputably grand white marble building at the southern side of Waterlooplein, directly across from the governor-general's palace.

That year the Concordia ranked marginally ahead of the Harmonie, which was admittedly the older of the city's two great social clubs and remains the better known today; even though the Harmonie's building was torn down in the 1960s, like that of the Concordia, the part of modern Jakarta in which it once stood retains its name. At the time of the Krakatoa events the fabric of the Harmonie was looking somewhat shabby, and the club was suffering from a temporarily declining membership. Far better

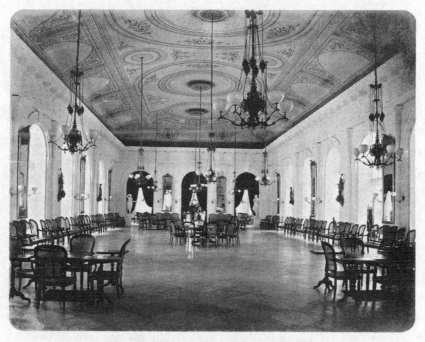

The grand ballroom of Batavia's Concordia Military Club.

balls and soirées were being held at the much swankier Concordia, where, to the delight of most Europeans who were invited, they were marked by a lavishness that often bordered on the decadent.

For example, a masked ball was held at the Concordia on Saturday, July 28, just before the arrival of the circus. Three hundred couples came, by horse-drawn carriage or in the back-to back conveyance known as a *dos-à-dos*. The gardens were illuminated with Chinese lanterns, there were obelisks lit with piped gas, and a Turkish kiosk with a sky-blue cupola had been built for the outdoor band.

And inside—under the ornate ceiling and gas chandeliers, within walls groaning with portraits, mirrors, and statuary, beside rare plants, flowers, and soft-colored veils of gauze, upon a floor of polished teak squares dusted with French chalk—they danced—

until the sun came up like thunder, as in the East it was wont to do. The women, the finest of all Batavian high society, were seen to be wearing dresses that, to the more matronly onlookers and chaperones, were positively outrageous. So short! they chorused next day. So delightful! the men reminisced. "If you want to enjoy the sight of beautiful pink satin shoes with fine ankles moving on the dance floor—go to the masked ball at the Concordia!"

The masks and costumes were as various as could be. One lady came dressed as a swallow, with a headdress and wings made of feathers from Anjer songbirds—birds that had lately been flying, in other words, in the turbulent volcanic airs of the Sunda Strait. Another, *La Madame la Diable,* had black wings and gilded horns and a silk dress in black and red adorned with images of Lucifer. There was a Carmen, a Louis Quinze escorting an Italian farmgirl, buxom in gingham and Genoa lace. There was a toreador, a consistory of monks, and a group of British sailors from a passing Royal Navy warship whom everyone thought were in fancy dress, though they were simply in full uniform, officially.

And as if this display were not enough—in the center of the ballroom was a fountain gushing not water but pure eau de Cologne. This was by way of an experiment: A flower vase was used as the centerpiece, and from deep in the lush enfoldments of its blooms gushed fountains of perfumed water that, when mingled with the scents of the assembled dancers, the cigar smoke, and the rich aromas of spices from the *rijsttafel,* made for *a symphony of olfactory delights* that . . . the newspapers, gushing too, found simply too overwhelmingly wonderful for words.

Then came August, and with it, at last, the circus. The company had come clear across the Pacific—one of their earlier performances had been in a small town outside San Francisco—and were in the East determined to make the best impression. And so every imaginable act was on hand: tightrope-walkers, fire eaters, a pigeon charmer, an American who could somersault over eight horses, the Nelson Family of Acrobats, Hector and Faue the Lords of the Trapeze, Fräulein Jeanette and Her Amazing Bareback Riders, William Gregory the Gymnast King, the Well-Known Sweetheart from Earlier Years, Miss Selma Troost—with *Troost* being the Dutch word for consolation, a commodity that many lonely bachelors thousands of miles from home would no doubt readily welcome.

There were a hundred acts in all, and twenty Arabian horses on which the performers might show their paces. There were dozens of clowns; and on August 22 the Batavia Cricket Club staged a match against the Clown's First Eleven, all of them wearing their circus costumes. The Cricket Club won, handsomely.

John Holtum, known around the world as the Cannonball King, was perhaps the stellar attraction. He was a thirty-eight-year-old Dane who had developed the skill of catching a cannonball fired at him from the far side of the circus ring. The first time he tried this, the ball, which weighed fifty pounds and when it struck him was moving at more than a hundred miles an hour, took off three of his fingers. He was not discouraged by this setback, however; nor was he put off when a rival in an Italian circus was sliced in two by a grenade he was attempting to deflect with what up to the moment of his demise was said to be a powerfully muscled chest.

Mr. Holtum persisted, and by the time he arrived in Batavia that August he was the world's acknowledged leader in this limited field of endeavor. He liked to challenge men in the audience to catch cannonballs: 161 volunteers in Europe and America had already tried but failed, and the few who tried in Batavia failed as well, including the promisingly named Mr. Thor, whose fingertips the ball grazed, doing him no harm but losing him his wager. John Holtum also liked large men to try to pull him off balance with a rope—they always failed in that too. But when the great Dane pitted his resilience against four horses harnessed to a rope, they pulled in so confusing and angry a way that he fell over, and was pulled around the ring, in an undignified pother, showing himself not to be quite as invincible as he liked his audiences to suppose.

For every single night throughout the first four weeks of August,

John and Anna Wilson's Great World Circus staged a performance—two on Saturdays. And, such were the limitations on social
life for Batavia's Dutch lower ranks, as well as for those members of
Javanese society who were permitted to attend, that the newspapers
reported each and every performance, in full detail, and with a tone
of breathless excitement as if such amazements had never been witnessed before. The results of the steeplechases were published; the
names of all those who challenged John Holtum, and the out-
comes; the number of runs scored and wickets taken in the cricket
matches with the clowns; and, of course, the outbreaks of bad
behavior and curious mishaps that seemed to attend this circus, in
just this particular month of 1883, as rarely before.

The bad behavior broke out within the first week. Without the
guiding influence of the circus owner, John Wilson—he was overseas, recruiting more performers—the competing rivalries of the
various artistes apparently spilled over into violence. The circus
members all stayed at the Hôtel des Indes, said to be the grandest
hotel in the entire Dutch empire, and the fighting first broke out in
the bar. They were drinking champagne, the newspapers said, and
arguing volubly over who was the funnier clown or the most adroit
performer on the trapeze, when one of them, scandalized by a
wounding remark, hurled a glass at another. From then a full-scale
donnybrook broke out, with wine, beer, food, and fists being
thrown in wild abandon. Mrs. Wilson was hit in the face; one of the
performers had his cheek badly bitten; the athletes battled the
gymnasts, the horsemen the jugglers; in the end the police had to
be called. It was a sorry affair, and all Batavians loved every minute
of it.

The fighting in the Hôtel des Indes coincided with a sudden
quickening of the eruptive mood across on Krakatoa. After Captain Ferzenaar's visit to the island, when he had recorded the three
huge craters, the fourteen new steam vents, and the vast clouds of
boiling dust, ships passing along the Sunda Strait reported new
and dangerous-looking activity. One master reported a "vast erup-

tive column" on the twenty-second. Another spoke of "shakes and heavy blows" on the twenty-fifth. There were "falls of ash," "a milk-white sea," and "dull explosion." Most ominously worryingly of all—in Sumatra on the twenty-sixth, the last Sunday in August—a villager noticed "hot ashes coming up through the crevices in the floor of his hut." Whatever was happening deep below the ground was clearly beginning to overwhelm the capacity of the surface to keep it under control. Something, somewhere, had to give.

It is said that animals are presciently aware of impending seismic doom. Catfish jump out of the water. Bees mysteriously evacuate their hives. Hens stop laying for no apparent cause. Mice appear dazed and can be caught by hand. Deep-sea fish are found at the ocean surface. Hibernating snakes suddenly appear on the surface, frozen to death if they do so in a harsh winter. Dogs begin to howl for no obvious reason. (A retired U.S. Geological Survey scientist collected lost-pet advertisements in American newspapers and claimed to find a correlation, such that within two weeks of any rise in the number of missing animals, there was an earthquake nearby.)

There is no firm scientific evidence that there is a connection, nor is there a true basis for a new pseudo-science called ethogeological prediction, which seeks to forecast earthquakes by observing carefully calibrated animal activity. Yet not a few geologists do feel it is at least reasonable to suppose that the infinitesimal subterranean shiftings and strainings that precede massive eruptions or earthquakes can be sensed by animals long before being experienced by man or his machines. But no one has yet measured the link, if indeed there is one.

But in Batavia in that August of 1883 there was one animal that, for some unexplained reason, did begin to behave in a most unusual fashion. It was a very small circus elephant, said by his keeper, a Miss Nanette Lochart, to be the smallest trained pachyderm in world history. He had been caught in Java, and she had trained him in a matter of weeks. The entire company adored him, and the Batavian public—especially the children, who got in for half price, with their servants admitted free—were captivated by the sight of this miniature monster juggling balls with his three-foot trunk, or stepping gaily from tub to tub as he negotiated a little obstacle course.

But halfway through his stay, both the elephant and his keeper began to behave most eccentrically. Miss Lochart began to believe that the other artistes in the troupe, their dander up after they had started fighting with one another, might for some reason turn on the elephant and try to harm him. Perhaps, she thought, they might try to break into his pen and give him poison. Despite there being no evidence, she decided to take evasive action.

And so, midway through August, when the ash falls and thunderings and pillars of flame were just beginning to be noticed once again in Batavia, Miss Lochart moved her little elephant into her room in the Hôtel des Indes. He was, after all, her only worldly possession. She may have surmised that the hotel owner, a stern Frenchman named M. Louis Cressonnier, might take a dim view of an elephant in the hotel room, though no sign specifically forbade it. So she settled the animal down, said her good nights, locked the door behind her, and left for an evening of dining with friends.

The elephant, denied the company of his mistress, clearly unaccustomed to the luxury and comfort on offer in the East's premier hotel, and perhaps—just perhaps—sensitive to what was going on in the earth beneath his feet, promptly went berserk. He trampled through and across all the furniture in Miss Lochart's room, smashing it to smithereens. He trumpeted. He roared. He stamped his not-yet-enormous feet so aggressively that other guests thought the entire hotel was about to fall down.

And in the end the Batavia constabulary were called. They found Miss Lochart; they demanded that she persuade her two-ton charge to leave the room without further ado; and M. Cressonnier further demanded that the entire troupe, and such animals that others among them might have sequestered in their chambers, leave the hotel forthwith and find other accommodation.

This they did. And no one thought any more about it.

Except that then, shortly afterward, on Monday, August 27, 1883, the mountain that had been grumbling and groaning for the previous ninety-nine days finally exploded itself into utter and complete oblivion.

For sixty million years the two tectonic plates that converge on Java had been grinding slowly and steadily toward each other, four inches every year. Now, on the very morning after the Great World Circus had given its best performance yet—with the artistes now all newly accommodated and the little errant elephant by all accounts behaving obediently and impeccably once again—the event that would be consequent upon all those years of subterranean shiftings and slidings was at last about to occur.

8

The Paroxysm, the Flood,
and the Crack of Doom

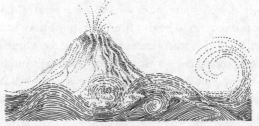

And I thought: I would give all these people's lives, once more,
to see something so beautiful again.

—Dutch pilot in Anjer, quoted in "Krakatau," a short story
by the novelist Jim Shepard, 1996

THE EVENT

The death throes of Krakatoa lasted for exactly twenty hours
and fifty-six minutes, culminating in the gigantic explosion
that all observers now agree happened at 10:02 A.M. on
Monday, August 27, 1883. The observers, as is often the way of
such things, agree on precious little else. Thousands of people, far
and wide, suddenly became aware of the events in the Sunda
Strait—but their accounts of it, like the accounts of any monstrous
and traumatizing event, present today a morass of conflict and
confusion.

The countdown to the final hours of the mountain's existence
properly began at 1:06 P.M. the previous day, Sunday. All across the
colony, Dutchmen and Javanese alike were looking innocently forward to the long lazy stretch of the afternoon of a much-needed
day of rest.

The first indications that all was not right became apparent more or less simultaneously to a number of people nearby. They were at the time almost all completing the last moments of their familiar Sunday ritual: pushing their chairs away from the luncheon-table, folding their napkins, draining the last dregs of the coffee, standing up and stretching their legs, picking up their cigars, dogs, and wives for the Dutch tradition of the afternoon family walk.

In Anjer, from where most of those very early reports originated, the relaxed mood of the afternoon must have seemed peculiarly suited to the place. Anjer was a sedate, pretty little port town, as pleasant a posting for a visiting Dutchman as it might be possible to find. It was situated in a shallow bowl in the volcanic coastal range, a place where the hills dipped steeply down to the sea and formed a cozily protective natural harbor. The beaches were wide and white, the fringing palm trees leaned into the trade winds, there were flowers and banyan trees and a paradise of birds, and everywhere a heavenly scent of spices.

The local people lived in *kampongs* of small thatched cottages, the colonials in neat white stucco houses with red roofs. The finer of these mansions—some, like that of the assistant resident,* sporting a Dutch flag flying from a staff in the middle of its magnificent lawns and a private dock with the official's impeccably maintained launch—could be seen to best advantage from out at sea, where they seemed separated from one another by acres of deep green jungle. There was usually a flag signal flying from the pilot station that spelled out: CALL HERE FOR MAIL; and the inbound oceangoing ships would indeed invariably call, their presence giving an air of vibrancy to the town; and, because they were stopping for orders and not for stevedores, Anjer was free of the slums and squalor of a cargo port.

The Europeans, in their white tropical suits and topis, would

---

*The resident, who at the time of the eruption was a Mr. van Spaan, administered the region from the headquarters of the residency in Bantam, thirty miles away. Anjer town itself merited only an assistant, a Mr. Thomas Buijs, who died in the disaster.

leave their houses—servants had Sundays off—and saunter of an afternoon along broad seafront avenues, under groves of tamarind trees. There would be crowds of local Javanese, children running everywhere, pye-dogs sleeping under upturned boxes, chickens, pigs, goats, creaking bullock carts, insistent pavement salesmen—all the carefree magic of an Eastern street, in other words, out for entertainment and fun on what seemed likely to be an easy, lazy summer Sunday.

Relations between colonizers and colonized in the East Indies were less than perfect—indeed much less than perfect, for the Dutch were not very kindly in the ways they wielded their imperial powers, and they are consequently remembered with much less affection today than are most of those other Europeans who ruled far-flung territories around the globe. Yet, by all accounts, on this particular Sunday any feelings of antipathy were soothed and muted by the holiday mood. People smiled at one another, Dutchmen muttered cordial greetings to Javans, everyone sauntered contentedly along in the broad heat of the early afternoon.

And then, without warning, from out to sea in the west—a sudden sound.

The first two accounts to be written were of a kind that was to be repeated, in essence, many dozens of times that day. "We plainly heard," wrote one, "the rumbling of an earthquake in the distance." "We didn't take much notice at first," wrote a second, "until the reports got very loud."

The newly appointed Anjer telegraph master, Mr. Schruit, was once again idling on the veranda of the Anjer Hotel, which was owned by his new friend, the confusingly named Mr. Schuit, the local Lloyd's agent. It was clearly Schruit's preferred place to spend his Sunday mornings. He was a young man, and cut rather a lonely figure. His wife and children were still in Batavia, and, after weeks of searching, he had only now found them a suitable house in Anjer, and was at long last now looking forward to a family reunion.

But until then, as for the past few months, his holiday Sundays were spent on the veranda, puffing on a cheroot, gazing out at the view. Three months earlier he had been there too, trying under Schuit's tutelage to decipher the flag message that was flying from the yards of the German warship *Elisabeth,* then sailing down the strait—a message that almost certainly told of the first eruption. He did nothing about it then: He was new to the area, he was not on duty, and in any case the local Lloyd's agent had the matter in hand, was well able to read the *Elisabeth's* flags, and was already sending his historic dispatch about what the telegraph operators would misread as "Krakatan" spewing out its great clouds of smoke and ash.

But on this occasion Schruit was on duty, and much more observant than before. He remembered later seeing a fully rigged barque heading north, a billow of white sails gliding along the blue mirror of the sea. Then he spied the much less pretty steamer *Gouverneur-Generaal Loudon,* the locally familiar government-chartered vessel that had taken the eighty-six tourists to Krakatoa back in May, heading into Anjer port.

(On this occasion she was undertaking more mundane and customary tasks: first picking up a hundred coolies who had been hired to help build a local lighthouse and ferrying them across the strait; then going on up the eastern Sumatran coast to the troublesome region of Aceh, in the north, and there delivering, among others, some three hundred miscreants, all members of a chain gang destined to be put to work on a variety of government building sites.)

It was at the very moment when Schruit was watching the *Loudon* steaming toward the safety of port that there came the first roar of an explosion.

It was an extraordinary sound, he thought: far, far louder than anything he recalled from before. He looked sharply over to his left and saw, instantly, the unforgettable sight of a tremendous eruption. To judge from the billows of white smoke that were now tumbling up from the mountain—"as if thousands of white balloons had been released from the crater"—it was a far, far larger eruption than anything he had witnessed when he was standing at this exact spot back in May.

Moreover, whatever was happening on the mountain was also having an immediate effect on the sea. It was rising and falling, strongly, irregularly, in bursts of sudden up-and-down movements of the seawater that seemed immediately unnatural and sinister. It wasn't tide or wave or wash: It was some terrific disturbance, and the water was slopping up and down, dangerous and unpredictable.

He ran down to the beach, where he had spied his deputy telegraphist watching the eruption, transfixed. He said he too was utterly perplexed by the movement of the sea. Perhaps the tide was on the turn, he said. But as he said so a furious rush of water roared toward them, sending them scurrying back up to the roadway. The two men ran to the small white stone building that was the Anjer telegraph office—and, as they did so, the enormous cloud from the volcano began to drift down on them. Within moments all Anjer was enveloped in dust and cloud and became strangely dark.

Some remember the cloud as black; others, like Schruit, are equally certain it was white. One of the pilots waiting for orders at the Anjer pilot station, a Mr. de Vries, swore it alternated in color, from white (when presumably it was largely made up of steam) to black (when it was composed largely of eruptive smoke). But no matter: It was so thick and heavy that within moments an artificial night had descended on Anjer port, and the two men who groped their way to the cable office then found they had to light lanterns, in the middle of the afternoon, in order to be able to send their first message. They timed it at 2:00 P.M. Krakatoa, they tapped out in urgent Morse to their head office in Batavia, was beginning a major eruption. It was "vomiting fire and smoke." It was so dark in town that it was now no longer possible to see one's hand before one's eyes. What were the instructions?

Batavia replied, with equal urgency. Yes, they had already become aware that something was taking place. The governor-general himself, his Sabbath ease disturbed, had been inquiring.

People were milling about in the streets, worried. Chinese traders in particular seemed to have a peculiar sense of unease: The cable office reported hearing wailing. So it was important for Anjer to keep its telegraph station open and the information coming in, the operator tapped out. For the next six hours the operators did just that, giving the Dutch officials in Weltevreden (the name, "Well Contented," must have prompted on this occasion a sardonic smile or two) a moment-by-moment chronicle, in the staccato language of the telegraph, of the unfolding events. "Detonations increasing in loudness." "Hails of pumice." "Rain of coarse ash." "First flooding." "Vessels breaking loose in harbor." "Unusual darkness." "Gathering gloom."

The Anjer harbormaster, who was by now aware that the crisis was frightening many of his friends and colleagues—"the Day of Judgment has come" was a common belief—tried to collect as many of the local expatriates together as he could, to reassure them. How he imagined he might accomplish this is left unsaid. But he did manage to assemble a fair number of the colonial establishment—the assistant resident, the public works supervisor, the lighthouse keeper, the registrar, the town clerk, a local doctor, and a prominent local widow—and tried to tell them that what they were witnessing would soon blow over, that it was, in his considered and experienced view, nothing to worry about. He could hardly have been more wrong.

At 2:45 P.M. the *Loudon,* all passengers aboard, set off for the forty-mile journey to the port of Telok Betong, at the head of Lampong Bay across in Sumatra. Her master, Captain Lindeman, steamed well to the east of the exploding island, trying as best he could to avoid the showers of rock and ash cascading down from the plumes of smoke. One British ship in the vicinity, the *Medea,* estimated that by midafternoon the column had risen to a height of seventeen miles, more than three times the height of Mount Everest; the *Medea*'s Captain Thomson said there were "electrical displays" in the cloud, and explosions every few minutes were shaking his ship—even though he was at the time at anchor off Batavia, more than eighty miles to the east.

In the center of the capital, meanwhile, people were very rapidly becoming aware that matters were getting out of hand. Two seasoned observers of Krakatoa's earlier throat-clearings—Dr. J. P. van der Stok down at the observatory in Batavia and the mining engineer Dr. Rogier Verbeek up in the hills above town—had already telegraphed each other to find out what was going on. Van der Stok—the man whose wife had lost her Delft dinner plate in the May eruption and who himself had noted with great precision the time of the very beginning of the earlier events—once again swung into official observatory mode, even though once again this was a Sunday. He checked his watch at the very moment he heard the first loud rumblings, dashed from his house to the observatory buildings and wrote the time down in the official log: 1:06 P.M. That time remains today, etched in official records, as the one known certain commencement of Krakatoa's final phase.

Confirmation of these figures then came from a totally unexpected source: the city's gasworks, to the south of Batavia old town. They proved, quite uncannily, to be of the greatest use to scientists who later studied the eruption. And they did so, quite simply, because of the way they had been built.

The most visible parts of any plant that produces gas from coke are the tall drum-shaped metal containers for the flammable gas—containers that are in essence telescopic, which "float" on enormous ponds of water or mercury and which grow taller or shorter, higher or lower, depending on the amount of gas pumped from the works to be stored inside them. (Today's gasometers tend to stand tall against a city skyline in the morning, fall gradually over the day as the gas inside them is consumed, and then are replenished as more gas is manufactured overnight. In those cities like Batavia in the 1880s that had gas street lamps, the profile of varia-

REDUCED COPY OF A PORTION OF THE
RECORD OF PRESSURE ON THE
BATAVIA GASOMETER
27ᵀᴴ AUGUST, 1883.

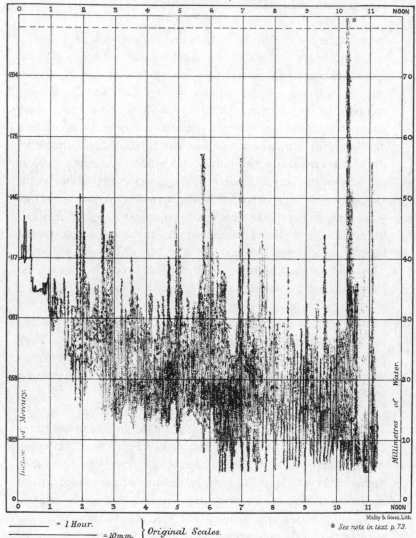

= 1 Hour.

= 10 mm. } *Original Scales.*

Malby & Sons, Lith.

* *See note in text p. 73.*

The Scale on the original diagram terminates at the point marked with a dotted line.

The invisible and inaudible pressure wave from Krakatoa's cataclysmic final explosion, measured—until it blows off scale—at the Batavia gasworks.

tions in pressure, and thus the times of the varying heights of these gas containers, would necessarily be rather different. They would stand tall in the early evening and fall away during the dark, during the gas-illumined hours.)

What superintendents of gasworks have long known is that these storage containers—commonly, but in fact wrongly, called *gasometers*—also act as gigantic barometers. The pressure inside them, and in the lines leading from them, goes up and down by infinitesimal amounts according to the rise and fall of atmospheric pressure outside them. Normally one would never notice these small amounts of movement, since they would be superimposed on the much larger movements resulting from the consumption of gas. But an eagle-eyed superintendent, in charge of arranging the pressure in the gas lines according to demand, could indeed notice. Moreover, a paper register of the gasometer pressure is always produced (and was, for most of the time, in Batavia), which would show all the movements with great precision.

The pressure records are being taken constantly, but they can in fact record the minute fluctuations in the atmospheric pressure (and hence also the fluctuations caused by an event like Krakatoa) only when the base pressure is low enough for the recording meter to be affected by them. The gasworks superintendent would increase this pressure each evening, when the street lights were illuminated; he would keep it high until the middle of the evening; and he would then lower it in hourly stages until dawn. So the best recordings of changes in the atmospheric pressure would be made during daylight hours, when the pressure in the gas lines was kept low because of the low demand.

Which is exactly what happened—after a hiccup. For there is no record, mysteriously, when the very first explosions occurred at lunchtime on Sunday. But then whatever problem existed vanished, and the recording trace begins properly at 3:34 P.M. Batavia time (which, since this was still some while before the formal international establishment of time zones, was a little more than

five minutes ahead of Krakatoa time).* From that moment on
until dusk, and then throughout all of Monday morning after
dawn, the Batavia gasworks pressure gauge provides an incredibly
accurate, minute-by-minute record of the massive air-pressure
waves that radiated out from the volcano, each and every time it
erupted. The paroxysmal eruption itself at 10:02 on Monday
morning blew right off the scale: It caused a pressure spike of
more than *two and a half inches* of mercury, unheard of in any
other circumstance.

By five o'clock on Sunday evening, when in normal circumstances
ordinary civil twilight would be only an hour away, it was, in fact,
nearly totally dark up and down the entire west Java coast, and was
becoming similarly so in the capital. At this point enormous
chunks of pumice began to rain down from the sky.

There were three European ships inside the narrowest part of
the Sunda Strait at the time—the *Loudon* with Captain Lindeman,
which, because of the bucking sea, failed to reach Telok Betong
and so anchored well out in Lampong Bay instead; the Danish
salt-carrying barque *Marie,* which also rode out the enormous and
ever growing waves in the same Sumatran bay; and the cargo-
carrying barque *Charles Bal,* close to reaching the end of her long
voyage from Belfast to Hong Kong. All three vessels were deluged
with pumice: dangerous, heavy, sharp, fast-moving masses of rock,
the larger pieces still warm to the touch.

The captain of the *Charles Bal,* W. J. Watson, found himself in
a more perilous situation yet—somehow embayed, horribly pin-

---

*It was the International Meridian Conference held in Washington, D.C., in 1884 that for-
mally established the system of twenty-four time zones, each of which were essentially fifteen
degrees wide (though there were deviations around countries, states, and islands, where necessary, for neatness and convenience). The International Date Line was also set to run from
pole to pole through the Pacific, though it had to be jiggled around because of various mid-
Pacific islands that turned out not to exist. The whole notion of time zones was essentially
the brainchild of a man named Charles Dowd, principal of a women's college (which eventually became Skidmore) in Saratoga Springs, New York, who wore his beard and hair in
perfect emulation of his hero, Abraham Lincoln.

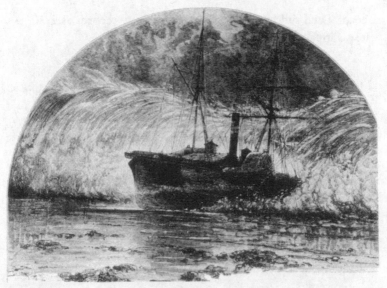

The Royal Dutch Navy's armed paddle steamer *Berouw* about to be picked up by one of the giant tsunamis generated by the eruption.

ioned in the sudden dark rain of rocks and compelled to beat around purposelessly, navigationally blinded. He was for a long while during the Sunday night just ten miles away from Krakatoa, closer to it than anyone else who survived. It enabled him to leave a record that was vivid in the extreme—except that chronologically (since it is thought that in the confusion he forgot to set his bridge chronometer to Batavia time) it was different by a single hour from everyone else's.

We first encounter Captain Watson* when he was beating northward, with Java Head and First Point, Welcome Bay and Pepper Bay, to his starboard, the great mountains of Sumatra to port, and the islands in the narrows of the Sunda Strait directly

---

*In an 1884 issue of the *Atlantic Monthly*, in a long interview conducted in San Francisco (where the *Charles Bal* sailed after visiting Hong Kong) by the journalist E. W. Sturdy.

ahead. Then suddenly, at what he incorrectly records as 2:30 P.M. (it was in fact only 1:30 P.M.):

. . . we noticed some agitation about the point of Krakatoa, clouds or something being propelled from the northeast point with great velocity. At 3:30 we heard above us and about the island a strange sound as of a mighty crackling fire, or the discharge of heavy artillery at one or two seconds' interval.

At five the roaring noise continued and was increasing [wind moderate from the SSW, notes Captain Watson here, his mariner's routines never quite deserting him]; darkness spread over the sky, and a hail of pumice-stone fell on us, of which many pieces were of a considerable size and quite warm. We were obliged to cover up the skylights to save the glass, while our feet and our heads had to be protected with boots and sou'westers.

. . . we sailed on our course, until at 7 P.M. we got what we thought was a sight of Fourth Point light; then brought the ship to the wind, SW, as we could not see any distance, and knew not what might be in the Strait.

The night was a fearful one; the blinding fall of sand and stones, the intense blackness above and around us, broken only by the incessant glare of varied kinds of lightning, and the continued explosive roars of Krakatoa made our situation a truly awful one.

At 11 P.M. . . . the island became visible. Chains of fire appeared to ascend and descend between it and the sky, while on the SW end there seemed to be a continued roll of balls of white fire. The wind, though strong, was hot and choking, sulphurous, with a smell as of burning cinders, some of the pieces falling on us being like iron cinders. The lead came up from the bottom at thirty fathoms, quite warm.

From midnight to 4 A.M. of the 27th . . . the same impenetrable darkness continued, while the roaring of Krakatoa less continuous, but more explosive in sound; the sky one second intensely black, the next a blaze of light. The mast-head and yard-arms were studded with corposants* and a peculiar pink flame came from fleecy clouds which seemed to touch the mast-head and the yard-arms.

At 6 A.M., being able to make out the Java shore, set sail, and passed the Fourth Point lighthouse. At 8 A.M., hoisted our signal letter, but got no answer. At 8:30 passed Anjer with our name still hoisted, and close enough in to make out the houses, but could see no movement of any kind; in fact, through the whole Strait we did not see a single moving thing of any kind on sea or land.

At 10:15 A.M. we passed the Button Island, one half to three quarters of a mile off; the sea being like glass all around it, and the weather much finer looking, with no ash or cinders falling; wind light, at SE.

At 11:15 A.M. there was a fearful explosion in the direction of Krakatoa, then over 30 miles distant. We saw a wave rush right on to the Button Island, apparently sweeping entirely over the southern part . . .

. . . by 11:30 we were enclosed in a darkness that might almost be felt, and then commenced a downpour of mud, sand, and I know not what . . . we set two men on the lookout for'ard, the mate and the second mate on either quarter, and one man washing the mud from the binnacle glass. We had seen two vessels to the N and NW of us before the sky closed in, adding not a little to the anxiety of our position.

At noon the darkness was so intense that we had to grope our way about the decks, and although speaking to each other on the poop, yet we could not see each other. This horrible state and the downpour of mud and debris continued until 1:30 P.M. the roaring of the volcano and the lightning from the

---

*This is the old name for those spheres of charged and luminous electrical clouds—known commonly to sailors as St. Elmo's fire—frequently seen about the masts of sailing ships during storms. The word comes from corpus sanctum, "holy body"—a reminder than in peril a sailor can do aught but put his trust in God.

volcano being something fearful. By 2 P.M. we could see some of the yards aloft, and the fall of mud ceased; by 5 P.M. the horizon showed out to the northward and eastward, and we saw West Island bearing E by N, just visible. Up to midnight the sky hung dark and heavy, a little sand falling at times, and the roaring of the volcano very distinct, although we were fully 75 miles from Krakatoa. Such darkness and such a time in general few would conceive and many, I daresay, would disbelieve. The ship, from truck to water-line, was as if cemented: spars, sails, blocks and ropes were in a terrible mess; but thank God!, nobody hurt nor was the ship damaged. But think of Anjer, Merak and other little villages on the Java coast!

Other ships, more distant, experienced even more drama. The *Berbice,* a German paraffin carrier bound from New York under the command of a Glaswegian, William Logan, found herself in a peculiarly exposed situation. When Logan saw the towering black clouds and lightning flashes ahead of him, from his position in the strait's western approaches, he supposed it to be no more than a tropical storm. But as soon as flaming ashes began to fall on deck—a wooden deck that was only inches from his highly flammable cargo—he recognized what was going on, understood the perils of his position, and promptly hove to in the lee of a protective island. He huddled there for the next two days—even though the island, by all accounts, afforded him precious little protection:

> The lightning and thunder became worse and worse. Lightning flashes shot around the ship. Fireballs continually fell on deck and burst into sparks. . . . The man at the rudder received heavy shocks on one arm. The copper sheathing of the rudder became glowing hot from the electric discharges.

The electricity in the air proved an even more serious problem, as Logan later recalled in an interview with an Australian newspaper:

Now and then when any sailor complained that he had been struck, I did my best to set his mind at ease, and endeavoured to talk the idea out of his head until I myself, holding fast to the rigging with one hand, and bending my head out of reach of a blinding ash shower which swept past my face, had to let go my hold, owing to a severe electric shock in the arm. I was unable to move the limb for several minutes afterwards.

Logan's crew were gripped with terror, volcanic dust covered the ship "at least eight English thumbs deep," his masts and sails were alive with fire and sparks, his barometer fell impossibly low, all the ships' chronometers mysteriously stopped, and the world beyond him was concealed in a frequently impenetrable miasma of whirling dust and smoke. His account does, however, conceal one note of incongruous optimism.

As well as his thousands of gallons of paraffin, Captain Logan carried in his cabin a small package wrapped in brown paper and tied with string, addressed to the curator of the Botanical Gardens at Buitenzorg. Inside were five specimen seedlings of a variety of spurge found in the forests of the Amazon, known as *Hevea brasiliensis*—wild rubber. There had already been numberless plans to harvest rubber commercially from such trees where they grew wild in Brazil, but, for a variety of reasons,* all the schemes failed: The *Berbice* was now bringing Amazonian plants to the Indies from which, it was hoped, plantation rubber could in due course be grown. The conditions of weather and soil made it likely that it would grow well in the East, the botanists predicted: Dozens of related plants, like tapioca, castor bean, and poinsettia already flourished there.

And this particular small story does have a happy ending. Despite the rigors of his passage through the strait, Captain Logan, together with his ship, his cargo of paraffin, and his infant

---

*Not the least of which was an endemic South American leaf blight caused by *Michocyclus ulei*. The five plants sent on the *Berbice* were free of blight, and the plantations of Southeast Asia have still not yet been infected, which many regard as akin to a miracle.

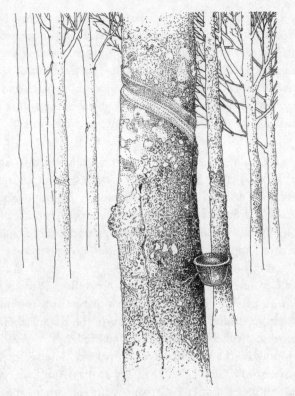

Hevea brasiliensis, Brazilian rubber.

rubber trees all survived Krakatoa; and the parent plants of what are now some of the most economically important rubber planta-tions in the world remain today in the Buitenzorg Botanical Gar-den, duly and safely delivered.

But otherwise the story of that long Sunday night makes for grisly reading. "Everything became worse," wrote an elderly Dutch pilot at Anjer. "The reports were deafening, the natives cowered panic-stricken, a red fiery glare was visible above the burning mountain."

At 6:00 P.M. the cable linking Anjer with Batavia finally broke—

the line going dead at the very moment Telegraph Master Schruit was telling government officials that yes, the eruption was continuing and indeed intensifying still. Schruit, tapping frantically at his Morse key, found he could not even make contact with the small town of Merak, seven miles up the coast. With his assistant telegraph operator in tow, he promptly dashed out into the gloom, ran through the old Dutch fort, fully intending to press on up the coast road to find and repair the rupture. He found it soon enough, just as he reached the drawbridge at the mouth of the harbor:

> . . . there, a fearful sight met my eyes: a schooner and twenty-five or thirty *prahus* were being carried up and down between the drawbridge and the ordinary bridge as the water rose and fell, and nothing remained unbroken, including the telegraph wires which had been snapped by the schooner's mast.
>
> But we felt no alarm as the water did not overflow its banks. Not entertaining any idea of danger, I sat down to table at about half-past 8. Of course I had made the necessary arrangements for beginning the repair of the broken line the first thing in the morning.

It was not to be. Anjer would not speak to Batavia again for the duration of this crisis—and Batavia would thus be wholly unaware of the terrible fate that would soon befall the town, and would befall all its neighbor villages up and down both the Java coast and across on the far side, in Sumatra.

The astronomical logs for August 26 note that civil twilight in Anjer port began that night at 6:22 P.M., half an hour after the sun had set, when artificial lights were first needed in the street; nautical twilight, the time when the horizon ceases to be the sharply delineated line that a navigator deems essential for working with his sextant, began at 6:47. Both periods would in normal circumstances endure for thirty minutes. This night there was no such thing. It had been dark in Anjer since midafternoon, and when the invisible sun did set, the darkness was Stygian indeed—the air a hot, ashy breath, filled with grit and sulfur, disorienting, confusing, and poisonous.

By late evening it was the turn of the ocean to take over as the more terrifying manifestation of Krakatoa's gathering power. As the great volcanic engine pumped and stoked more and more explosive energy into the atmosphere, so the waters surrounding the dying mountain became progressively more and more disturbed—and communities that were already huddling, frightened, along the low coastline of the strait began to experience ever greater waves, ever more dangerous seas.

The reports from after sunset speak continually of smashed boats and inundations of low-lying land, of ruined houses and of bystanders pulled off their feet into the raging waters. The first and most melancholy accounts were those later given to a number of newspapers in Java by the colonial *contrôleur* in the south Sumatran town of Ketimbang, Willem Beyerinck—the man who had given the first official news of the impending eruption back in mid May, when he telegraphed the resident of Lampong to say he had felt an outbreak of ominous tremors.

From that first Sunday afternoon, Mr. and Mrs. Beyerinck and their three children were to endure a week of the most exquisite agony—much of which they remembered well; and by so doing they provided one of the more reliable chronicles of this very complicated series of events.

The Sunday had begun innocently enough, with the opening of a new village market. There was the ritual slaughter of a baby buffalo, the playing of a gamelan orchestra, perhaps a *wayang kulit* puppet show—the kind of ceremonial that the Beyerincks had seen countless times during their tour. But these were strange times. Krakatoa had begun to rumble again that afternoon, and it somehow cast a shadow over an opening that was not, in consequence, an entirely happy affair. There had been a local outbreak of cholera too—a housemaid had just died, and Mrs. Beyerinck was worried about the health of her children. The children's *ayah* had seemed agitated for other reasons—complaining among other things that the birds that normally flocked around the family house had lately seemed restless, and that the auguries were not good. And Mrs. Beyerinck, warily watching the smoke roiling about the summit of Rakata, was wise and prescient enough to accept that one ignored the superstitions of the local people at one's peril.

When they returned from the market, Mrs. Beyerinck made what at the time seemed a strange request—that the family not go home, but make right away for a tiny village in the hills, where they rented a holiday cottage. Her husband, however, wouldn't at first hear of it. The locals, he said, would wreak havoc in Ketimbang town if he did so; and the wilder elements who had only lately been hired to pick the summer pepper harvest up in the highlands would soon hear that the Dutch *contrôleur* had run away, and would descend on the town in short order. No, he declared; the family would stay. Mrs. Beyerinck went off to her room to sulk, only remembering later the distant tinklings of the gamelan and the rhythm of a great drum sounding what she thought was a threnody.

But then everything changed, very suddenly. Her husband sauntered down to the shore to see what effects the new eruption might be having—and came on a scene that astonished him. While in the distance the mountain was roaring and boiling from behind an immense pillar of clouds, here enormous waves were breaking on the beach, and the level of the sea was piling up, rising and falling alarmingly, crashing with a weird randomness against anything solid on the shore. There was no wind, no storm. But the surface of the sea had a terrible, writhing, coiling awfulness about it.

He could see the *Loudon,* with its cargo of Anjer coolies bound for the pepper fields of Telok Betong, beating up Lampong Bay toward him, then trying desperately to dock. It was being tossed every which way, one moment corkscrewing high on the crest of a huge mass of water, the next being twisted as if by an unseen hand and plunged deep down into a trough. The master then evidently lost his nerve and gave up the struggle, for, as Beyerinck watched in horror, the boat, now looking so vulnerable and fragile, suddenly turned away, presumably to try to ride out the fury in midchannel.

The *contrôleur* stared, now momentarily dumbstruck, as the thrashing waters rose higher and higher up the shore, soon reaching the outbuildings of his own residency. Water began crashing against the stucco, breaking hard against what looked like increasingly insubstantial structures. It was this very sight that finally made up his mind for him. His earlier decision had been wrong, he told his servants: Mrs. Beyerinck and the children should leave right away. Everyone should flee to their summer cottage up in the hills.

For a few moments, it seemed unlikely that they would make it. For at 8:00 P.M., as a hail of pumice began to rain down, the waves began their first orgy of destruction. They were eventually to reach well over a hundred feet in height, and right from the start even the precursors of the mighty waves, even the first tentacle feelers of water, did the most amazing damage. In an instant Beyerinck's office suddenly came crashing down, along with a clutch of outbuildings. The family and their servants escaped drowning only by shinnying up coconut palms and waiting until the waters receded for a few moments' respite. Then they climbed back down, gathered up valuables, set the horses and their other animals loose, and ran, as far as they could manage, inland.

Their flight was the stuff of cinema epic. There was a dreadful roaring behind them as they stumbled, half blind, frightened, soaking wet, through miles of paddy, sinking into thick mud while trying desperately to outrun the ever pursuing monster. At one stage Mrs. Beyerinck, by now covered with mud from head to toe, tried to shout, but her throat was horribly sore and she couldn't utter a sound. She felt her neck—it was thick with a collar of leeches. They ran on and on, getting lost, from time to time joining forces with other local people, who themselves were fleeing in great crowds from the thundering, roaring floods behind them. Pieces of pumice hurtled down from the sky, burning fiercely like jagged meteorites.

The family and such servants as had run with them reached the hilltop cottage at midnight. They broke out supplies, fed their terrified

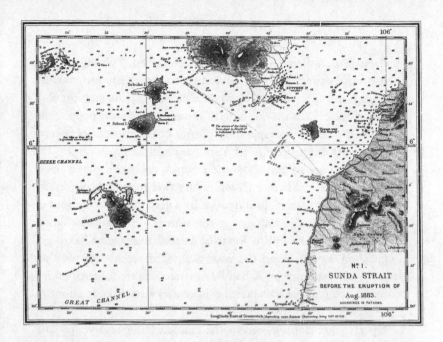

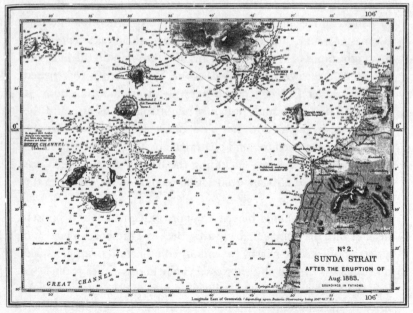

Admiralty charts show the islands of Krakatoa before and after the 1883 disaster.

children, settled them to some kind of fitful sleep. The adults then knelt on the slatted floor and peered through the window toward the raging volcano, which they could see distinctly through the fog of falling rock. Outside the hut lay thousands of local people, all crying and wailing in desperation. Some of the more sober were praying to Allah for some relief from the nightmare.

But it was not to end for some hours yet. One of Beyerinck's servants arrived just before dawn, saying that the entire residency had been ripped from its foundations by a gigantic wave at about 2 A.M. All the signs suggested that the waters were getting higher and higher, and the entire town of Ketimbang would likely go under. And indeed when the *contrôleur* sent scouts downhill at dawn to see the damage, it had been destroyed, totally. An enormous series of waves had flooded over every rooftop at about 6 A.M.: Nothing was left standing.

Up at Telok Betong, where the resident later said water had come within ten yards of his house, which was perched at the top of a hill 120 feet high, there was a mass of destruction. The harbormaster, waiting in vain for the *Loudon,* said he was swept off his feet eight times before he ran for his life. From up on the hills surrounding the town he watched as the Dutch navy's well-armed paddle steamer *Berouw* strained mightily at her mooring buoy. It looked to him as though not only might the chains give and the vessel founder, along with her crew of twenty-eight, but that the buoy might free itself too and be hurled about the harbor, destroying all other ships there—including the barque *Marie,* which was waiting in the roads. Mass destruction seemed to be lurking in the wings.

And on the *Loudon* herself Captain Lindeman found himself dealing with the unimaginable waves, the close presence of land and a host of other ships nearby (a perilous combination that sailors fear most in a storm, denying the vessel vitally necessary sea room)—as well as a terrified crew. The locally hired men in particular were petrified by the eruption of St. Elmo's fire in the riggings, and they left their posts in droves to try desperately to beat them out, to extinguish what was in fact inextinguishable. The phosphorescence was, they insisted, evidence of ghostly spirits: If these phenomena found their way below they would eat their way through the hull and the vessel would sink like a stone.

There was an almost endless succession of other, very similar reports. There were those from nine other nearby ships—like the magnificent American barque *W. H. Besse*, forty miles north of Krakatoa; the British vessels *Sir Robert Sale* and the *Norham Castle*, both of which were hove to off Sumatra; the Norwegian barque *Borjild*, eighty miles northeast of the volcano; the Welsh cargo vessel *Bay of Naples*, under way to Singapore, and 120 miles south of Java Head; the Rotterdam Lloyd's steamship *Batavia*, well to the southwest of the entrance to the strait; the steamer *Prins Frederik*, which had passed Krakatoa on August 25 and was by the time of the explosion also well out into the Indian Ocean; the *Annerley*, southbound, and at the time of the eruption standing to the north of the strait; and the British-flagged *Medea*, whose Captain Thomson had managed to measure with some accuracy the height of the volcanic cloud at the outset of the eruption, at seventeen miles.

There were all manner of curious survivals. One man had fallen asleep at home and awoke to find that the wave had lifted him and his bed to the top of a hill, depositing him there in perfect safety. Another grabbed the corpse of a cow and floated to high ground. More bizarre still—and barely credible—was the man who reportedly found himself being swept inland next to a crocodile: He clambered on to its back and hung on for grim death with his thumbs dug deep into the creature's eye sockets.

And there were in addition statements and newspaper interviews and private letters home from lighthouse keepers and residents, assistant residents, *contrôleurs*, agents of Lloyd's, telegraph operators, harbormasters and sharp-eyed civilians of all stripes, as well as an enterprising member of the Catholic clergy named

Julian Tenison-Woods,* who wrote an exceptionally long letter about the events to the *Sydney Morning Herald*. From this mass of information a broad summary can be distilled.

Krakatoa's final twenty hours and fifty-six minutes were marked by a number of phases. First, from early afternoon on Sunday until about 7 P.M. there was a series of explosions and eruptions of steadily increasing frequency and vigor. From early evening the ash falls and the deluge of pumice began. By 8 P.M. the water had become the next medium of transmission of the volcanic energy, and as night fell the temper of the sea in the Sunda Strait became one of unbridled ferocity.

Then, just before midnight, a series of air waves—fast-moving, low-frequency shocks sent out invisibly and inaudibly by the detonations—began arriving in Batavia. The time-ball on the astronomical clock down at Batavia's harbor stopped dead at eighteen seconds after 11:32 P.M. because of the ceaseless vibrations. Audible evidence of the explosions began to radiate outward too, and there was a report from Singapore and Penang† that thudding sounds could be heard at about the same time. In Batavia a large number of people, kept awake by the explosions and for want of something better to do, were walking around the Koningsplein; they noticed that the gas lanterns suddenly dimmed at about

---

*This was the kind of man whose like we do not see today. A London-born Passionist minister, amateur geologist, and naturalist, he was compelled by ill health to move abroad when he was in his twenties, and became an expert on the conchology and paleontology of Tasmania during the time he worked as a traveling missionary for the Catholic archbishop of Sydney. He left the priesthood in 1883, when he was fifty-one, and traveled to Malaya, Singapore, and the Dutch East Indies—where he became fascinated by the eruption of Krakatoa—and then on to the Philippines, China, and Japan. He returned to Australia and died in Sydney six years later, in 1889—"a man of wide culture, a musician, an artist, and something of a poet," as well as a writer of hymns, the author of a book on the fish of New South Wales, another on the history of Australia, and more than 150 other contributions to scientific literature.

†These reports were also collected by Tenison-Woods, a reliable and properly disinterested observer.

1:55 A.M. Along Rijswijk, the main shopping street, several shop-windows suddenly and inexplicably shattered at about the same time.

Then at about 4 A.M. the nature of the explosions reportedly changed, very slightly, becoming less continuous but more explosive. Someone described the sounds as like a steam-engine, emitting full-throated *whoomphs* as it gathered speed. At about 4:56 A.M. an enormously powerful air wave was detected at the Batavia gas-works—suggesting, if travel time over the ninety miles to the volcano is allowed for, that something else had just happened deep within Krakatoa's heart. The culminating explosion—though no one on the ground at the time knew it—was soon about to happen.

There were four gigantic explosions still to come. The first was noted at 5:30 A.M. The Sumatran town of Ketimbang was then destroyed at 6:15 A.M., and Anjer, its Javan sister port across the strait—according to the few who survived to tell the tale—was inundated and wrecked very shortly thereafter. The second mighty explosion came at 6:44 A.M.—forty-one minutes after a dawn that, to those in all of western Java, never arrived that day. Ashes began to fall on Batavia at 7:00 A.M.—although Oscar Hatfield, the American consul, reported seeing them falling in the consulate grounds two hours later.* At 8:20 A.M. a third, quite terrible explosion was felt in Batavia, and many of the buildings started to make what were described as "crackling" noises. And then finally, at 10:02 A.M., came the culminating, terrifying majesty of it all.

Two minutes to go, and according to simultaneous reports: The sky was completely darkened in all of southern Sumatra; the *Loudon* was weathering heavy ash falls in Lampong Bay; the nearby *Marie* reported "three heavy seas came after each other; at once a

---

*For some unexplained reason, Mr. Hatfield left little by way of an interesting record of the Krakatoa events—and his report of the ash fall, which he timed much later than did Tenison-Woods's very detailed and certain description ("pale yellow . . . gloomy . . . very dense"), suggests that certain of his consular faculties might have been wanting.

fearful detonation; sky in fire; damp." The *Annerley* lit all her lights, noted that it was raining pumice stone, that the barometer was rising and falling half an inch a minute. In Batavia it became eerily dark again, and—most significantly—it started to get cold. From 10:00 A.M. the temperature began to fall—as many as fifteen degrees Fahrenheit over the coming four noontide hours.

Explosions like a battery of guns are heard across in Telok Betong. Lightning strikes the lighthouse at Vlakke Hoek in southern Sumatra. The lighthouse at Fourth Point, just to the south of Anjer, is hit by a vast wave and destroyed, ripped off at its base, leaving only an amputated stump of jagged masonry. An immense wave then leaves Krakatoa at almost exactly 10:00 A.M.—and then, two minutes later, according to all the instruments that record it, came the fourth and greatest explosion of them all, a detonation that was heard thousands of miles away and that is still said to be the most violent explosion ever recorded and experienced by modern man. The cloud of gas and white-hot pumice, fire, and smoke is believed to have risen—been hurled, more probably, blasted as though from a gigantic cannon—as many as twenty-four miles into the air.

"A fearful explosion." "A frightful sound." Captain Sampson of the British vessel *Norham Castle* wrote simply in his official log: "I am writing this blind in pitch darkness. We are under a continual rain of pumice-stone and dust. So violent are the explosions that the ear-drums of over half my crew have been shattered. My last thoughts are with my dear wife. *I am convinced that the Day of Judgment has come.*"

The British consul in Batavia at the time was one Alexander Patrick Cameron; and five days later he sat down in his study and wrote out, in the usual impeccable sweeps of fine Victorian copperplate, his summary of what he then knew of the disaster. The document remains today in the Public Record Office in London, largely unread and unconsulted because of a confusion that has led those

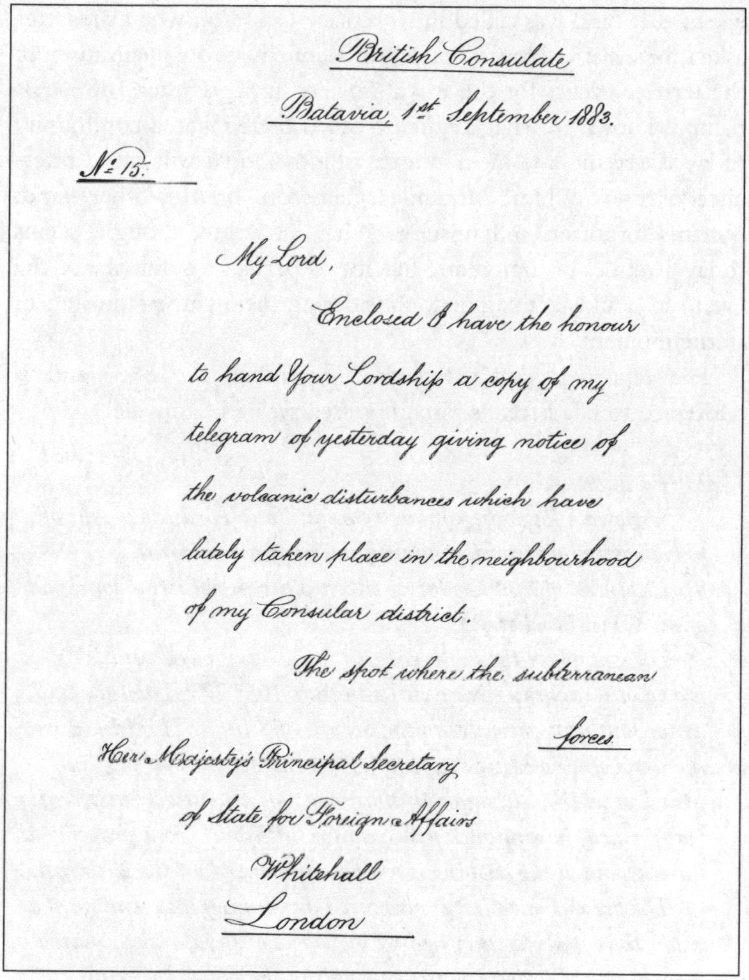

British Consulate

Batavia 1st September 1883.

No 15.

My Lord,

Enclosed I have the honour to hand Your Lordship a copy of my telegram of yesterday giving notice of the volcanic disturbances which have lately taken place in the neighbourhood of my Consular district.

The spot where the subterranean

forces

Her Majesty's Principal Secretary
of State for Foreign Affairs
Whitehall
London

The elegant copperplate and exquisitely courteous tone of Consul Cameron's lengthy Krakatoa dispatch to Lord Granville, in London.

who have chronicled the Krakatoa events to believe that the British consul was in fact a man named Henry George Kennedy.

The error is understandable. Kennedy had in fact been consul

in Sumatra, and was called in to replace Cameron when the latter
asked for leave in November 1883. Kennedy wrote a summary of
the terrible events for the Royal Society in September 1883. His
name is known to what might be called the Krakatoa community
today as a result, and most indexes of most books will have a reference or two to him. Alexander Cameron, on the other hand,
remains forgotten and unsung. What he wrote, though, seems
today a model of diplomatic felicity, as perfect a summary of the
events as could be imagined, considering the awful circumstances
of the moment.

His report is dated Batavia, September 1, 1883, and is
addressed to Gladstone's foreign secretary, Earl Granville:

> My Lord:
>   Enclosed I have the honour to hand Your Lordship a copy of
> my telegram of yesterday, giving notice of the volcanic
> disturbances which have lately taken place in the neighbourhood
> of my Consular district.
>   The spot where the subterranean forces have found vent is the
> island of Krakatau* lying in Longitude 105° 27E, Latitude 6° 7S,
> at the southern entrance to the Straits of Sunda. This island was
> the scene of a volcanic eruption of less importance on the 20th
> May last which, although on that occasion an entirely new crater
> was formed, had no such disastrous results to life and property as
> have attended the explosions which commenced on the 27th inst.
>   The present outburst commenced on Sunday last, and on that
> night the inhabitants of nearly the whole of Java and Sumatra
> were alarmed by loud noises resembling the reports of heavy
> artillery, which continued throughout the night and at rarer
> intervals during Monday 28th inst. It soon became known that

---

*Note that the consul spells the volcano's name correctly: As already suggested, the mis-
spelling *Krakatoa* seems more a consequence of the carelessness of journalism than diplomacy. However, in the same paragraph Cameron makes an error with his dates. *Even Homer
nods.*

these noises were produced by a fresh eruption of Krakatau and since Monday intelligence has been slowly reaching Batavia from various quarters apprising us of the extent of damage done, and proving by the loss of life and property that this is one of the greatest calamities of this century.

The residencies of Bantam and Batavia were darkened throughout the early hours of last Monday by a thick cloud of grey ashes, the light diminishing gradually, as the cloud progressed from west to east, from twilight to almost total darkness at midday, and a continuous shower of ash fell during the forenoon giving the ground an appearance as if covered by snow. At about 11.30 a.m. at Batavia and at earlier periods of the day in the more immediate vicinity of Krakatau the sea suddenly rose, presumably owing to the subsidence of part of Krakatau and other islands or to a submarine upheaval, and a wave of considerable height advanced with great rapidity on the shores of western Java and southern Sumatra, causing greater or less damage according to its distance from the centre of disturbance. A second wave higher than the previous one followed the first at an interval of about an hour with even more serious results. It is now reported that part of Krakatau island, the island of Poeloe Tempora and other small islands in Sunda Straits have disappeared, and that a reef has been formed between Krakatau and Sibesie islands, the channel usually taken by steamers. Dwars-in-den-weg/Athwart-the-Way, an island at the northern entrance to the Straits, is reported split into five pieces, while numerous small islands are said to have been raised which had no existence previously.

These reports however still require verification and with a view to ascertaining the extent and nature of the changes caused by the volcanic action a Government survey-steamer has been dispatched to the neighbourhood to take a new survey of the Straits.

The destruction caused by the waves on shore both to life and property, although known from reports already to hand to be very

widespread, can hardly yet be estimated with any degree of
certainty, as owing to the action of the sea and the heavy rain of
ashes, telegraph and road communication has been either entirely
interrupted or is much delayed.

It appears beyond a doubt however that the whole of the
southeastern coast of Sumatra must have suffered severely from
the effects of the sudden influx of the sea, and thousands of
natives inhabiting the villages on the coast must have almost
certainly perished.

The west coast of Java from Merak to Tjeringin [has] been
laid waste. Anjer, the port where vessels bound for the Java and
China Seas call for orders and a thriving town of several
thousand inhabitants (natives), no longer exists, its former site
now being a swamp.

The lighthouse at Anjer (Java's Fourth Point) has also been
much damaged.

Many Europeans, including numerous officials, and many
thousand of natives have been drowned, in the district of
Tjeringin alone on the southeast coast of Java it is reported that
no less than ten thousand persons have lost their lives. The result
to agriculture in west Java [is] not yet officially known. The fact
however that owing to the covering of ashes which spreads over the
whole country, the cattle are deprived of their ordinary
nourishment, is in itself a very serious consideration and
measures have already been taken to supply the afflicted districts
with food for man and beast. It is to be feared that the natives
will be greatly impoverished by the damage done to fruit and
palm trees which form a source of wealth, while coffee and tea
gardens and standing crops of all descriptions must have suffered
severely.

With a view to rendering safe the navigation of the Sunda
Straits the Rear Admiral, Commander in Chief of the
Netherlands Indian Navy, has stationed one man-of-war to
cruise off the southern and another to cruise off the northern

entrance to the Straits to warn vessels to proceed with caution.

In view of the quantity of shipping (principally British) which daily passes through Sunda Straits and the important nature of the circumstances above related I have thought it my duty to dispatch the telegrams mentioned in the accompanying memorandum, and trust my action will meet Your Lordship's approval.

> *I have the honour to be,*
> *My Lord,*
> *Your Lordship's Most obedient, Humble Servant,*
> *A. P. Cameron*
> *Her Britannic Majesty's Consul, Batavia*

The island of Krakatoa, meanwhile, had in essence disappeared. Six cubic miles of rock had been blasted out of existence, had been turned into pumice and ash and uncountable billions of particles of dust. The rumblings and roarings continued for some while, then on Monday afternoon became ever fainter. By dawn on Tuesday they had stopped completely. That last great detonation at two minutes past ten on that Monday morning had blown the island apart, and sent most of it to kingdom come.

Now it was time for those who could, together with those whose duty it was, to venture out to see just what damage the eruption had caused.

THE EFFECTS

It was just before dawn on the Monday, and an elderly Dutch harbor pilot, one of those stationed in Anjer to guide ships to and from the Batavia roads, was walking on the beach. He couldn't sleep; besides, staying inside was perilous, not least because the intermittent hails of pumice stones, many of them too hot to touch, threatened either to set ablaze the *atap* thatch with which his house was roofed, or to smash holes in it and wreak who knows what damage inside. Much better, he thought, to watch the great events from the comparative safety of the shore.

There was not much visible through the gloom. The clocks said that the sun ought to be ready to rise; but the falling, swirling ash had effectively dimmed the view for more than a few yards in any direction. Krakatoa itself, thundering away angrily to the west, was quite invisible—except that there was a dark orange glare to the ash clouds in the mountain's direction: It was like the view of a very distant furnace glimpsed, only half seen, through the dark clouds of its smoke.

But then, all of a sudden, the image shifted. Suddenly the old pilot, who had spent a lifetime guiding vessels through dangerous and unpredictable waters, became aware of something that was just barely visible, something that shouldn't have been there at all. He related to the Reverend Philip Neale, the British chaplain in Batavia, who later in the year set about collecting the stories of eyewitnesses, exactly what it was:

> Looking out to sea, I noticed a dark black object through the gloom, traveling toward the shore. At first sight it seemed like a low range of hills rising out of the water—but I knew there was nothing of the kind in that part of the Sunda Strait. A second glance—and a very hurried one at that—convinced me that it was a lofty ridge of water many feet high.

In the aftermath of Krakatoa's eruption, 165 villages were devastated, 36,417 people died, and uncountable thousands were injured—and almost all of them, villages and inhabitants, were victims not of the eruption directly but of the immense sea-waves*

---

*The term *sea-wave*, or its Japanese equivalent, *tsunami*, is now generally preferred by the scientific community; *tidal wave*, a term that yet survives, is condemned as inaccurate, since the waves caused by earthquakes or volcanoes are in no sense tidal.

that were propelled outward from the volcano by that last night of detonations.

It was in this one respect—the production of a number of massive and highly destructive sea-waves—that Krakatoa was then and remains today so very unlike almost all the other of the world's great volcanic disasters. Its scale was phenomenal. The number it killed was unimaginably vast. But it was the way that it killed all those people that still sets Krakatoa apart.

Other volcanic eruptions around the world kill people in more direct and predictable ways—and they kill and injure, it should be remembered, a not insignificant number of people, since one in ten of the world's population is currently reckoned to live near volcanoes that are either active or have the potential to become so. So far as volcanoes are concerned, a great number of people—in the Philippines, in Mexico, on Java, in Italy even—currently live in harm's way.

The types of hazards to which such people are likely to fall prey, or to which their forebears fell victim in the past, are many and manifest. Erupted boulders and lumps of partly congealed lava—generally known by the term *tephra*, from the Greek word for "ash"—scream back down from the skies and flatten anything in their path. Perhaps a relatively small number of people, fewer than a thousand, died in this way from the Krakatoa eruption. All of them were in southern Sumatra, in the path of the prevailing wind: The hot ash that burned them alive had sped westward from Krakatoa on top of a cushion of superheated steam.

Most of the other means with which volcanoes kill their victims were not experienced here. In other eruptions lava flows surround and trap victims and sear them to death. Earthquakes associated with volcanoes destroy buildings, and huge seismically caused cracks in the earth swallow people and the buildings in which they live. The terrifyingly fast-moving clouds of hot lava, ash pumice, and incandescent volcanic gases, known to the French as *nuées ardentes* and to the rest of the world as *pyroclastic flows*, sweep people up and incinerate them in seconds—as with, for example, almost every one of the 28,000 inhabitants of St. Pierre, in Martinique,* who in May 1902 had been persuaded to stay in town for a supposedly important election, but were burned and suffocated by the sudden pyroclastic flows coursing down from the eruption of Mt. Pelée.

Clouds of sulfur-dioxide gas, usually released during eruptions, choke and poison their victims. Clouds of carbon dioxide suffocate them. Clouds of hydrochloric acid gnaw away at their lungs. The torrents of volcanic mud and water slurry that course down the sides of certain volcanoes and that have the Javanese name *lahars* (since so many such flows run down the sides of Javanese volcanoes—though not, as it happens, Krakatoa) carry victims miles away, and drown and bury them.

Sometimes secondary events can prove fatal. In 1985 a small eruption of a Colombian volcano called Nevado del Ruiz melted a glacier near the summit: The resulting river coursed down a valley that was quite unused to such huge flows, and the mud sea that was eventually created drowned an entire village below, killing 23,000 people. There are still more obscure risks: For example, volcanoes that erupt beneath glaciers—which tend not to have too many people living near them—produce sudden floods of melting ice, which have recently been given the exotic Icelandic name *jökulhlaups*. These can also prove fatal.

However, of all the victims whose deaths can be attributed directly to volcanic activity during the last 250 years, fully a quarter are now believed to have died—drowned or smashed to pieces—as a result of the gigantic waves that were created by the

---

*There were precious few survivors: The best known was a prisoner named Louis Augusté Ciparis (or Sylbaris), who was in solitary confinement in an evidently well-insulated and nearly airtight cell. When he was found to be a survivor, a miracle was promptly declared and he was set free. Barnum & Bailey had him perform in their traveling circus for some years before he got into trouble again and ended up in an American jail. He was fired from the circus when people lost interest, and died a pauper in Panama in 1955. His cell in St. Pierre still stands as a tourist attraction; visitors are taken there in a bus called the Ciparis Express.

eruptions. The entire Minoan civilization on Crete was supposedly wiped out in 1648 B.C. when volcanic tephra from the eruption of Santorini—or, much more probably, the tsunamis thrown up by the eruption—destroyed the palaces at Knossos. More than ten thousand people died in 1782 in the waves that were created by an avalanche of volcanic debris that hurtled into the sea from Japan's Mt. Unzen. In 1815 a similar number of Javanese died when Tambora exploded, sending pyroclastic flows raging into the ocean, with tsunamis radiating out in all directions and inundating the coast.

Careful study of the records for the last two and a half centuries has come up with a total of some ninety tsunamis for which volcanoes alone can be held responsible—and the greatest of these by far was the 1883 eruption of Krakatoa. About 35,500 men, women, and children died as victims of the two gigantic waves that accompanied or were caused by the death throes of this island-mountain, and they account for more than half of all those in the world who are known ever to have died from waves caused by an erupting volcano. So this should be remembered well: It was neither fire nor gas nor flowing lava that killed most of the victims of Krakatoa. All but the thousand who were burned in Sumatra by the immolating heat of newly made ash and pumice and scalding gases died by the primary agency of water.

During the eruptive days back in late May, the state of the sea was certainly noticed, but was never once reported to be the cause of any undue alarm. The hopper *Samarang* noticed a swell powerful enough to lift her screw out of the water; the lighthouse keeper saw the surface of the strait turn suddenly white; the rudder of the *Bintaing*, another small hopper, swung around and hit her own hull with a mighty clang when a freak wave caught her. But that was about all: The eruption in its opening stages was about ash falls and noise and that seven-mile-high column of coiling smoke. The ocean seemed to prefer not to become involved.

But three months later matters were very different. The way

that Krakatoa's immense outpouring of thermal energy was converted into mechanical energy—for this conversion is what essentially determines both the immensity and the enormity of any volcanic eruption—was altered. The noise was there, on an extraordinary scale. The expulsion of material high into the sky went on, both in gigantic amounts and for a very long time. But most of the mechanical energy went into the enormously difficult task of moving the ocean—movement that, once started and given additional shoves from behind, can become one of the most powerful natural forces imaginable.

In August the state of the sea was something noticed by all. Right from the beginning, when Telegraph Master Schruit took his lunch and strolled out onto the hotel veranda to first see the column of smoke, it was the strangely erratic motion of the sea that most alarmed him. On the far side of the strait, in Ketimbang,

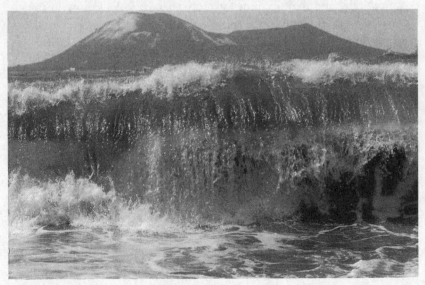

A classic wall-of-water tsunami, generated by only the most moderate of earthquakes on Krakatoa.

Contrôeur Beyerinck too was astonished by the punishment his town's little dock was having to take from the curiously restless waters. The ships out in the strait—the *Loudon,* the *Marie,* the *Charles Bal*—all reported on the state of the sea. For them it was not so serious, as waves at sea are less dangerous to a ship than waves close to land. The electricity in the air and the rain of flaming rocks from the sky were quite dangerous enough.

As darkness fell, so the sea became ever more furious. At 7:00 P.M. on Sunday, Beyerinck saw small boats being tossed around. At the same time on the Javan side, Schruit found his telegraph cable had been snapped by the mast of a schooner tossing on the waves. Between 7:00 P.M. and 9:00 P.M. several houses close to the seafront in the small town of Tyringin, well to the south of Anjer, were reported destroyed and swept away.

At about 7:30 P.M. a quarry near Merak, where dozens of Chinese laborers were hewing out stone* for the new Batavia docks, was then inundated, and the camp where the workers slept was washed away: they may have been the first casualties of what was to be a long and mortally expensive night. But now there was a lull: though a village five miles out of Anjer was reported to be submerged at 10:00 P.M., by midnight the sea was glass-smooth once again; and at 1:00 A.M. on Monday morning Schruit, still trying feverishly to repair his severed cable (he eventually failed), noticed only small oscillations in the surface of the sea close to where the Anjer canal debouched into it.

Then at 1:30 A.M. one almighty wave is reported to have rushed up the long funnel of Lampong Bay to Telok Betong, where it ripped through and ruined several houses. Although it was clearly highly destructive, and though the time of its occurrence seems to be accurate, having been cross-checked with other

---

*It seems now a measure of the Chinese laborer's legendary tolerance for appalling working conditions that so many were still hewing stone in Merak, despite the terrifying nearby concussions, to say nothing of the flames and the clouds of ash.

witnesses (not least the servants of Beyerinck, sheltering with his family in their hilltop cottage), this one wave appears to be something of an aberration—far larger than its predecessors but unrelated to any particular event at the volcano. It was indicative, however, of what was to come.

The greatest and most terrifying volumes of water began moving in concert with Krakatoa's four culminating explosions—the first eruptive paroxysm being timed, as we have seen, at 5:30 A.M. It was as though something deep within the mountain had begun a series of low-frequency pulsations, the sea moving back and forth in time with each pulse, and the amplitude of these movements becoming greater and greater, the *volume,* as it were, of the waves becoming stronger and stronger with each sequence of pulses. The four major tsunamis that were caused by, or were coincident with, these giant volcanic explosions then hit the shores like planet-size wrecking balls, the effects all unimaginably and fatally destructive.

The destructive capacity of a great wave can be calculated, with difficulty, from a mess of competing and combining features, including the configuration of the shoreline, the funneling effect of cliffs and headlands, and the depth of the coastal waters. It seems from the various eyewitness reports that what was most impressive about the waves that struck the shores of Java and Sumatra that morning was their sheer size—the high and unstoppable moving walls, the majestic volume of hundreds of billions of tons of roiling, thundering, foaming green water.

The last four great explosions of Krakatoa's life took place at 5:30 A.M., 6:44 A.M., 8:20 A.M., and, finally and most terrifically of all, at 10:02 A.M.—all these well-chronicled moments being recorded in Krakatoa Time, which (because each local Dutch administrator still set his official watch according to when the sun rose and set and reached its noontime peak *in his own district*) was in those days five minutes and forty-two seconds behind what the capital's civil servants regarded as Batavia Standard Time. The energy that was released in these eruptions was transformed into a variety of violent effects. There were massive expulsions of rock and ash and gas. There were torrents of heat, searing and welding together everything around them. There were sounds—bangs, cracks, thunderous roars, shattering low- and high-frequency noises—that were so loud they could be heard thousands of miles away. Seismic shocks were triggered that caused buildings five hundred miles away to rock on their foundations.

And the eruptions also produced two kinds of shock waves. One was a wave that passed invisibly through the air, a sudden burst of pressure that bounced around the world, and was recorded as doing so, moreover, a remarkable *seven* times. These air waves—which recorded as pressure spikes at the Batavia gasworks, ninety miles to the east—radiated outward from Krakatoa very fast, at what was an easily calculated velocity of about 675 mph. They were recorded as reaching Batavia at 5:43 A.M., 6:57 A.M. and (there seems, curiously, to be no firm record of any airwave resulting from the third explosion) at 10:15 A.M. respectively, Batavia Standard Time. (As noted, these events took place before the invention of time zones, either in the East Indies or anywhere else in the world. Taken together with the dubious accuracy of many of the mechanical clocks of the day, the absence of the coordinating abilities of radio, which had of course not quite yet been invented, and the wide range of anecdotal reports from frequently panicky eyewitnesses, this makes it tricky, though not entirely impossible, to construct a firm chronology of what took place in the aftermath of the eruptions.)

The other shocks, considerably more complex in the way they moved, of much shorter duration but of equally extraordinary geographical spread, involved the disruption of the surrounding sea-water. Sea-borne waves in general move much more slowly: in the relatively shallow waters of the Sunda Strait probably at an average speed of about 60 mph.* However the Krakatoa tsunamis

---

*The speed of a tsunami is directly proportional to the square root of the depth of sea through which it travels. In midocean, where the depths are measured in thousands or tens

were forged, it would take one of them about thirty minutes from the moment of eruption to travel to the closest point on the mainland.* And it would be thirty-seven minutes before the wave was close enough to the town of Anjer for the people there to see it, to recognize just what it was that was bearing down on them, and to start—a fairly but not entirely futile gesture—to try to outrun it.

It would take a further fifteen minutes for a great wave like this to seek out and destroy the quarries in Merak in the north, and drown all the Chinese workers there (as it did). It would take seven minutes fewer to flatten and wipe out all of Tyringin in the south (as it did also). And it would be one hour and one minute† before the same wave, slowing itself down but building itself up all the while, would reach all the way up to the head of Lampong Bay and, as it was equally sure to do, wreak havoc in the attractive little south Sumatran town of Telok Betong.

The coastlines of Sumatra and Java are, like any coastlines, made hugely complex by all the inlets and island-shadowed estuaries, bays and peninsulas, rocks and reefs. The way that an inrushing wave behaves as it courses toward the shore is only vaguely understood—making it somewhat challenging to try to work out from the survivors' tales which wave actually struck and destroyed each

---

of thousands of feet, it can move at speeds of up to 500 mph; where the water is 900 feet deep it will slow to 115 mph. In the Sunda Strait, where the water depth varies from about 500 feet in midchannel to twenty feet and less at the edges, the speed of travel appears to have averaged about 60 mph. When a fast-moving deep-sea wave encounters the shore, it changes shape drastically: out in midocean it will be fast moving but only a few feet high; as it reaches shore it will slow down, pile up on top of itself and very swiftly become enormous. A wave generated by Krakatoa might have started out with a height of ten feet and a speed of 100 mph; when it got to Telok Betong, or Anjer, or Merak, it may then have slowed to only 20 mph—but it could have been as tall as a ten-story house, with frightening consequences for anything or anyone in its path.

*The nearest land is Point Tikus in southern Sumatra, on the west side of Lampong Bay— but happily it is in the shelter of a small island, which would slow the wave's moment.

†The accuracy of these figures comes from the Royal Society report, which calculates the speed of the onrushing waves as the square root of the product of gravity and the water depth, $V = \sqrt{gh}$. The Dutch mining engineer Verbeek came up with slightly different figures and used a more complicated equation to find V, the wave speed: $V = \sqrt{g/2h}(h + \in)(2h + \in)$, where g is gravity, h the water depth and $\in$ the height of the wave crest above normal.

affected town, village, kampong, and home on the edges of the two great East Indian islands.

Which wave was it that killed the vast majority of those thirty-six thousand who were lost?

Was it "the low range of hills rising out of the sea" that was seen, chillingly, by that elderly Dutch pilot in Anjer at dawn? "The sight of those receding waters haunts me still," he was to write later, since for him this *was* the killer wave, without doubt. "As I clung to the palm tree . . . there floated past the dead bodies of many a friend and neighbour. Only a mere handful of the population escaped. Houses and trees were completely destroyed, and scarcely a trace remains of where the once busy, thriving town originally stood."

Or was it the climax of all that terrible agitation of the sea that compelled Mrs. Beyerinck across in Ketimbang to demand that her husband and family flee for the hills and the safety of high ground? Was it the "giant black wall of water" that roared into Telok Betong at 7:45 A.M., picked up the gunboat *Berouw* as though it were a child's bathroom toy and dropped it in the middle of the Chinese quarter of town? The same wave that stranded the government's revenue cutter, and smashed all the local *prahus* and scattered the fragments of their hulls about like so much confetti?

Could it have been one of the "four waves" supposedly seen that morning by an engineer named R. A. van Sandick? He was a passenger aboard the *Gouverneur-Generaal Loudon*—the steamer that, it will be remembered, was unable to dock at any of the quays in Lampong Bay because of the raging surf. The waves, which came in at tremendous speed sometime between 7:30 and 8:30 A.M.

> . . . destroyed all of Telok Betong before our eyes. The light tower could be seen to tumble; the houses disappeared; the steamer *Berouw* was lifted and got stuck, apparently at the height of the cocoanut trees; and everything had become sea in front of our eyes, where a few minutes before Telok Betong beach had been. The impressiveness of this scene is difficult to

describe. The unexpectedness of what is seen and the tremendous dimensions of the destruction, in front of one's eyes, make it difficult to describe what has been viewed. The best comparison is a sudden change of scenery, which in a fairy tale occurs by a fairy's magic wand, but on a colossal scale and with the conscious knowledge that it is reality and that thousands of people have perished in an indivisible moment, that destruction without equal has been wrought.

Or was it perhaps the wave that struck Merak at 9:00 A.M.—the wave that drowned all but two of the town's 2,700 inhabitants? An accountant named Pechler who somehow survived by running before it, climbing farther and farther uphill until he was beyond range, certainly would imagine this tsunami to be immeasurably vast: It destroyed stone buildings that stood on top of a hill later measured at 115 feet high; it drowned all thirteen Europeans who lived there and who had had good reason to feel secure, surrounded as they were by walls of heavy masonry on the summit of a good high hill. But the wave displayed all the insouciance of its great power; and at the time it roared over, submerged, and then wrecked these mansions it was towering above them by a good twenty feet—meaning that whether what Pechler saw was *the* wave or not, it was at least 135 feet high, formidable in its terror. It drowned everyone in the town below, and when the waters receded almost everything in the town was either smashed beyond recognition or swept clear away.

Or yet again, might the great wave have been the one that was recorded savaging Merak an hour later, at 10:30 A.M., when a Dutch *contrôleur* named Abell, on the road to Batavia with his *wedono** to report to his superiors details of tragic happenings yet

---

*A *wedono* was an indigenous colonial official of almost the same rank as the European *contrôleur*, who acted as his district officer. Dual administrative rule was a permanent feature of Dutch colonialism—a native official called a regent supposedly worked alongside the Dutch resident, for example—and, providing both sides acknowledged where the power truly lay, the system generally worked well.

farther down the coast, looked around to see "a colossal wave" roaring up the shore? It was, he said later, taller than the tallest palm tree he could see—a wall of water that no one caught by it could possibly have survived, something so dreadful it was quite beyond nightmares. Might this have been the one?

The answer on this occasion is probably yes. In fact, almost without a doubt, however compelling and awful the accounts of eyewitnesses to the other tsunamis of that dreadful morning may be, this last was indeed the one, the real killer wave. It happened at what seems to have been the correct time—with a travel speed of sixty miles per hour, its arrival at Merak at 10:30 would put its time of origin at Krakatoa at almost exactly ten o'clock, which is the moment of the culminating, self-destroying explosion.

Most crucially, this one wave is recorded as having hit with extraordinary destructive power, a short while either before or later than the Merak 10:30 A.M. arrival, at all the population centers of the west Java and south Sumatra coast. "An immense wave inundated the whole of the foreshores of Java and Sumatra bordering the Strait of Sunda," reported a contemporary study, "and carried away the remaining portions of the towns of Tjiringin, Merak and Telok Betong, as well as many other hamlets and villages near the shore."

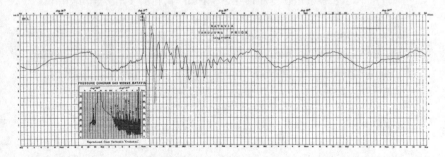

The tide meter at Batavia registers a sudden swell at 12:36 P.M., two and a half hours after the eruption—showing how relatively slowly tides move, compared to the fast-spreading barometric pressure wave recorded at the gasworks.

Its arrival was also recorded on the well-armored tide meter in Batavia Harbor—at 12:36 P.M. A wave so powerful as to give an almighty jolt to that tide recorder would have to have been enormous indeed. It would also have had, if traveling at sixty miles per hour, to have begun its journey some two and a half hours beforehand. This means, in other words, that it would have originated at a few minutes past ten in the morning. Undoubtedly, from all the evidence, this too was Krakatoa's most colossal wave, the biggest consequence of the biggest and final explosion. It was a wave so enormous and so powerful that it turned out to be the grimmest of grim reapers, the terrible climax to a long and deadly day.

"Everyone was frozen with horror," wrote the resident of Lampong, Mr. Altheer, of the moment when he heard the almighty explosion, just after ten on that Monday morning. He well knew, from what had already happened disastrously three or four times before during the previous twenty hours, just what to expect: Another tidal wave, probably much larger than before since this was so great an explosion, would now come racing out from the island, and it would arrive within minutes. That is, of course, had there been an island: Altheer had no means of knowing that Krakatoa was no more, having just been blown to oblivion.

The wave reached Telok Betong at 11:03. One anonymous European, writing some days later in a Batavia newspaper, was down on the town's shore, helping the local people who had already had their houses wrecked by the morning's earlier onslaughts from the sea. He was just lifting a huge wooden beam from on top of a trapped man, when he heard a scream. He looked up and saw a tall front of water rearing up and rushing toward him at a barely believable speed. There was a thunderous noise as it hit the beach and began rushing, crashing upward, through the town.

At this point the man's specific recollections become confused, his sudden panicked flight jumbling all conscious experience together into one amorphous mass. He is not alone in his confusion. The event must have been unforgettably dreadful, but in its details liable to the highly selective amnesia of those caught up by it.

Each of those snared by the Telok Betong wave speaks of running, wildly, panicked, trying madly to stay ahead of the wave, following natives running wildly too; and, in the particular case of the anonymous European writing in the Java *Bode*, of running behind a woman who stumbled and dropped her baby and could not abandon it and so was swept away, of running behind another woman who was—somewhat incredibly, it must be said—in the very process of delivery as she raced on, screaming and bloody, of seeing a man desperately trying to avoid the wall of water by climbing up as high as possible, by running up every slope that he found, of snatching hurried looks behind him to see, horrible in its immensity, the ever pursuing wall, which from time to time smashed against some obstacle and broke, disintegrating into huge and dirty gray piles of spray and wreckage-filled foam, but then regrouping and following him always with a roaring relentlessness, with an unstoppable energy, with a dogged and seemingly murderous resolution such that he could only continue to run, despite being so leaden-legged and air-starved and exhausted, run ever onward, always impelled by the frenzied gale that howled ahead of the wave, and by the certain knowledge that if he stopped or took a wrong turn that sent him downhill rather than up, he would be brought down drowned and his body crushed and hurled against the broken walls and jagged edges of spars and smashed glass and masonry that was rising up all around him.

Any doubts about the power of this single awful wave would be dispelled later by the discovery of a single compelling piece of evidence: the position of the Dutch steam gunship, the *Berouw*. The brief fame of this doughty little craft—four guns, a draft of six feet, a thirty-horsepower reciprocating steam engine, paddle wheels, and a crew of four European officers and twenty-four native seamen—provides a singular measure of the ocean's ferocity.

The captain of the *Berouw* had been the first to warn the mate of the *Loudon* that too strong a sea was running for him to risk a landing. That was at about 6:00 P.M. on the Sunday. The Telok

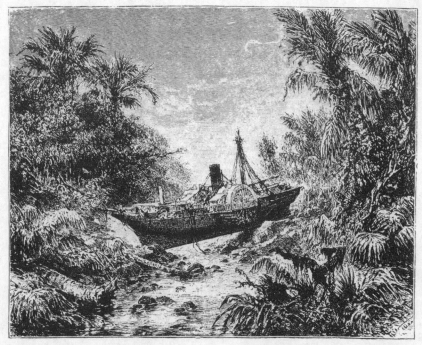

The *Berouw*, well and truly stranded—but very little damaged—a mile and a half up the Koeripan River. Hunks of rusting iron remained in the jungle until the 1980s.

Betong harbormaster then spotted her in difficulties about five hours later, in the middle of the night—she was well lit, and her plight was clearly visible through the gloom. Exceptionally strong waves were breaking about her, and the official feared that not only might she break her mooring lines, but that the heavy chains holding down the two-ton conical steel mooring buoy to which she was attached might snap as well.

Then early next morning, disaster struck. There were two eye-witnesses: Both the anonymous European in Telok Betong and the *Loudon* passenger N. H. van Sandick saw her lifted up on high by one of the 7:45 A.M. waves, then saw the mooring springs part one by one. The ship broke free of her buoy and was transported high on the crest of the mighty wall of green water. She was swept westward for a quarter of a mile until, as the wave broke, she was crashed down precipitously on the shore, at the mouth of the Koeripan River.

It is thought that this fearful crash—in which the vessel remained upright—killed all of the crew. But it was not the end of the ship's own nightmare. When the great wave of 11:03 A.M. hit, the ship was picked up once again and carried westward another two miles. She was driven all the way up the Koeripan River valley, along which the tsunami sped, and crashed down when the wave was spent, about sixty feet above the level of the sea from which she had been plucked. She lay askew across the river, forming a bridge. She was upright once again, a macabre tomb for the twenty-eight members of her crew.

She was found and inspected the following month by the crew of a rescue ship: "She lies almost completely intact, only the front of the ship is twisted a little to port, the back of the ship a little to starboard. The engine room is full of mud and ash. The engines themselves were not damaged very much, but the flywheels were bent by the repeated shocks. It might be possible to float her once again."

But whether possible to float her or no, no one ever seemed interested in trying to slide her back all that way to sea. This was no *Fitzcarraldo*. And so the *Berouw* remained where she was thrown, lying athwart the river for the better part of the next century, picked apart by scavengers over the years, like a carcass, or rotting quietly in the steam and sun.

The hulk was more or less intact when it was visited in 1939: It was rusting and swathed in vines, and had become home to a colony of monkeys. Pieces of her were last seen in the 1980s. Nowadays she is all gone. The Koeripan River trickles uninterrupted past where she lay, and the only memorial is her great mooring buoy, sitting on a plinth at the site where it was washed up, two miles from where it had last floated, and fifty feet higher than the level of the sea. The name *Berouw* is the Dutch word for "remorse."

The devastation in Sumatra was fully matched by that across the strait in Java. The tales from the survivors are every bit as memorable and dismaying. In the main towns—Anjer especially—the ruination was near total. There were only momentary lapses into levity: A telegram received in Batavia reported tersely: "Fish dizzy and Caught with Glee by Natives." For the rest, all was melancholy. And for the most melancholy memorial of all, a symbol on the scale of the *Berouw* over on the far side, one need only look at the great granite lighthouse at what was called Java's Fourth Point, a little way to the south of Anjer.

It survived the first onslaught, as had the gunship; it survived the wave that drove the *Berouw* up on to the beach; but when the wave that hit Telok Betong at 11:03 struck Anjer—about fifteen minutes earlier, since Anjer is closer to the volcano—it picked up an immense piece of coral rock, weighing perhaps six hundred tons, and dashed it against the column. Despite its iron cage of reinforcing ribs, the light crashed down, extinguishing one of the most important navigation beacons of the entire Sunda Strait. And although his wife and child were drowned, the keeper himself survived. With the phlegmatic way of both the well-trained lighthouse keeper and the fatalistic acceptance of a true Javanese, he returned to his duties as soon as was physically possible, and had a temporary light erected, and lit, within a matter of hours.

The stone stump of his lighthouse can still be seen, standing like an old and rotten tooth, rising no more than ten feet above the ever grinding waves of today's more peaceful sea. A replacement, built by the Dutch government three years after the eruption, is close beside it—except that it has been placed a prudent distance, probably about a hundred feet, back from the shore. And it has been constructed entirely of iron, just in case.

*   *   *

The stump of the Fourth Point light remains. And the carcass of the paddle-steamer gunship high in the river valley. But that is about all. The town of Anjer did not survive. Nor did Ketimbang. Nor Telok Betong. Nor Merak. Nor Tyringin. The Anjer Hotel, from whose veranda the first signs had been spotted weeks before, was no more than foundations and twisted banyan roots. The massive Dutch fort walls, which had survived the depredations of centuries, were cracked and tumbled into no more than a shapeless mass of weathered stone. Railway tracks were twisted and scattered across the ground like so many yards of iron ribbon. Iron gearwheels, shards of broken iron, and fractured lumps of machinery seemed to be everywhere. Giant boulders sat in entirely improbable places, picked up and smashed down as though they had been pebbles. Thousands upon thousands of houses and settlements up and down the coasts of Java and Sumatra were ruined, flattened, everyone in or near them crushed, drowned, or never to be found again.

And what also did not remain was the volcano that had caused it all. To everyone's astonishment, it was seen, once the dust had cleared and the gloom had been swept from the sky, to have totally vanished. Krakatoa, after the final majestic concatenation of seismic and tectonic climaxes that occurred just after ten on the Monday morning, had simply and finally exploded itself out of existence.

Lloyd's agent in Batavia, the Scotsman Mr. McColl, was able to send the following message within the week to his colleagues back in London, as concise a summary of the reality as that from his diplomatic colleague Consul Cameron down the road, and only a little less elegant:

> We shall probably not be in possession of full particulars for some days yet, as telegraph lines are damaged and roads destroyed, but so far we can give the following particulars. The island of Krakatoa, the summit of which peak was 2,600 feet

above water level, has totally disappeared beneath the sea, and the neighbouring island of Dwaisindeweg* is split in five parts. Sixteen new volcanic islands have been formed between Krakatoa and Sibesie,† and the sea bottom in the Straits of Sunda has completely changed. In fact the Admiral Commanding-in-Chief has issued a circular stating that till new soundings have been taken the navigation of the Straits of Sunda is likely to be extremely dangerous. Anjer and lighthouse and the other lights of southwest Java have all been destroyed. The subsidences and upheavals we have alluded to caused a large wave about 100 feet in height to sweep down on the southwest coast of Java and south of Sumatra. This was swept in for a great distance, thereby doing great injury both to life and property. We are here only twelve miles away from one of the points on which the wave spent its fury. The whole coastline to the southwest has changed its configuration. The inhabitants of the island of Onrust were only saved from the flood which swept over the island by taking refuge on board two steamers. At Merak government establishment the inhabitants took refuge on a knoll, fifty feet high, but were all swept off and drowned, with the exception of one European and two Malays, who were saved. Mauk and Kramat, on the west side of Batavia roads, have been laid waste, and about 300 lives lost. In Tjeringin only one house has been left standing. Both the native and European officials have perished. A rain of mud also fell at the above place, which is situated opposite to where Krakatoa once lay.

Anjer seems to have been completely destroyed. Lloyd's sub-agent there wires from Serang: "All gone. Plenty lives lost."

---

*This is the skerry that British Admiralty charts were wont to call Thwart-the-Way. Mr. McColl, just like Consul Cameron, was wrong to suggest it had split: It retains its insular integrity to this day, sitting smugly as a large navigational hazard at the northern entrance to the strait.

†This also turned out to be untrue.

THE EXPERIENCES

Not much of great excitement seems ever to have happened on the island of Rodriguez, which in the late nineteenth century was one of Britain's more remotely idyllic tropical possessions. According to the 1881 census some five thousand people lived there, contentedly farming the forty square miles of agreeable farmland and happily fishing the 200 miles of sandy coastline (the relic of an old volcano itself) that had been set down in a lonely corner of the western Indian Ocean. Mauritius, the notional mother ship of which Rodriguez was then and still remains a dependency, lies 350 miles away to the west. There is "a regular steamer service" today, and an occasional plane; in the latter part of the nineteenth century there were infrequent supply calls by a chartered sailing ship. A telegraph cable connecting the capital village of Port Mathurin with the capital of Mauritius, Port Louis, was not built until the beginning of the twentieth century.

The people of Rodriguez, who were Creole speaking and (according to a short book written in 1923 by a civil servant who claimed "no pretence of fine writing") possessed of "a deep brown velvety skin . . . hair of a very deep black, woolly and curly . . . protruding thick red lips . . . and magnificent snow-white teeth," were descendants of slaves imported by the French to cultivate their sugar plantations. They had been left behind when the French were thrown out by the British at the end of the Napoleonic Wars.

Their lives were run under the genial superintendency of four British imperial administrators, a quartet who might well have found their way into a Gilbert and Sullivan operetta—a magistrate, a medical officer, a chief of police, and a "First Class" priest (the last paid an annual government stipend of a thousand Mauritian rupees to remind the woolly-haired local people that God was, most naturally, an Englishman and, in this corner of the world, a Catholic Englishman to boot).

Placidly unexciting though Rodriguez may have been through its three centuries of inhabited existence, it did make an appearance in the history books, courtesy of the volcano in the far-off East Indies. In August 1883 the chief of police on Rodriguez was a man named James Wallis, and in his official report of the dependency for the month he noted:

> On Sunday the 26th the weather was stormy, with heavy rain and squalls; the wind was from SE, blowing with a force of 7 to 10, Beaufort scale. Several times during the night (26th–27th) reports were heard coming from the eastward, like the distant roar of heavy guns. These reports continued at intervals of between three and four hours, until 3 pm on the 27th, and the last two were heard in the directions of Oyster Bay and Port Mathurie [*sic*].

This was not the roar of heavy guns, however. It was the sound of Krakatoa—busily destroying itself fully 2,968 miles away to the east. By hearing it that night and day, and by noting it down as any good public servant should, Chief Wallis was unknowingly making for himself two quite separate entries in the record books of the future. For Rodriguez Island was the place furthest from Krakatoa where its eruptions could be clearly heard. And the 2,968-mile span that separates Krakatoa and Rodriguez remains to this day the most prodigious distance recorded between the place where unamplified and electrically unenhanced natural sound was heard and the place where that same sound originated.

A popular Victorian science writer, Eugene Murray Aaron,* explained to his readers why this figure of 2,968 miles should amaze them one and all:

> If a man were to meet a resident of Philadelphia and tell him that he had heard an explosion in Trenton [New Jersey], thirty miles away, he might be believed, although there would be

---

*He was also a world expert on the poisonous bites of the tarantula.

some doubt as to his powers of imagination. If however he should make the same assertion of an explosion in Wheeling, West Virginia, three hundred miles away, all doubts of his accuracy would vanish. But if, with every sign of sincerity and a desire to be believed, he should earnestly insist upon his having heard an explosion in San Francisco, three thousand miles away, he would receive a pitying smile, and his listener would silently walk away.

Yet just this last marvelous thing was true of those . . . on the island of Rodriguez.

It was heard in a score of other equally exotic places besides. No sound was heard in Rodriguez or anywhere else before Sunday, the twenty-sixth, nor any after the night of the twenty-seventh. And there is general agreement (although among the usual welter of confusions, not the least of them caused by the same lack of time zones that frustrates attempts to make a chronology of the eruption and the sea waves) that the loudest sounds occurred everywhere in the middle of the day on the Monday, suggesting that they originated on Java some short while before noon.

So, for example, the detonations were clearly heard in early afternoon, local time, on what is now the notorious British-owned American base-island of Diego Garcia.* In those days it was also a Mauritian dependency, where local farmers pressed palm oil and made copra, and where there was a coaling station for steamers crossing the Indian Ocean. The plantation supervisors plainly heard the explosion while they were taking their lunch. "Nous avons cru tellement à l'appel d'un navire en détresse," they later reported ("We thought it was a ship in distress firing its guns").

---

*Diego Garcia has an exceptionally unhappy recent history. To help the United States create what is presently its most strategically important overseas base, the British in 1970 leased the Americans the islands, assuring the Pentagon that they were uninhabited—when officials in Whitehall knew this to be palpably untrue. The two thousand islanders were forcibly removed to Mauritius to make way for the base. In 2001 the High Court in London ruled their exile illegal and suggested they be allowed to reclaim their former lands.

There were scores of broadly similar reports. The explosions were heard in Saigon and Bangkok, Manila and Perth, and at a lonely cable station south of Darwin called Daly Waters.* From Port Blair, the capital of the Indian prison-islands of the Andamans, came news that someone heard a sound "as of a distant signal-gun." No fewer than eighteen different sets of witnesses in what was then Ceylon came forward with stories ("Captain Walker and Mr. Fielder were puzzled at various times . . . by hearing noises as if blasting was going on," "Sounds as of firing of cannon at Trincomalee," "Mr Christie of the Public Works depart-ment . . . presumed some man-of-war was practising with her big guns, out of sight of land, as he could see no ships").

His highness the raja of Salwatty Island, in New Guinea, said he had heard strange sounds and demanded of a local doctor why the white men were firing their cannons. Stockmen driving cattle across the Hammersley Range† in western Australia heard what they thought was artillery fire to the north-west. In Aceh, at the northern tip of Sumatra, the location then (as still now) of a fierce proindependence rebellion, the Dutch garrison commander assumed that a local fort had been blown up by insurgents, and ordered all his men to battle stations. The Honorable Foley Vereker, who, despite belonging to one of Ireland's leading mili-tary families, had been cast out to the remoter colonies as com-mander of HMS *Magpie,* off Banquey Island, near Borneo, recorded in his log how he and his crew all heard the sound. And close by them, those Dayak islanders who had recently murdered a local official named Francis Witti (and, according to lore, had eaten his torso and limbs and had shrunk his head as a keepsake) heard the extraordinary sound too, assumed it to be the authori-

---

*Daly Waters, with a current population of about thirty-five, once had Australia's first inter-national airfield, since Qantas used it as a refueling stop. The telegraph cable from Singapore and Batavia once terminated here, and a pony express would, until 1872, take messages far-ther south across the desert to where the cable to Sydney began once again.

†The Hammersleys are now the site of one of the world's largest iron-ore mines, at Mt. Newman.

ties coming to get them and fled deep into the jungle.

Ships were launched from scores of places by eager would-be rescuers and salvagers, convinced there was an unseen vessel in trouble. A pair of fast ships set out from Macassar,* for instance, because they assumed another was in dire straits; two more set out from Singapore; a government boat went out to search off Timor; and when the sounds continued in Port Blair, the British authorities on the Andamans sent out a lifeboat too. In Singapore it became impossible, on one set of telephone lines, to hear yourself speak, since "a perfect roar, as of a waterfall, was heard, and by shouting at the top of one's voice the clerk at the other end heard the voice, but not a single sentence was understood. The same noise . . . was noticed on every line here."

There was one aberrant report—from a man also named Foley on Cayman Brac, now a small nodule of highly expensive Caribbean real estate, but then a forlorn sliver of a tropical sandspit, a day's sailing south of Cuba. Mr. Foley insisted that he heard the banging cannonades, and that they happened in rapid succession on the Sunday. But there is neither corroboration nor a common-sense explanation. There was no eruption anywhere in the Caribbean at the time (it would be another nineteen years before Mount Pelée exploded with classically Plinian intensity); and though freak atmospheric phenomena might be able to explain away this report of the explosion being heard *twelve thousand* miles away, the fact that Mr. Foley also claimed to have heard it twelve hours *before* Krakatoa exploded suggests that his memory, or hearing, was indeed at fault.

Yet there were some oddities about the dissemination of the sounds—not the least being that a great number of people in Batavia, Buitenzorg, and west Java generally heard nothing. Others simply felt strangely deaf, or heard a curious buzzing in their

---

*From whence came the eponymous sweet-smelling hair oil, the bane of countless English chairbacks and the cause of the creation of the protective lace furniture shroud called, somewhat unoriginally, an *antimacassar*.

ears, or else were aware of wild swings in the pressure all about them, as if they had been caught in some silent atmospheric hypertension. Experts in the field have taken many a stab at explaining why this might be. Some point out that the behavior of sound waves in the upper and lower atmospheres is very different, and that in the speeding up and slowing down that any wave experiences on passing through them a focuslike phenomenon may be brought into play that would make one locality receive a lot of sound, another very little. Others, less sophisticated, explain it by reminding us how snowfall muffles sound—and that ash falls, which covered Batavia and its suburbs at the time, are likely to do much the same.

Nonetheless, the overall conclusion remains inescapable, both to historical anecdote and to science: The sound that was generated by the explosion of Krakatoa was enormous, almost certainly the greatest sound ever experienced by man on the face of the earth. No manmade explosion, certainly, can begin to rival the sound of Krakatoa—not even those made at the height of the Cold War's atomic testing years. Those other volcanoes that have exploded catastrophically in the years since decibel meters were invented—Mount St. Helens, Pinatubo, Unzen, Mayon—have not come close: no one suggests that the explosion of Mount St. Helens in May 1980 was heard much beyond the very mountain ranges in which it was sited.

Dr. Verbeek, with the modest assurance of one who had seen and heard and experienced the vastness of the eruption, stated in his report of 1885 that "the exceptionally loud noises require our attention . . . the large explosions that in loudness have far exceeded all known noises. At no earlier event was a noise heard over such a large part of the earth's surface." Under the impact of Krakatoa's explosion, 13 percent of the earth's surface vibrated audibly, and millions who lived there heard it, and when told what it was were amazed.

The inaudible waves, it was soon discovered, had ventured even farther afield. The thousands of Europeans and Americans who noticed and recorded them—without in most cases ever realizing just what they were—did so, all around the world, at more or less the same time. The fact that they did so at all points up, quite unexpectedly, one of the newly formed habits of middle-class Victorian society—a set of habits that never anticipated such a catastrophe as Krakatoa, but nonetheless in due course took full advantage of its effects.

The newfangled habits of the late nineteenth century included, among the time's multitudes of other scientific advances, the development of increasingly precise means for forecasting the weather. And while cost and complication limited popular access to most other sciences, it did become rapidly possible—and indeed rather popular—for people to buy and use scientific instruments to help them understand the daily fluctuations in their climate. Consequently, as Victorian homes, clubs, and hotels filled up their halls and vestibules with the ever newer and more handsome barometers, recording thermometers, sun gauges and rain gauges, so the middle classes became an unwitting army of amateur meteorologists, faithfully tapping the glass each day the better to predict whether it would be Fine or Stormy, Changeable or Fair.

The most costly and sophisticated of these instruments was the recording barograph. Because of its price, it was reserved most commonly for the mantel in the club, rather than for the hall at home. The task of this small machine is to record, with an ink trace on a sheet of graph paper secured around the circumference of a clockwork-driven drum, the slight hourly variations in atmospheric pressure over the week that it takes for the drum to rotate a single time.

A well-made barograph is a joy to behold—an elegant confection of brass and steel, mahogany and glass, its mechanicals visible inside its crystal case, its clockwork heart ticking away happily as it takes the pulse of the day. And the ink trace—flowing gently up and down, sometimes more steeply downward if bad weather is on the way, arching up again if the storm passes over at speed—has a seductively sinuous beauty to it also: The records of the passing week's atmospheric alterations are stored in a small drawer under the instrument, to be studied later, whenever the weather becomes the subject for reminiscence or chatter.

One very noticeable aspect of a barograph's recorded trace is, however, just how smoothly the weather changes. The line invariably curves up or down slowly, steadily. It does not jump erratically, like a seismic record during an earthquake or a lie detector revealing a falsehood. Atmospheric pressure, by contrast, changes with measured deliberation—a feature reflected by the curves on a barograph with their steady and considered moves across and along the ever unrolling snake of recording paper.

But as scores of people around the world came to change their paper for the week that ended on Sunday, September 2, 1883, and as they smoothed the records out to place them away in the drawer, they all, almost simultaneously, noticed something. On the trace for the Monday of the week just gone, August 27, there was a sudden and unanticipated jerk. A hiccup. A notch, an interruption—an altogether most puzzling thing.

The pen that had been so smoothly and seamlessly noting the pressure on the instrument's vacuum chamber had suddenly been flicked up, and then equally violently snapped down again. When looked at more closely, the oscillation was even more peculiar than that: First there was a sudden recorded rise in pressure, then two or three minor oscillations, then a very deep depression, followed by a less steep rise, then more small oscillations, and then finally, after an interruption that lasted for the better part of two hours, back to the smooth and gently changing trace of normal times. In summary, it seemed as though, and for some inexplicable reason, there had been the quite impossible occurrence of *an earthquake in the air*.

It took only a matter of hours of excited discussion between observatories and weather-fascinated members of the lay public to draw the inescapable conclusion: This, faraway though it may have been, was all Krakatoa's doing.

As soon as news of the extraordinary degree of the eruption

had become widely known, it remained a simple matter to check the times of the strange two-hour-long blips on the barograph traces, to make an allowance for the probable approximate-speed-of-sound rate of travel of the shock wave, and to figure in the time difference between Krakatoa and the various club mantels around the world. And lo! they all matched. The eruption that had sent out flame and ash and tidal waves and an incredible explosive sound had also sent an invisible, inaudible shock wave that passed cleanly through the atmosphere, and had been recorded, quite unexpectedly, on scores of machines designed for the much more prosaic task of suggesting to middle-class Victorian gentlemen in Birmingham and Boston and beyond whether they should take their umbrellas to luncheon.

Except that it was, as is much that relates to Krakatoa's eruption, much more complicated than it seemed at first sight. When the traces were examined more closely and compared with other barograph traces from more distant cities around Europe and then around the world, it appeared that the shock wave from Krakatoa's final cataclysmic explosion had traveled around the earth not once but *seven times*.

The barographs, barometers, and weather stations all recorded the signature two-hour wave, with the amplitude of the oscillations diminishing at each pass; it had apparently reverberated, flown back and forth and around the planet in a manner that seemed quite out of proportion to the magnitude of the original event itself.

All this provoked much fluttering in the scientific dovecotes. Every weather expert in the world suddenly wanted to know what was going on—why a pressure wave like this would behave in this peculiar way. In scientific London the degree of interest was particularly intense—and it prompted what was an entirely unanticipated response: that, despite Krakatoa being on Dutch sovereign territory, it would be best if it were to be left to a distinguished *and entirely British* body to investigate its eruption.

The high-handedness, seen from today's perspective, quite

boggles the mind. Perhaps it was the vast reach of British influence of the day that made this seem desirable to some; perhaps, more specifically, it was the existence of all those records from all those British-owned and British-designed barographs (for all those instruments, in places as far-flung as Melbourne and Mauritius and Bombay, turned out to have been manufactured in England) that provoked this entirely unwarranted—but, it has to be said, quite unresented—example of British imperial busybodyness.

The timetable went like this: The explosion occurred in late August. In early September the British barograph paper records— initially from the leather-bound and entirely masculine fastnesses of London clubland, but later from weather observatories at Greenwich, Kew, Stonyhurst, Glasgow, Aberdeen, Oxford, and Falmouth—were changed, and the blips noticed. Suspecting that something of absorbing scientific interest was afoot, a senior British government official named Robert Scott—secretary to the Meteorological Council—promptly sent telegrams to his colleagues in observatories around Europe, asking if they would examine their traces too.

And back came the reports, from Vienna, Berlin, Leipzig, Magdeburg, Rome, Paris, Brussels, Coimbra, Lisbon, Modena, and Palermo and other places besides, confirming in all respects what Scott suspected: that a wave of sudden pressure had swept around and around the planet from its birthplace in the Sunda Strait; that the passage of the wave had been a remarkable event; and that, moreover, it had lasted, echoing around the globe, for no fewer than fifteen days following the eruption.

Scott found this quite extraordinary, and told his superior, an old India hand and engineer named General Richard Strachey. In December—only four months after the eruption, and so with quite unusual dispatch—the pair presented a brief paper to the Royal Society, that most estimable and ancient of British scientific institutions. It was entitled "Notes on a Series of Barometrical Disturbances which Passed over Europe between the 27th and 31st of August 1883." It caused an immediate sensation.

And it did so because here was one of the first provable instances in which a natural event occurring in one corner of the planet had effects that spread over the entire world (or what would be the entire world, if further records could be sought from the Americas and Asia and elsewhere, for they would show the same evidence). Here was the event that presaged all the debates that continue to this day: about global warming, greenhouse gases, acid rain, ecological interdependence. Few in Victorian times had begun to think truly globally—even though exploration was proceeding apace, the previously unknown interiors of continents were being opened for inspection, and the developing telegraph system, allowing people to communicate globally, was having its effects. Krakatoa, however, began to change all that.

The world was now suddenly seen to be much more than an immense collection of unrelated peoples and isolated happenings: It was, rather, an almost infinitely large association of interconnected individuals and perpetually intersecting events. Krakatoa, an event that intersected so much and affected so many, seemed all of a sudden to be an example of this newly recognized phenomenon. And so it was up to a British scientific society—most decidedly a British one, given the imperial mood of the day, like it or not—to investigate it.

A decision to do just this was taken in January, following the reading of two more short papers presented to the Royal Society, both describing the scene on the ground in and around the Sunda Strait. The first paper was by the British consul in Batavia—now a Mr. Kennedy from Sumatra, since Consul Cameron had fallen ill—and the other by the socially well-connected Captain Vereker of HMS *Magpie,* who reported from Borneo. Both papers amply confirmed the view in London that this extraordinary event off Java had been so huge and so *world affecting* that a body had to be set up right away to investigate it. And so on February 12, 1884, an advertisement, in the form of a letter to the editor, was placed in *The Times:*

THE KRAKATOA ERUPTION

Sir—The Council of the Royal Society has appointed a committee for the purpose of collecting the various accounts of the volcanic eruption at Krakatoa, and attendant phenomena, in such form as shall best provide for their preservation and promote their usefulness. The committee invite the communication of authenticated facts respecting the fall of pumice and dust, the position and extent of floating pumice, the date of exceptional quantities of pumice reaching various shores, observations of unusual disturbances of barometric pressure and of sea level, the presence of sulphurous vapours, the distances at which the explosions were heard, and exceptional effects of light and colour in the atmosphere. The committee will be glad to receive also copies of published papers, articles and letters bearing upon the subject. Correspondents are asked to be very particular in giving the date, exact time (stating whether Greenwich or local), and position whence all recorded facts were observed. The greatest practicable precision in all these respects is essential. All communications are to be addressed to—

Your obedient servant,

G. J. Symons, Chairman, Krakatoa Committee,*

Royal Society, Burlington House

It took five years for what at first might have seemed a somewhat pumice-obsessed committee to study all of the complex and new types of information that flowed from the event. Their final report, issued in 1888, had 494 pages of text, as well as countless drawings, graphs, and exquisite color prints. It stood as a lasting masterpiece of determined study, elegantly composed style, and

---

*The thirteen-member committee was esteemed in the extreme: Among the members were the hydrographer of the navy; the great geologists Archibald Geikie and Thomas Bonney; the physicist who discovered fluorescence, Sir George Stokes; and the aforementioned General Richard Strachey, who was such a towering figure of Indian engineering that he had a bridge over the river Jumna named after him.

splendid Victorian brio—and it collated, among other things, all the pressure-wave observations that had led to the establishment of the committee in the first place.

And what was found out about them was simple but remarkably beautiful: that the shock waves from Krakatoa had radiated across the world like ripples on the surface of a pond—the surface of a vast and slightly flattened sphere in the case of the earth, of course, rather than the merely circular surface one would find on a

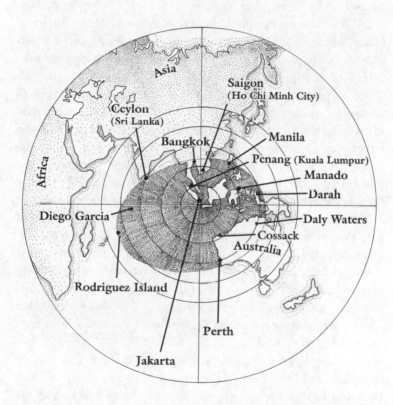

The sound of the great explosion could be heard as far off as Rodriguez Island, nearly three thousand miles from its origin. Records suggest that people living within the dotted line were well able to hear the rumbling and banging, most thinking it the noise of a naval bombardment.

pond. They had radiated outward at a speed approximating that of audible sound: between 674 and 726 miles per hour.

The waves had traveled outward, expanding their width until they were halfway to their goal and then contracting again until they reached their precise antipode—which, in the case of Krakatoa at 6° 06'S, 105° 25'E, is a point (at 6° 06'N, 105° 25'W) in the Pacific Ocean off Bogota, Colombia. Then, having taken nineteen hours to reach this nameless and watery antipode, they headed off back again and returned to Krakatoa (with the passage noted on all the barographs in between, at a bewildering variety of places that included St. Petersburg, Toronto, the Antarctic island of South Georgia, and a village that is now a pretty New York suburb called Hastings-on-Hudson).

Every time the wave was noticed, it was found that the time of its passing agreed with the eruption time that had been marked by observers back in the Sunda Strait. The wave's passage above the Greenwich Observatory, for example, was recorded on all the barographs' registers—with the sharp upward tic of pressure, the minor ruffles on the record, the sudden downward blip, further ripples, the slow upward rise to the resumption of normal state—at 1:23 P.M. on the Monday. Krakatoa Time was seven hours ahead of London—meaning that at 10:02 A.M., the local time when the volcano exploded, it was 3:02 A.M. in Greenwich. Subtracting that from the time that the observatory barographs recorded the blip gives a figure for the wave's travel time of ten hours, twenty-one minutes—the precise figure that could be calculated for inaudible shock waves moving across the 7220 miles of a great circle separating London from Krakatoa.

The Greenwich Observatory corroborated the figures for the first pass—and for the six further passes of the shock as it moved back and forth across the capital before it had weakened to the point where it could no longer be detected. And as it did so, two things became even clearer: that the eruption time was exactly right, 10:02 A.M., and that the world now knew a great deal more about the transmission of atmospheric shock waves than ever

before. Meteorology in general profited mightily from the findings; and in the mid–twentieth century, when large atmospheric
explosive tests were conducted, the way in which shock waves were
propagated through the atmosphere was well understood also.
The Krakatoa Committee had, even if only in these respects, fully
justified its existence.

The tsunamis that killed so many on the shores of Java and Sumatra crossed the world as well. It could be seen from the beginning
that up close to the volcano the waves were enormous, and killed
thousands. That they then became ever smaller in proportion to
their distance from Krakatoa was to be expected. But the discovery
that they were in fact so deeply powerful, and radiated away from
the volcano so aggressively that they could still be detected in the
sea as far away as the English Channel, was the cause of general
astonishment.

The first to receive firm news of the appearance of some faraway
Krakatoa-induced waves was Charles Darwin's son George, shortly
after he was elected (by the narrowest of votes) the professor of
astronomy at Cambridge.* His friend Major A. W. Baird, who had
the post of chief of the Tidal Survey of India, wrote to him in
Cambridge to say that "the wave caused by the volcanic eruption
at Java is distinctly traceable on all the tidal diagrams hitherto
received, and I am informed of a great tidal disturbance at Aden
on August 27; but the daily reports are always meager in information. Kurrachee and Bombay also show the disturbance, and as far
as I have examined the wave reached halfway up to Calcutta on the
Hooghly."

This last—the news that waters had rushed up the Hooghly
River almost to the city that was then the capital of British India—

---

*Sir George Darwin's contribution to science was mostly theories that were later to be dis-
proved. Most egregiously he calculated mathematically that the moon was created by being
torn away from the cooling earth—an idea now universally discredited. He also wrote papers
on contemporary fashion, and claimed never to do more than three hours' work a day.

did it. The Royal Society promptly asked for an immediate report. Major Baird swiftly obliged with a six-page summary of his survey stations' reports from across that immense swath of British imperial territory that stretched between Aden and Rangoon. And the Krakatoa Committee, recognizing a significance that went beyond the simple threat to Calcutta, immediately commissioned a senior Royal Navy captain to investigate the phenomenon worldwide.

Records from tide gauges at ports all across the world were speedily gathered. An early analysis showed a fascinating, but not entirely unanticipated, trend: Almost all the stations that recorded sudden and unexpected waves that could be positively linked to the eruption (by their timing and their type) lay to the west and south of the island.

Almost all: not Batavia itself, which lies eighty-three miles away to the east as the crow flies, and considerably farther so far as a tidal wave might pass. Despite where it lay, the capital city did indeed witness what even the Royal Society saw fit to call "a wall of water," when the wave hit its gauge at 12:36 P.M. on the Monday afternoon—two hours and thirty-two minutes after the explosion. According to the Reverend Neale, the water rushed into the Batavian canal system, rising suddenly by several feet, forcing hundreds of merchants and residents to flee for their lives.

The day—unusually cold, half dark and drab, the air still filled with gray, gritty ash that got in everyone's hair and eyes and teeth—had begun, surprisingly, with a fair sense of stoic normality. The steam trams were filled with people setting off for work, the markets were thronged, the private horse-drawn carriages were trotting around the Koningsplein, their occupants talking excitedly about the events of the night before, confident that the worst was over. Then came the arrival of what was swiftly understood to be the huge relic of the great tsunami—the remains of a wave that somewhere had been much, much worse—and it made all these good burghers of Batavia realize, very suddenly, that in fact the worst was still to come.

The maximum height of this bore (the needle on the Batavia

tide meter shot up vertically, clear off the scale) was at least seven feet and six inches—a fraction of the height of the devastating waves that destroyed Anjer and Telok Betong maybe, but an impressive enough display. The waters promptly fell back again, to ten feet *below* normal sea level, and then rose back up again, then sloshed back down—oscillating a total of fourteen times over the next twenty-eight and a half hours, the height of the successive waves diminishing all the while. Finally, after what was no more than a three-inch ripple hit the Batavia tide meter at 5:05 on the afternoon of the next day, Tuesday, they vanished completely.

But Batavia alone in the area experienced the great wave. Almost no other places to the north and east of the volcano experienced anything at all—Singapore's tide meters registered nothing, nor was there any discernible blip in the records of Hong Kong, Yokohama, or Shanghai; and even at Surabaya, at the eastern end of Java, the disturbance that was picked up on the port's three tide gauges was only ten inches, "too insignificant to be otherwise noticed." There is a very simple reason for the lack of any dramatic effects on this side of the volcano, as a glance at the map will show.

To the east of Krakatoa the two sides of the Sunda Strait pinch inward like the jaws of a nutcracker. There are islands blocking the way too—Thwart-the-Way Island being one such, notorious in its nuisance value—and before a wave has any chance of touching Batavia port itself it reaches long fingers of shallows and sandbanks and further islets and reefs, all of them conspiring to slow down and frustrate the eastward movement of any wave. Nothing would stand in the way of a sound wave or a shock wave; but, faced with the dissipating influence of the shoals and headlands, a water wave would essentially not move eastward at all, as the recorders everywhere confirm.

However, to the west of Krakatoa, barring the presence of a small headland called Vlakke Hoek in southern Sumatra that acts as a small chicane on a westbound wave's right-hand side, there is only the wide-open sea of the Indian Ocean. Any tsunami moving out from the eruption in this direction would be free to go wherever it wished, without maritime let or hindrance. And in August 1883 the great ten o'clock wave did indeed fan out westward entirely untrammeled, and managed to go just about anywhere and everywhere it wanted. Two types of wave were detected: what were called long waves, which reverberated back and forth at periods of as much as two hours; and the short waves, which were steeper and with a much less regular and more frequent repeat.

The old Dutch port of Galle, close to the southern tip of the island of Ceylon, is where the arrival of these short waves—or more precisely, a sequence of fourteen waves, each separated by just a few minutes—was first noticed. The *Ceylon Observer* correspondent filed on August 27 that

> . . . an extraordinary occurrence was witnessed at the wharf at about 1.30pm today. The sea receded as far as the landing stage on the jetty. The boats and canoes moored along the shore were left high and dry for about three minutes. A great number of prawns and fishes were taken up by the coolies and stragglers about the place before the water returned.

A woman was killed at the port of Panama—a town in Ceylon, not on the isthmus—when she was swept from the harbor bar by an immense influx of water. Both the Panama harbormaster and the local ruler, the splendidly titled Ratamahatmaya, said later that ships had suddenly sunk downward and were then drawn backward to be left stuck in the drying mud, their anchors exposed—and just as suddenly were borne up by an inrushing surge of water. The local streams, hitherto sweet water, all promptly turned salty for at least a mile and a half upriver. The woman who died, from the injuries she sustained in falling while she was carrying a sheaf of paddy from the fields, is thought to have been the most distant casualty of the eruption that took place nearly two thousand miles away.

At Hambantota, farther south still, Ceylon government officials estimated the height of the wave to be twelve feet, and said that, as at Panama, its currents were irresistible, taking small craft back out to sea, and then sweeping them back and dashing them to pieces on shore. But however violent, there were no casualties here; nor were there any farther afield.

The long waves tend to be the ones that were recorded by the automatic tide gauges around the world, and it is these that make up the bulk of the formal record; the short waves tended to be more the stuff of anecdote, and, because they oscillated so swiftly, rarely made an appearance on the recorders. By the time the long waves reached India they were diminishing, fast—fourteen inches high in Madras, a series of ten or so six-inchers at Calcutta, a foot high in Karachi, half that in Aden. They spread southwestward toward the African coast as well: They broke a hawser of a boat moored in Port Louis, Mauritius; and in the rarely visited Indian Ocean reef harbor known as Cargados Carajos, the captain of the *Evelina* reported huge, smooth oscillations in the sea, breaking only when they came into contact with coral heads. Already the wave was 2,662 miles from its point of origin, and racing steadily along at a calculated 370 miles each hour.

A four-foot-high wave was noticed at Port Elizabeth, on the bleak and generally unlit east coast of South Africa,* and undulations were picked up at Cape Town. A visiting German South Polar expedition (which failed to reach its goal) on the island of South Georgia saw the icebergs and brash in the harbor of the whaling station at Grytviken lifted fifteen inches in a series of a dozen recorded and remarkable swells.

And so the waters progressed ever outward—with numerous waves of different types and styles, heights and frequencies, and

---

*That this particular wave was so unusually large is hardly surprising: The coastline between Durban and Port Elizabeth is notorious for giant waves, caused when northbound Antarctic storm systems slam into the southwesterly-flowing Agulhas Current, concentrating the confused water into shallows above a peculiarly narrow continental shelf. It is one of the most hazardous coastlines in the world—and the arrival in 1883 of a four-foot Krakatoa-induced tidal wave would have served only to remind local mariners about sorrows, single spies, and battalions.

deriving from what oceanographers now reckon were a number of different causes too. Eventually they ran out of steam and reached as far as they could go, in the farther recesses of northwestern Europe. By the time they reached the North Atlantic, and then the Bay of Biscay, the oscillations were small indeed—such that the tide diagrams had to be photographed and blown up in order to be able to measure the fluctuations in the record.

But they are there all right, tiny but still distinct. In Socoa, a French harbor near the more celebrated resort of Biarritz, and 10,729 nautical miles from Krakatoa, there were seven undulations, each of them three inches high—barely enough to be noticed by promenaders on the beach, though I like to fancy pomaded young men and their lady friends skipping amusingly out of the way when the unusual small swells suddenly threatened their boots and their trouser cuffs. Further north at Rochefort, a town on the Charente a little north of the Gironde and Bordeaux, the magnifying power of the estuary pushed the waves up to five inches—they had sped there from the volcano at a calculated (and now slightly faster) rate of 414 miles per hour, barely slowing as they did so.

And finally, turning the corner into the English Channel, only the slightest trace is noticeable. There is a ripple recorded at Cherbourg, another at Le Havre, and an irregular but discernible series of undulations at Devonport, close to where all Royal Navy cadets are now trained to be officers. But closer in—at Portland, Plymouth, and Dover—nothing. Maybe there is just the vaguest sign of it on the gauge on the inner side of the Portland breakwater—though the Royal Society report admits, "the indications of disturbance are not very conclusive, as no regularity of period is traceable in the small indentations which do appear."

I confess I would have derived no small pleasure from discovering that on the Tuesday morning, when waves were seen in France and in the West Country, the tide gauge in Dover port also suddenly startled its keeper, with a splash, or a slop, or a curious and inexplicable swell. But there was to be no such luck. With Dover

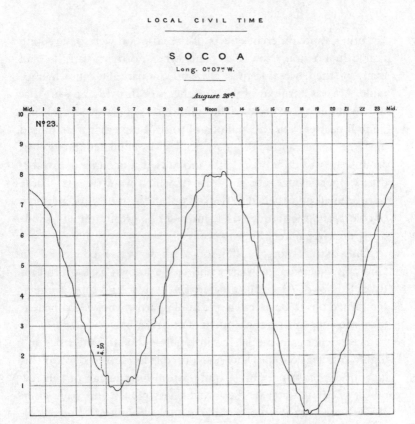

LOCAL CIVIL TIME

SOCOA
Long. 0ʰ 07ᵐ W.

August 28ᵗʰ

Tiny oscillations in the tides were noticed nearly 11,000 miles away from Krakatoa, such as here at Socoa, a small port near the French resort of Biarritz.

being no less than 11,800 nautical miles from the Sunda Strait, and lying at the distant end of a channel in which the water shallows to a few hundred feet and less, which would kill or dampen the kind of long waves that were produced by the eruption, it is perhaps not surprising to learn that the gauges here showed no deflection at all. The airwave may have found its way to Greenwich, seven times; but the sea wave was brought up short, 500 miles from home, and never arrived near the British capital even once.

*   *   *

Yet other, quite different effects of the eruption were before long visible in London, New York, and other northern capitals, and many of them were aesthetically and dramatically quite memorable. Art was born out of the aftereffects of this volcano—art that was then quite unexpected. Though not so popular today, it is not quite forgotten. For the millions of tons of dust that were hurled into the upper air in the East Indies disseminated themselves around the world for many years and caused all manner of extraordinary phenomena—not the least of which were sunsets. These were seen all over the world decked out in the most lurid rainbow of colors, and they attracted the interest of a great number of suddenly excited painters.*

One of the more prominent of these was Frederic Edwin Church, a member of what came to be known as the Hudson River School of American nineteenth-century landscape painters. He turned out to be precisely the kind of artist who would benefit from the atmospheric consequences of Krakatoa.

Frederic Church specialized in highly dramatic landscapes and highly colored skyscapes—he had a predilection for the grand (his gigantic *Niagara* presents a quite astonishing image of the raw power of falling water) and the excessively radiant (his *Twilight in the Wilderness* has an unforgettable richness about its evening color). Both of these artistic preferences came together in December 1883 when Church—supposedly well aware of the remarkable effects of Krakatoa's spreading trail of dust on the world's sunsets—traveled north from his ornate Moorish castle in the riverside town of Hudson, to the very tip of upstate New York, on the Canadian border. There he would try to capture an image of what he suspected would be an especially vivid northern twilight.

---

*It has long been supposed, but never proved, that J. M. W. Turner, whose impressions of sunsets made him an international institution, was painting evening skies colored by the aftermath of the 1815 eruption of Tambora. The link between Krakatoa and the sunsets of the late 1880s is well established; the similar effects of Tambora can only be surmised, with Turner's paintings usually offered as evidence.

He chose Chaumont Bay, at the very eastern end of Lake Ontario. The early-winter ice would be piling up in the westerly winds; there would be newly undraped trees on the wispy peninsulas; and above all there would be the vast expanse of the lake into which the setting sun would appear to sink. He chose to render the image in watercolor. The resulting painting—not surprisingly called *Sunset over the Ice on Chaumont Bay, Lake Ontario*—has a range of colors in the sky, a gently intermingling play of pinks, mauves, orange, salmon, and purple that is quite astonishing, and unusual, and suggestive of *something* happening, something inexplicable, high in the evening atmosphere. It is the only major painting to be created in the immediate aftermath of Krakatoa: and even if Church had not set out with the deliberate intention of making posteruptive art, the piece stands

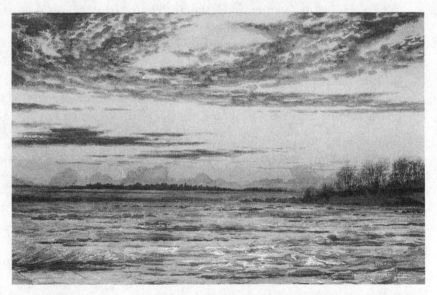

Frederic Edwin Church's *Sunset over the Ice on Chaumont Bay, Lake Ontario*— painted on December 28, 1883, and showing a sunset tinted with the vivid crepuscular colors known to have been caused by Krakatoan dust in the upper atmosphere. (*Courtesy of Powerstock*)

now as vivid documentary testimony to the giant volcano's effects.*

Lesser artists had a field day. The most notable was William Ascroft, who lived beside the river Thames in Chelsea. Early in September, two weeks after the eruption, he noticed that London was suddenly being gifted with a series of memorably setting suns and, more interesting, unusually strong afterglows. He painted them at a furious rate—creating a total of no fewer than 533 watercolors over the months of his fascination.

On especially gorgeous evenings he might make several paintings, one every few minutes, creating as he did so a kind of time-lapse image of the entire process. On November 26, for example, he painted the aftermath of the sun's disappearance once every ten minutes between 4:10 P.M. and 5:15 P.M., catching a sequence of fiery purples and oranges with all the fleet-footed accuracy of a film camera. He wrote lengthy notes and analyses of what he saw— "Blood Afterglows" and "Amber Afterglows" among them—and examples of the beaming coronas that often surrounded the setting sun itself and which were named, after the Hawaiian naturalist[†] who first spotted them, "Bishop's Rings." All 500 of his paintings later went on show at an exhibition in a museum in South Kensington. They remain today inside what is now the Natural History Museum, locked away and half forgotten.

Dust of all grades and compositions was thrown into the air by the eruption. Much of it, too heavy to be kept on high for long, fell down as drifting veils of gray, and was widely reported as having done so. Ships at sea experienced dust falls for a fortnight after the eruption: the *Brani* and the *British Empire* came under a slow rain of a white ash that one master said "looked like Portland cement" when they were sailing in the Indian Ocean within a two-

---

*It is privately owned by a collector in Philadelphia.

[†]This was the Reverend Sereno Bishop, who while sporting a name well known in Hawaii, where he worked as a missionary, was not related to the founder of Honolulu's famous Bishop Museum.

thousand-mile range of the volcano; the *Scotia* experienced falling dust until September 8, when she was off the Horn of Africa, 3,700 miles away.

But the lighter material, the finest particles of all, was thrown up right through the troposphere until, almost defying the pull of gravity, it was caught up for months in the lower reaches of the stratosphere* itself. Modern estimates suggest the Krakatoa eruption hurled material at least 120,000 feet into the air—some say 160,000 feet, or thirty miles. Tests have since shown that up so high, material can hover in a kind of weightless stasis.

A dust particle with a diameter of a micron—whether it is an aerosol droplet or a tiny fragment of volcanic silicate mineral makes no difference—has recently been shown to take many weeks to descend vertically through just half a mile of stratosphere. A particle half a micron in diameter will take many months to fall, so slight is the tug of gravity upon it. Yet horizontal movement was apparently no problem for the fragments of Krakatoa material. The strong globe-circling winds spread them far and wide. And the rains that would tend to flush out those particles that might have been tempted to stay in the lower reaches were of course nonexistent up high.

And so the particles stayed, undisturbed, for very long periods. And by refracting and filtering and in a myriad of other ways altering so vividly the colors of the sunlight that passed by them, and by staining the crepuscular sky with vermilions and passion fruits and carmines and royal mauves, so they ensured more potently than any other effect that Krakatoa would soon become the most famous volcano in world history.

Krakatoa killed more people than any other eruption, and for

---

*The stratosphere, a band of near-vacuum that lies between about eleven and thirty miles above the earth's surface, differs from the lower band of the troposphere in one important way: While the temperature declines steadily with height in the troposphere (as any climber or flier well knows), it does not do so in the lower reaches of the stratosphere. And in its upper few miles the temperature, so long as the sunlight is not cut off by the earth's shadow, actually begins to climb.

that it remains notorious, but it became much more widely known to hundreds of millions of people around the world for a more benign and beautiful reason, one that all could readily see for themselves each time they looked westward at eventide.

Poets were inspired in much the same way as painters. Tennyson is widely believed to have been thinking of Krakatoa when, in his almost wholly forgotten epic poem *St Telemachus* (published some nine years after the eruption), he wondered out loud:

> *Had the fierce ashes of some fiery peak*
> *Been hurl'd so high they ranged about the globe?*
> *For day by day, thro' many a blood-red eve, . . .*
> *The wrathful sunset glared . . .*

The Royal Society's Krakatoa Committee, displaying the almost obsessive need for the complete and the comprehensive that was so much a signature of Victorian studies like this, invited responses from the general public. They received wagonloads of material,* and painstakingly cataloged every single report of every atmospheric phenomenon, no matter how trivial, of which they were made aware. Two-thirds of the society's eventual 494-page report is devoted to "the unusual optical phenomena of the atmosphere, 1883–1886, including twilight effects, coronal appearances, sky haze, coloured suns, moons &c."

Forty-eight further pages give the details of the eight hundred places where unusual phenomena had been seen, organized into chronological order. The list actually starts some months *before* the cataclysm, presumably to give some kind of observational baseline from which the deviations that follow can be measured. So the first report comes from a town called Graaff-Reinet, in the

---

*When the editors of a centenary study of the eruption, published in Washington by the Smithsonian Institution in 1983, similarly asked members of the public to send reports of anything to do with the eruption, they received not a single response.

center of the South African tablelands, beginning during the southern winter—with someone noticing the presence of "fine sunsets, gradually increasing from February to June." The very first eruption of the final paroxysmal cycle, it will be recalled, took place toward the end of May.

And thereafter came a cascade of responses from people who had seen or been told about the Royal Society's announcements in the newspapers. A seemingly endless string of reports is included, from a bewildering variety of places, people, ships, and lighthouses—all with news of strange phenomena, most of them seen in the skies. A blue moon was spotted by a Mr. Haughton in Kokkulai, Ceylon. The lighthouse keeper at the Chinese city then called Chefoo—now Yantai—saw a pale-red glow, like fire, in the west. There was blue sun seen by a Dr. Earl Flint in Rivas, Nicaragua. Captain Faircloth of the Caribbean Signal Service saw "a lurid glare" emanating from the sun setting over Nassau in the Bahamas. The scientific journal *Nature*—which collected dozens of such reports in tandem with the society—had first a report of a green sun seen in Colombo, and then a letter from a Reverend W. R. Manley, a missionary in Ongole, south India, who saw "splendid twilight glows . . . deep red more than one hour after sunset."

And Gerard Manley Hopkins, the Victorian poet and Jesuit priest, then teaching classics at Stonyhurst College, wrote a lengthy essay for *Nature*, outlining in detail what he had seen—in style unusual for the normally arid magazine, but easily recognizable to those who know their *Wreck of the Deutschland:*

> . . . the glowing vapour above this was as yet colourless; then this took a beautiful olive or celadon green; not so vivid as the previous day's, and delicately fluted; the green belt was broader than the orange, and pressed down on and contracted it. Above the green in turn appeared a red glow, broader and burlier in make; it was softly brindled, and in the ribs or bars the colour was rosier, in the channels where the sky shone through it was a mallow colour.

There were four main kinds of phenomena, all amply evident from these pages and pages of reports. There were the sunsets themselves; there were the vivid and highly unusual colorations of the moon (often blue,* sometimes green), the occasional colorations of the sun and, very rarely, of some of the larger planets; the whitish solar coronas that were frequently seen just before sunset; and the monstrously flaming afterglows.

The afterglows in fact turned out generally to be a more dominant and widely noticed feature than the actual sunsets themselves. They are brilliantly hot-looking glows that appear from time to time some way up in the sky above where the sun has set some minutes before. The geometry of their origin is well known—having been studied assiduously after the eruption. They are caused when the rays of the sun, at an ever increasing distance from the observer, pass tangentially to an optically unusual layer in the atmosphere. In the case of the afterglow sightings that were made after Krakatoa, the optically unusual layer was the drift of ash—the ash particles absorbing and reflecting and causing a fiery reddening of the red light that is the last to be bent from the source, before, thanks to the disappearance of the sun and all its light now coming from well below the horizon, they eventually vanish altogether.

A pattern also emerged swiftly from the Royal Society's massive catalog of observations. There could be no doubt that the immense cloud of stratospheric ash that spread out from Krakatoa wound itself around the planet in a westerly direction—as one might expect, with the world turning eastward underneath it—and spread both northward and southward as it did so. At first, in other words, the blue suns and moons and the fiery afterglows and the Bishop's Rings and the extraordinary sunsets appeared in the low latitudes, close to the latitude of the volcano: during all of September in places no further north than Honolulu, no further south than Santiago, Chile.

---

*Though the phrase "blue moon" dates back to the sixteenth century at least, and has no connection.

But then the cloud spread itself farther afield. By the beginning of October the phenomena were seen in the Gulf of Mexico, then Nashville, Buenos Aires, the Canary Islands, Shanghai. By six weeks after the eruption, the particles were refracting, reflecting, dissipating, and dispersing light over fully sixty-two degrees of latitude and had spread, quite literally, half a world away from their birthplace.

And so the trend continued: By late October the amazing sunsets were causing bystanders to gasp and write poetry, to send letters to newspaper editors, and to paint vivid pictures in places like Tasmania, South Africa, and the southern cities of Chile. After which the cloud of volcanic aerosols, still wafting and widening north and south, performed a strange shiver and apparently began to move backward and outward at the same time, so that sometime around November 23, after touching western Canada and California, it became apparent not in Alaska, the Aleutians, or Hawaii but in England, Denmark, Turkey, Russia, and (coming from the west) Siberia.

Examination of all of the Northern Hemisphere records—including the five hundred sunsets of the Chelsea artist Mr. Ascroft, who started to paint in earnest in November what he saw each evening on the Thames—appears to show that countries lying to the east were to be affected later. It looks, in other words, as though the cloud that had in its early days been sliding ever westward was now moving in the opposite direction, as though it were trying to describe a long and lazy spiral across the surface of the earth.

(Close study of these upper-air movements later turned out to be quite crucial for modern meteorology. Analysis of the effects of the Krakatoa eruption—which still continues today, particularly at universities in Hawaii, Rhode Island, Oxford, Auckland, and Melbourne—has informed very many areas of science, but in particular has quite revolutionized the business of weather forecasting, helping it to nudge from its earlier and rather dubious standing as a mere drawing-room fancy into something that approaches the modern science of today. Each time, for instance, that one sees a map of the unfolding patterns of the jetstream, it is well worth remembering that it was the study of the stratospheric movement of the Krakatoa aerosols that led to the understanding of this particular weather-making phenomenon. Reverend Bishop in Hawaii seems to have been the first to notice, and he called the initial spread of ash the Equatorial Smoke Stream.)

The eruption particles continued to move "at about 73 mph"—such again was the scientists' exactitude—until, by the end of the year, they settled wherever they were and stayed, moving downward under gravity's pull with an infinite slowness. And so they seemed to hover permanently in the sky and, though having effects not quite as radiantly beautiful as in those first months, gave the greatest of pleasure to all below who saw them, for the next two or three years. Mr. Ascroft chronicled the entire long life of these sunsets from his Chelsea studio: They did not, he wrote, "entirely fade from view until the early part of 1886."

And the spectacle was only the half of it. The cloud reached the environs of New York in December. In the city itself they saw one thing. "The entire island seemed ablaze," said the World.* "Great tongues of flame shot up from across the water, reddening the southwestern sky and tinting the Jersey shore with delicate shell colors." The Times, then a less restrained newspaper than today, was only marginally more circumspect: "The clouds gradually deepened to a bloody red hue, and a sanguinary flush was on the sea; the brilliant colours finally faded to a soft roseate [Peter Mark Roget's Thesaurus had been available since 1852], then into pale pink, and finally died away upon the darkening horizon."

Out in the countryside, the results were very different, and the florid sunsets caused no end of confusion. In Poughkeepsie, seventy miles north along the Hudson Valley, the Daily Eagle noted with wry pleasure on Wednesday, November 28, Thanksgiving Eve, that:

---

*Of baseball's World Series fame.

Po'keepsie firemen have always been noted for their zeal and promptitude in going to fires, and their efficiency in putting them out, but the effort last night was a little too much for them. The light of an immense conflagration was visible, the bells pealed out vehemently and the boys ran vigorously down Market to Montgomery, down Montgomery to Riverview Academy they rushed—and when that point was reached it became evident that the fire was on the other side of the river, and a few moments of cooler reflection convinced them it was too far to reach, although it could probably last until they got there. Still, there was no means of transport available, and no likelihood of water. The fire being located 91,000,000 miles off, it would have taken the boys somewhere in the neighborhood of seven or eight million years to get there, let them run their best, and their terms would have expired before getting back. The sun is rather a large customer to tackle in the ways of fire, it beats a barn to death. We think the boys did well not to run out of hose.

Various causes are assigned for the peculiar light in the sky, the main one being the reflection of the rays of the declining sun upon the haze in the horizon. For several days past also there has been a peculiar state of atmosphere, which may or may not have something to do with it.

The no-doubt-embarrassed "boys" who were duped so nicely by the sunset belonged to the town's Hose Company No. 6 (of seven), better known as the Young America Hose Company, and they were based close enough to the river to be invariably the first called out to deal with a fire in the western end of town. A few days later, trying to assuage their pain, a meteorologist wrote helpfully to the Rochester *Democrat and Chronicle,* to say that the same confusing light had been experienced at his end of the state as well; and in his opinion whatever it was that had prompted the volunteers down in Poughkeepsie to dash off on their horse-drawn pumping engines was caused by "a stratum of decomposed vapours in the upper atmosphere."

The Poughkeepsie *Sunday Courier* went further, suggesting with some prescience that "the apparent reflection of a large conflagration, which called out our fire department" was caused by the sun's rays bending through "extremely small water droplets . . . perhaps mixed with dust and smoke, sifting out the blue and yellow rays." But another correspondent, probably taking down the office copy of *Old Moore,* offered only that the light show, all *lumière* and no *son,* was simply an augury of fine, anticyclonic weather in the days to come. Given the coldness of the coming season—made much worse by the dust—he could hardly have been more wrong.

Certainly the month of November 1883 was a time of unforgettable light shows throughout the northern world. In addition to the fire-spotters believing they had seen a blaze in the western suburbs of Poughkeepsie, there were reports that the fire-engines from New Haven, Connecticut, had been sent out too, for much the same reason. There was for a while a curious, half-panicked mood about people who had to see these ghastly skies night after night: To some they seemed almost apocalyptic, often unnerving; and it was only when they were explained away as being caused by dust from a distant volcano that people began to relax, and to bask in the simple sight of a terrible beauty they would long remember. But it was not to be a permanent fixture, there or anywhere.

And then there was the matter of temperature. Batavians noticed the chill immediately. At dawn on the very Monday morning of the eruption—though it was less a dawn, more a vague lightening of the drab ash-filled gloom—it was colder than records had shown for years—sixty-five degrees, fifteen Fahrenheit degrees lower than normal. People were seen shivering in the streets—though perhaps as much from fear as from a need to keep warm. Dense clouds hung in the air for days afterward, enveloping the city and an area perhaps 150 miles in diameter in a gray shroud the sun's rays could not penetrate.

Logic would suggest that a veil of dust particles spreading

around the world would in time produce much the same effect, with those places that languished under the sunset-producing dust clouds feeling chillier than normal—though on a lesser scale than places nearby. And to an extent they did. Oddly, the Royal Society, so assiduous in cataloguing the sound and light shows put out by Krakatoa, never bothered to consider the idea that the world might be cooled down by all the particles in the upper air. The society's editors present a catalog of the world's barometric pressures but none of the world's ambient temperatures. Those studies had to wait until much, much later—a rather puzzling and never satisfactorily explained omission, considering the time's enthusiasm for climate-related studies.

And when these studies of temperature came—the first carried out in 1913, the second in 1982—both found that, as expected, there had indeed been a worldwide drop in temperature. It had amounted on average to about one Fahrenheit degree and it had occurred, according to all the surviving records, at a time that appeared to be coincident with the eruption of Krakatoa. What has not been established, and what still concerns the scientific community, is which came first: Did the eruption lower the world's temperature? Or did a lowering of the world temperature because of some other reason perhaps—unthinkable though it seems—somehow prompt the crust to undergo stress and strain and crack, and a rash of volcanoes to explode?

There is no doubt that there is a correlation, a definite link. Benjamin Franklin* was the first to notice it, and told an audience at the Manchester Literary and Philosophical Society that the "dry fogs" that had seemed to cool the European summer of 1783 and made the ensuing winter exceptionally bitter were almost certainly the work of dust in the air. He had examined the volcanic records and found a fissure volcano called Lakagígar, or Hekla, in Iceland that had erupted earlier in the year (curiously, exactly a century

---

*He was at the time—1784—the first American ambassador to France.

before Krakatoa). It had produced great clouds of dust, for week after week, which tumbled up high into the atmosphere. This, he declared, must surely be the culprit.*

The infamous eruption in 1815 of Tambora, on the Indonesian island of Sumbawa, seven hundred miles east of Krakatoa, ejected twice the volume of material into the atmosphere (eleven cubic miles of rock, ash, and dust, compared with Krakatoa's six). The devastation it caused locally was profound—supposedly fifty thousand dead, an entire language (Tambora) extinguished, an entire island rendered uninhabitable for years. But its climatic effects were astounding too. For it lowered the world's temperature by almost one Celsius degree, on average: for every day when the normal temperature might be thirty-three, just above freezing, the temperature in the year after Tambora would be thirty-one degrees Fahrenheit, and ice would have formed on every pond and, more fatally, on every newborn crop, flower, and hatching egg.

So in New England the farmers claimed that 1816 was "the year without summer." There were frosts as far south as New Jersey in late May, in upper New England in June and July, and the growing season was slashed from the usual 160 days to seventy. Soup kitchens opened in Manhattan. Livestock had to be fed on fish carried over from the Atlantic seaports—1816 is also still remembered as "the mackerel year." There were crop failures— "the last great subsistence crisis of the Western world"—and, as a result, there was emigration to the Western states. No small number of today's Californians can rightly lay responsibility for their *being* Californians squarely at the door of the proximate cause of that year's ruinous cold—Tambora, a volcano unknown to most of them, and ten thousand miles away. (Although there was migration into California from Europe, in Newfoundland quite the

---

*Gilbert White's *Natural History and Antiquities of Selborne*, an essential bedtime companion for many an English reader, mentions the weather during the summer of 1783 as being unusually cold, with his Hampshire village experiencing "thick ice on 5th May."

reverse took place: Migrants were sent back east across the ocean, because there was not enough for them to eat.)

And yet back in Europe it was just as bad. The weather for 1816 is the worst recorded, with low temperatures stretching as far south as Tunisia. French grapes could not be harvested until November. The German wheat crop failed entirely, and prices for flour had doubled in a year. In some places there were reports of famine, and in others there were riots and mass migrations. The diaries and newspapers of the day present a litany of miseries. It is said that Byron composed his most miserable poem, "Darkness"—*Morn came and went—and came, and brought no day*—under the influence of that dismal year; and Mary Shelley may have written *Frankenstein* while gripped by a similarly unseasonable melancholy.

Nowadays sophisticated instruments and the measurements they take have usurped the role of anecdote and diary. Ice cores show minute layers of ash, or increases in sulfuric acid, as indicators of eruptive material in the atmosphere. And there is stunting of tree rings—the creation of a "frost ring" in trees that have suffered through an exceptionally cold winter. The examinations of deep ice cores, and of such still-extant trees as were living in the nineteenth century, have confirmed what the stories had long suggested: that eruptions of any of the world's larger volcanoes tend to coincide with periods of a cooling of the earth, some of the periods longer and with a very much lowered temperature, others shorter and with less of a fall in the mercury (the precise decisive factors are still not quite agreed on). Tambora's eruption of clouds of ash coincided with a cooling of the world in 1815; and so, in 1883, did the appearance of the clouds of ash from Krakatoa.

The most tragic of cargoes to move out from the volcano happened also to be the slowest. The audible sounds and the unhearable shock waves may have sped away at more than seven hundred miles per hour, and the dust may have wandered across the globe at more than seventy. The immense rafts of floating pumice that drifted away from where they splashed into the seas around Krakatoa made it as far as the southeast coast of Africa—but they did not make landfall for more than a year, traveling half a mile each hour, at best.

And when they did arrive, and were found washed up on the shores, they were discovered in some horrifying cases to have transported skeletons along with them, bringing as passengers the unidentifiable remains of some of the unfortunate thousands of Javanese and Sumatrans, Dutchmen and Chinese, who had perished.

Pumice is one of the better-known by-products of vulcanism. The most widely used products are, of course, volcanically produced rocks, used for building stone. These range in type from the dark basalts and gabbros, such as were used for making the spires of Cologne Cathedral, to the paler andesitic rocks, as are to be found in the temple lanterns all around Kyoto. Volcanic soils, especially a group called andosols, are uncommonly rich in minerals, and, in the attractive phrasing of the *Encyclopedia of Volcanoes,* "have nourished many ancient civilizations." And scores of the most ordinary, everyday items have incorporated volcanic products—particularly a highly absorbent mineral derived from weather ash called bentonite—in their manufacture: batteries, surfboards, refrigerators, and air conditioners generally make use of ash invisibly, while the cheap building material known as breeze block makes no secret of its origins: crushed-up coke, cinders, furnace clinker, and, in places where they are readily available, finely ground pumice and volcanic ash.

Pumice resides also in many an old-fashioned bathroom, sitting beside the loofah and the scrubbing brush. It is pleasantly abrasive and, because of the high proportion of gas bubbles that are caught in it before it solidifies in flight, it is of such low density that it floats quite readily. Makers of distressed denim fabrics also like to use it in their giant washing machines, where it brushes softly against the cloth, whitening and aging it in a way that finds favor with youngsters.

But these more benign uses of pumice tend to shroud the awful truth about the immense tonnage of it that was released by Krakatoa. The headmistress of a mission school in Zanzibar,* off the east coast of Africa, wrote to the Royal Society in response to the appeal, to report that

> . . . about the third week in July 1884, the boys . . . were much amused by finding on the beach stones which would float, evidently pumice-stone. The lady who was with them . . . also noticed that there were a quantity of human skulls and bones "all along the beach at high water-mark"; these were quite clean and had no flesh remaining on them, and were found at intervals of a few yards, two or three lying close together.

The closer to the volcano, the thicker the rafts of pumice, of course. On the west coast of the island of Kosrae, in what is now Pacific Micronesia, huge plates of pumice sixteen inches thick were hauled from the beach early in 1884: They were covered with barnacles, and many were accompanied by the roots of huge trees, with extra pumice lumps caught up in their roots, helping them stay afloat. These trees, torn up and floated three thousand miles to the east, were presumably parts of Krakatoa's old forests—the same forests that had been noted and painted by Captain Cook's homebound expedition in 1780, and those that the Sundanese boatbuilders had been cutting when they were forced to flee for their lives from the first eruptions of May.

The crews of ships moving through fields of pumice—like those, "acres in extent," that were encountered by one vessel coming into the Sunda Strait from Australia in January—were struck by the peculiar sound of the bow slicing through the rock. There was no real noise, "just a soft sort of crushing sound." And all the passing ships did their best to avoid desecrating the terrible cargo the pumice rafts all too often carried. A crewman on the vessel

---

*Zanzibar, a British Crown protectorate, became the empire's leading producer of cloves, after seeds were brought from the Spice Islands of the Dutch East Indies in 1818.

Samoa, which was heading southwestward, off into the Indian Ocean, wrote of the nightmarish unreality of such encounters:

> For two days after passing Anjer we passed through masses of dead bodies, hundreds and hundreds of them striking the ships on both sides—groups of 50 and 100 all packed together, most of them naked. We passed a great deal of wreckage, but of course we cannot tell if any vessels were lost. We also passed bedding chests and a number of white bodies, all dressed like sailors, with sheath knives on them. For ten days, we went through fields of pumice stone.

What was witnessed by the seamen on the *Samoa,* the *Bothwell Castle,* the *Loudon,* the *Berbice,* the *Charles Bal,* the *Kedirie,* and a score of other ships that scoured the Sunda Strait during those weeks in late August, September, and October does not bear too much repeating, so awful is what they have to say. Most reports were much more dreadful than

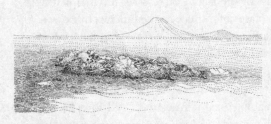

Rafts of pumice, many laden with the remains of victims, drifted as far away as Zanzibar.

the following account, which was published in a letter to *The Times,* from a correspondent in Batavia in October:

> The British ship *Bay of Naples* had called at these islands and had reported that on the same day, when 120 miles from Java's First Point, during the volcanic disturbances, she encountered carcasses of animals including even those of tigers, and about 150 human corpses, of which 40 were those of Europeans, besides enormous trunks of trees borne along by the current.

Yet it is what they did *not* see that remains, to take the longer view, of greater significance. For what they did not see was this: the half-mile-high pointed peak of Krakatoa. They did not see it smoldering menacingly in the aftermath of what it had done, as most volcanoes are prone to do.

Usually—whether named Vesuvius, or St. Helens, Pinatubo, Unzen, or Etna—a volcano explodes, and in so doing causes manifold devastation and death. Then it simply stands there, raging and smoking ever less fiercely, presiding with a titanic smugness over the ruin it has so lately made. But, uniquely in much of volcanic history, Krakatoa did no such thing.

Krakatoa had gone. Six cubic miles of rock of it, most of the island's great bulk, had just vanished, either blown into the sky or collapsed into the sea, and with the most thunderous roar and the greatest loss of life ever recorded.

For so long before that Krakatoa had been an island of no consequence. It had been little more than a genial, easily recognized sailors' companion spotted off the bow of an approaching ship, a seamark that would always help guide any navigator who was feeling his way up or down this most vital waterway between Java and Sumatra. It was that old "island with a pointed mountain," and nothing more.

Now, in the middle of the summer of 1883, it had suddenly and without much warning gone totally berserk, sent the sea berserk as well, and then, essentially, vanished. Not vanished in name, perhaps; not vanished from memory either; and recently, and in much the same form, reborn (as we shall see). But in August 1883 this inconsequential little island went mad and disappeared. The reasons why this happened—as it did, and when it did—have occupied the minds of a great community of geologists around the world for all the long years since.

THE EXPLANATIONS

Why did Krakatoa happen? Why indeed, and more generally, do volcanoes do what they do? why does the terra firma upon which we so confidently and innocently secure all our lives, sometimes and so capriciously tear itself open, causing such fearful destruction?

To those caught up in such a moment of appalling terror, such as the thousands whose lives were wrecked in 1883, it all must seem a most monstrous injustice, a terrible cheek perpetrated by the earth and its presiding deities. Krakatoa is a stark reminder of the truth of Will Durant's famous aphorism "Civilization exists by geologic consent, subject to change without notice." Yet geology, which is an unemotional and rational science, allows us to step back from our shock and dismay at such events, to accept a longer view—and to be awed by something rather different: that despite her seemingly cruel caprices, this planet in fact enjoys by and large an extraordinarily fortunate situation.

The simple, very obvious features of the earth—its location in space, its size, the processes that led to its creation, processes that include the very volcanic events that took all the lives west of Java—happen to have been suited perfectly, when taking the long view, to the sustenance and maintenance of organic life.

To victims of a volcanic eruption like this, of course, the very reverse must seem true. But consider location, for instance. Planet Earth is sited just close enough to the star around which it orbits to derive only benefits from the latter's infernal solar heat. It is neither so close as to risk the boiling of its oceans and the loss of its water into outer space by photodissociation in the upper atmosphere nor so far away that all its present liquid water remains uselessly and inconsumably frozen.

The size of the earth is ideal too. Thanks to its moderate size its gravitational pull is just right. It is strong enough to overcome in particular the escape velocities of the molecules both of water and carbon dioxide, which means that we have a sheltering canopy—a benevolently situated *greenhouse,* even though this is a word with more negative associations today—that first allowed life's building blocks to be assembled, and then ensured that the fragile living entities so made could be protected from the perilous radiations from outer space.

And then there are the volcanoes—just the right number, of just the right size, for our own good. The deep heat reservoir inside the earth is not so hot, for instance, as to cause ceaseless and unbearable volcanic activity on the surface. The amount of heat and thermal decay within the earth happens to be just perfect for allowing convection currents to form and to turn over and over in the earth's mantle, and for the solid continents that lie above them to slide about according to the complicated and beautiful mechanisms of plate tectonics.

Plate movement and convection and the volcanic activity that is their constant handmaid may not seem, to the victims of eruptions and tidal waves, to be in any way benign, or to be good for the planet as a whole. And yet, taking the long view once again, they most certainly are: The water, carbon dioxide, carbon, and sulfur that are so central to the making and maintenance of organic life are all being constantly recycled by the world's volcanoes—which were also the probable origins of the earth's atmosphere in the very first place. It is not merely that volcanoes bring fertile volcanic soils or useful minerals to the surface; what is more crucial is their role in the process of bringing from the secret storehouses of the inner earth the elements that allow the outer earth, the biosphere and the lithosphere, to be so vibrantly alive.

Almost all our neighbor planets are, so far as is known, volcanically lifeless. They are also, on all the available evidence, more or less biologically lifeless—and that is quite probably at least in part because they *are* so volcanically dead. Only Io, one of Jupiter's many moons, seems to sport a significant number of volcanoes: spectacular sulfur-rich fountains of magma have been seen spouting on its surface. But there is no suggestion of plates or of any movement of the solid crust, either on Io or on any planet or moon known to exist between Mars and Pluto. The vigorous business of plate movement apparently does not occur on planets that are hotter than our own; nor does it on those that are much more frozen and more deeply dead.

But it is the movement of the plates, and the internal storms that rage below and cause them to slip beneath or alongside one another or tear themselves apart along their suture lines, that is the driving force behind our earth's highly unusual degree of vulcanism. Plate movement, as well as shaping the planet's topography, also creates most of the very vulcanism that is central to its life. Plate tectonics, in other words, is the key to it all—and any examination of just why Krakatoa happened as it did, and how it did, must inevitably now refer to this newly minted catalog of knowledge about the workings of the earth.

It was of course not always so. In the distant past, whenever the earth behaved with terrible and unanticipated violence, mankind could do little more than wonder, horror-struck, at the sheer effrontery of it. In very early times this wonderment was answered, inevitably, mainly by religion and the making of myths. Volcanoes were hills occupied by temperamental gods: They could be appeased by frequent sacrifice. The appeasing flesh could be that of a young human (a small child thrown every twenty-five years into the crater of a particular Nicaraguan volcano, for instance, would guarantee its quietude) or an animal (Javanese today toss chickens into the crater of Mount Bromo—superstition still plays an important role in East Indian attitudes toward their volcanoes).

The ancient Romans and Greeks then hammered some kind of order into their beliefs, as might be expected: The idea of the existence of Hades, the nature of such gods as Pluto and Vulcan, the character of Titanic monsters like the fearsome, wild-eyed, and flaming-tongued Typhon were all connected with the wayward behavior of an earth that all then knew had a terrible and dangerously hot interior. It was no coincidence that the gateway to Hades—believed by the ancients to be in the earth's center—was the Romans' most notorious local volcano, Mount Etna, and the phrase "sailing to Sicily" was for a while a euphemism for entering the fiery furnaces of the devil's domain.

The seers of the classical world were on rather shakier ground when it came to deciding just why, other than for divine reasons, there was just so much heat inside the earth. The Greeks—the philosophers Anaxagoras and Aristotle in particular—favored the human analogy of *trapped wind,* with the friction of the escaping wind causing the generation of heat, a sort of volcanic vindaloo. The Romans, on the other hand, and among them most notably Lucius Seneca, favored the notion that the heat came from the combustion of a vast inner-earth storehouse of sulfur—and in some Roman poetry of the time this idea extended to the burning of deeply buried reservoirs of alum, coal, and tar.

This idea, that volcanoes were the consequence of the steady burning of a finite store of earthly combustibles, exerted a grip on the scientific mind for centuries. Then, as chemistry developed as a science, so its innumerable secrets offered themselves as the favored sources for all the necessary heat, and were widely accepted as doing so. During the seventeenth and eighteenth centuries a great many seers—Isaac Newton among them—believed that so-called exothermic chemical reactions were the answer. By 1807, when the Geological Society of London, the world's first such body, was founded, the oxidation of newly discovered alkaline metals, such as sodium and potassium, was thought to be an answer.

Even as late as the 1920s there were two now notoriously blinkered scientists who clung to what might seem today quite fatuous chemical theories. One of them, Arthur Louis Day, proposed in 1925 that volcanic heat was due to a series of complex chemical reactions between gases, and he won support from the redoubtable and influential Sir Harold Jeffreys,* while at the same

---

*Sir Harold was one of the last holdouts against the theory of continental drift, writing into the late 1950s that the earth's crust was simply too rigid to permit it. He was widely regarded as a brilliant figure—an expert on the monsoon, and on the physics of cyclones

time dismissing vulcanism generally as a phenomenon that was merely "local and occasional, not perpetual and worldwide."

However, in tandem with all those chemists and physicists who for so long had such an influence on geophysical thinking, there were also other natural philosophers—René Descartes most notable among them—who started out on what would prove to be the right track. In the mid–seventeenth century Descartes—better known for his *cogito, ergo sum,* and for his legacy of Cartesian coordinates—came up with a quite revolutionary idea: that the earth originated by way of gravitational attraction and gaseous condensation, that heat was an essential primordial component of this process, and that its slow decay resulted in the earth having three internal concentric parts: a highly dense and incandescent liquid core, a half-cooled plastic central region, and a cold, solid, and comparatively light crust. Moreover, there was ample primordial heat left over from the creation process to power all known volcanoes for a very, very long time.

The subsequent advent of the science of field geology, the furious debates that went on between Neptunists, who believed all rocks to have precipitated from a primeval ocean, and Plutonists, who saw countless of them as having their origins in melting and magma, belongs to another story, temptingly diverting though the various interlocking sagas may be. In essence, though, the mystery that occupied most minds for most of the late nineteenth and early twentieth centuries was simply *why rocks melt*—what combination of physics and chemistry, of depth, of heat, and of the presence or absence of water in the mix of minerals would lead a rock to become plastic and mobile and molten, and then to emerge on the surface and cool and harden and solidify back into rock once again.

The chemists and the chemistry that sought to answer ques-

—but was a memorably bad lecturer, devoting much time to the dynamics of boiling porridge and confiding its secrets not to his Cambridge classes but to passersby outside the open window to which he customarily spoke. Like George Darwin, mentioned earlier, he also held the Plumian Chair in Astronomy.

tions about the makeup of the earth in earlier times had been over-taken by the physics and the physicists seeking to do much the same in recent years; and though the physics answered much of the detail involved, many of the fundamental questions remained doggedly, in essence, unresolved.

Or at least they did until that memorable July day in 1965 when, as I have explained in an earlier chapter, the soft-spoken and self-effacing Canadian geologist J. Tuzo Wilson managed to combine both chemistry and physics of the earth into one, inaugurating the science of plate tectonics. In doing so he launched a brand-new and all-encompassing global theory that would offer the answer to almost everything volcanic that had ever been wondered at.

Uniquely in the solar system this planet sports a crust that is, by virtue of this process, being constantly destroyed and regenerated—an ever mobile chemical factory where materials that exist in solid, liquid, and gaseous states are being recycled endlessly. They are burst out from the middle of oceanic plates by a process, newly understood, that allows upwelling materials to melt *without heat being added to them*—to melt simply because the pressure on them is relieved by their being convected upward and outward toward the atmosphere.* There are volcanoes here in these mid-crustal ridges, big but not especially explosive volcanoes, mountains that ooze basalt, like those in Hawaii and Iceland, the Azores and the rift valleys of East Africa.† They are the stuff of research and fasci-

---

*This is a complex subject, well beyond the scope of this story: Since it involves phrases and concepts such as enthalpy, adiabatics, isobaric entropy, and mole fractions, this is perhaps just as well. However, it is vitally important to a proper understanding of plate tectonics, and I have included references to useful books on the topic in the "Recommendations for Further Reading."

†Mount Nyiragongo, in the East African Rift, is one of the more malign of the volcanoes to be found at a place where plates are pulling apart from one another. The lava flows are immense and unpredictable (though the calamitous eruption of 2002 happened to have been predicted, accurately, by an unpaid and heroic Congolese vulcanologist named Dieudonné Wafula, whose reports were widely ignored), and its gas emissions kill people and wildlife, including large numbers of elephants, who are suffocated and then covered with a thin coat of lava.

nation in their own right. But they are the distaff side of the volcano of this account, the Alpha to Krakatoa's Omega, the midplate
reciprocal to all that goes on at the plates' edges, *the other side of
the story.*

For the materials that rise up in the middle, along with whatever
they sweep before them on their way, are in due course swept down
again at the peripheries of plates. They are swept down by the
process that is most crucial of all, which, though an essential part of
earthly regeneration, also leads directly to the making of highly
explosive, dramatic, and deadly arc volcanoes like Krakatoa. Colloquially the phenomenon that exists at these plate edges is known
among geophysicists and vulcanologists as *the subduction factory*—
and Krakatoa stands front and center in one of the largest and most
complicated of these extraordinary, world-shaping entities.

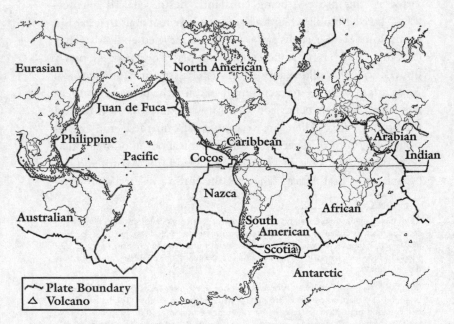

The world and its pattern of tectonic plates. Where oceanic and continental
plates meet, there is volcanic and seismic activity in great—and often terrible—
abundance.

The factories and the subduction zones that underpin them are essentially coextensive. It probably bears repeating that each of the zones is where one of the world's many heavy oceanic plates slowly collides with one of the many lighter and thicker continental plates, and slides, buckling as it does so, underneath. The zones are very long, and very thin. If unraveled they would extend for about nineteen thousand miles. But they are rarely more than sixty miles wide. The total area of the subducting world's assembly lines amounts thus to about a million square miles—about the size of Greenland, or the American Confederate South, or Argentina.

And enclosed within the zones, and formed, allowed to grow, and then destroyed or mutated or otherwise dramatically affected by the processes going on inside them, are about fourteen hundred of the world's fifteen hundred historically active land volcanoes. Of all visible volcanoes, 94 percent, in other words, stand within subduction zones. A mere handful of countries—Indonesia, Japan, the United States, Russia, Chile, the Philippines, New Guinea, New Zealand, Nicaragua chief among them, *and in that order,* play host to most of them: these nine countries are home to more than nine out of every ten volcanoes that are liable to erupt today or have done so in recent history.

The most readily recognizable subduction zones are those that enfold the Pacific Ocean. As a reasonably familiar example, consider that which runs along the western edge of South America, and which has created the chain we known as the Andes.* This is where the heavy basaltic Nazca Plate collides with the lighter granitic-and-sedimentary-rock South American Plate. (It does so simply because it is at the same time splitting itself away from its neighbor, the Pacific Plate, along what is called the East Pacific Rise—close to where the Isla de Pascua, Easter Island, is hoisted above the ocean surface.)

The subduction factory that results is a classic of its kind, creat-

---

*From which comes the word *andesite,* one of the more common indicator-rocks found in a typical subduction-zone volcano.

ing dozens of volcanoes running from the Andean peaks of Ruiz and Galeras in the north, in Colombia, via Chacana, Cotopaxi, and Sangay in Ecuador, Huaynaputina* in Peru, Lascar in Chile, Llaima and Villarica on the frontier between Argentina and Chile, and, at the southern tip of the continent, Monte Burney and Cerro Hudson, this last volcano erupting massively in 1991. All told there are sixty-seven volcanoes that have been manufactured by the processes of this one subduction zone—and since four thousand miles separate northern Colombia from southern Chile, and since there is a sort of serrated regularity to the Andes, that means there is more or less one volcano piercing the sky every sixty miles.

Much the same number of volcanoes, with similar intervals between them, is to be found at the other subduction factories around the Pacific—in Alaska and the Aleutian Islands, in the Kamchatka Peninsula, in Japan and the Kurile Islands; and an even greater number is to be found in the most volcanic part of the world, the great subduction zone that stretches three thousand miles from the northern tip of Sumatra to what is called the Bird's Head on the northwestern tip (the West Irian side) of the island of New Guinea.

In this immense factory, there are at least eighty-seven volcanoes that make up much of the archipelago that politics has lately chosen to call Indonesia and the Philippines. Indonesia itself has and has had more volcanoes and more volcanic activity than any other political entity on the earth, in all recorded history. It is a country that is defined by its place at the heart of a subduction zone, is essentially made up of volcanoes and precious little else. On the island of Java alone today there are twenty-one volcanoes that remain fully active. Their eruptions are invariably spectacular and terribly dangerous. And because a great many people live and

---

*Despite its generally unfamiliar name, this rather small Peruvian volcano had one of the biggest eruptions in world history, in February 1600—creating a "spike" in Greenland ice cores rarely equaled, even by Tambora and Krakatoa.

work near the volcanoes (not least because volcanic soil, thanks to the recycling mentioned earlier, is nutritious and ideally suited for farming), they are the cause of a dismaying number of deaths.

The earth fashioned three of the five greatest volcanoes of historic time in this one gigantic factory. The largest the world has ever known was made there: Mount Toba, which erupted 74,000 years ago in what is now northern Sumatra. It had a Volcanic Explosivity Index, or VEI, of 8—the highest on a scale that is now universally used to classify all eruptions (save for those that merely ooze lava, without exploding). Toba's *humongous* explosion—the curious adjective is now officially used to describe giant volcanoes, the equivalent of the cyclonic sea state that is these days termed *phenomenal*—left behind an immense lake, fifty miles long and fifteen wide, with the sheer caldera cliffs rising eight hundred feet straight out of the water. The eruption left layers of dust eighteen inches thick on ocean floor fifteen hundred miles away, and must have placed a severe crimp on the development of such ur-humans as were struggling for existence in those times: it must have lowered the ambient temperature by many degrees, and made even harsher a climate that was already in the midst of changing into yet another Ice Age.

The eruption of Tambora in 1815, in this same subduction zone, is reckoned the second greatest in history, with an Explosivity Index of 7. (This index, which was first created at the Smithsonian Institution in Washington, is based on two features: the amount of material that is ejected in an explosion, and the height to which it is hurled through the atmosphere. These two factors are clearly observable in modern eruptions; they are also deducible from the records of the past. Even though there were no literate eyewitnesses to Toba, and very few to Tambora— which has to be the principal reason neither has lingered in the public consciousness, while Krakatoa clearly has—the total mass that was ejected from each eruption can be calculated with some precision from an examination of the local geological record, and the distribution of the ash falls faraway on the seabed can suggest with fair accuracy the heights to which the columns rose into the sky.

Third in the list is Taupo in New Zealand, which erupted in A.D. 180, hugely; and fourth is Novarupta in Alaska—better known as Katmai, on the landward end of the Aleutian chain—which did so in 1912. This last was the largest recent eruption on the North American mainland, but, because of its remoteness, it was little noticed except by what—in terms of calderas and domes and frozen lakes—it left behind.

And then, fifth in the list of all volcanoes known, with a VEI of 6.5, with more than six cubic miles of rock and ash and pumice and dust hurled dozens of miles into the lower stratosphere, with sounds heard three thousand miles away, with tidal waves of enormous force and height, with shock waves that ran four times to the far side of the world and almost three times back, and with more people killed and more livelihoods ruined than by any other eruption in world history, comes Krakatoa.

Seven weeks after the eruption, when the dust had cleared, the Netherlands government ordered Dr. Verbeek and his colleagues to investigate exactly what had happened. The team of four took off in the government hopper-barge the *Kedirie*, on October 11, and spent the following two weeks examining every possible aspect of the now seemingly dead remnants of the mountain. The orgy of self-destruction was more than amply evident. It was, Verbeek wrote later, "the most interesting eruption witnessed by the human race until now." But very little of the original mountain was left to see.

The southern quarter of the island remained but was sliced open, as if with a vertical carving knife—so that the original peak of Rakata looked, from the south, almost the same, but with everything to its north missing. The exposed northern face of Rakata was almost perfectly vertical, and in cross-section, when viewed from the north, quite perfectly triangular; it was pierced with vertical lines and radiating systems of lava-filled dikes and sills and plugs of newly formed rock, all covered with several feet of gray pumice dust, so that it looked from afar exactly as it was—a perfect cross-section, as one might see in a teaching chart, of a onetime volcano that had been blasted in half and into oblivion.

But at least the summit of Rakata was, more or less, still there. The two northern peaks of Krakatoa, the summits known as Danan and Perboewatan, were, to the expedition's awestruck fascination, no longer anywhere to be seen. Nor was the little skerry of andesite known (because of its shape) as the Polish Hat: It had quite vanished, presumably vaporized in that one paroxysmal instant.

The very opposite had happened to the two small islands, Lang and Verlaten, which had once enfolded Krakatoa like a pair of parentheses. Rather than vanishing into thin air, these two now appeared to be very much larger than they had before. Their beaches, it turned out, had since the eruption become choked and swollen with enormous amounts of stranded pumice. Larger they may have been—but the essential difference about their appearance as maritime parentheses was that now they were enfolding and bracketing *nothing*—between them was just a huge expanse of empty, lifeless sea, with the immense broken fang of Rakata peak rising alone straight out of the ocean as a reminder of what had once been there.

The sea to the immediate north of the cliff was very deep—nearly a thousand feet. Clearly an enormous new caldera had been created—almost the entire volcano had collapsed into an immense void below, and the cliff of Rakata, neatly bisected where it had been shorn off, was all that remained to the south of the collapse. Off to the northeast two entirely new little islands had risen from the waves, and were christened Steers and Calmeyer Islands: because they were composed of little more than stranded rafts of soft pumice they were eroded back to sea level in very short order; on today's charts there are just warnings of "patches of discoloured water," fifteen feet deep, suggesting where they used to be.

Back in 1885, when Verbeek wrote his official report, there were only the vaguest explanations of just why all of this might have happened. It was easy enough to describe what had happened—the science of descriptive vulcanology was in any case well advanced, and had been for many years. But when the vulcanologists of the day came to explain the reasons for the violent behavior of their charges—as true for every volcano in the world as it was for Krakatoa—there was very little understanding of the processes of the world to offer them a basis for coming up with a theory.

After all, only a few decades before many believed that basalt and flows of lava were simply precipitates from the sea. Until 1857 many geologists thought that volcanoes were caused by the bulging upward of horizontal flows of lava—and not that they had been built vertically by the discharge of their own products. And at the time of Krakatoa's eruption Alfred Wegener—who first came up with the idea of continental drift, which was to lead to the theory of plate tectonics, and who might well have set the bewildered community of vulcanologists off in the right direction—was only three years old.

And so while in all the official reports and learned papers about the event there was plenty of description of the ruin and dismay that Krakatoa had caused, and though there was much speculation about why the volcano exploded with the violence it displayed, there was next to nothing by way of sensible wondering about the larger mechanisms that triggered it.

This was true for Verbeek in his report of 1886, for instance. He spent countless pages describing in detail clogged pipes, steam vents, and collapses of central parts of the main volcano. What he concluded did display a remarkable prescience: He said that a good deal of the vanished volcano had foundered into the sea, and had not been blasted into the atmosphere. He suggested that the Plinian violence of the explosion was a result of seawater mixing suddenly with the magma, and flashing over, turning into superheated steam, in a gigantic and uncontrollable explosion that is these days given the somewhat less than attractive name of a *phreatomagmatic* eruption. But he never tried to step back and wonder why Krakatoa was where it was, and why it did what it did in the first place.

The same was true also for John Judd, president of the Geological Society of London and author in 1881 of a then classic work, *Volcanoes*. He too wrote eloquently of the way in which hot magma and seawater mixed, and of how pumice was created by a lowering of the magma's melting point by the addition of water—but, once again, he missed the central point. He never even tried to grapple with the central issue: Why Krakatoa?

The last popular book on the subject* was written in 1964. Even then, still lacking any solid theory that might account for the world's inner processes, the author could only really describe what a volcano was ("a hole in the ground through which hot gas, molten material and fragmentary products rise to the surface"), say where volcanoes were to be found, and name the kinds of material that came from them. And when he arrived at the specific case of Krakatoa, which was described over many pages and with a quite magical literary skill, the author turned desperate. The book begins to speak in terms of "The Demon" going in to "press the attack," his "searching fingers boring into the defences," and the "pent-up energies of time" and "primeval forces" readying themselves to do battle. One can hardly blame him. Neither he nor anyone else had an inkling of what really caused Krakatoa. And that was hardly his fault: he was simply a very few years too early.

But once plate tectonic theory was in place, all that changed. Nowadays there is a ready explanation for what happened and why. Essentially the same explanation accounts for the eruption of Toba at the northwestern end of the subduction zone, and for Tambora's at its eastern end, and for those of all the other volcanoes in between.

Krakatoa erupted because of what happens when two plates

---

*Krakatoa, by Rupert Furneaux, had the unfortunate distinction of being published in 1965, almost on the eve of the announcement of the discovery of plate tectonics.

collide—specifically, because of what happens when the north-bound Australian Oceanic Plate collides, as it has been doing for many millions of years past and as it continues to do today, with that part of the Asian Plate that, for the sake of simplicity, we will call by the name it enjoys today, Sumatra.

The oceanic plate is cold and made of the heavier, darker, less acidic suite rocks that underlie all oceans, so as it hits it begins to sink below the warmer and lighter rocks of which Sumatra and all other continents are made. As it sinks it takes with it, downward, the small wedge-shaped sliver of continental rock that it either scratched or smeared off the Sumatran edge; it also takes along with it some of the sands and clays that had accumulated on the Sumatran coast, the water that was trapped chemically in these, and a fair amount of atmospheric air and seawater besides. This entire geological cocktail—an amalgam of cold and heavy basalt from the

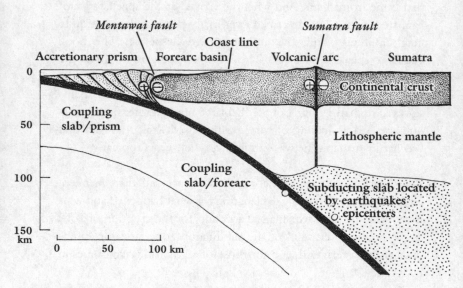

A cross-section showing the basic elements of an oceanic plate—upwelling new material from the center, spreading outward and then sliding beneath the lighter continental plate it then encounters. Volcanoes and earthquakes are an inevitable feature of this last process of *subduction*.

plate; granitelike rocks from the Sumatran crust; sands, clays, limestones, and vast quantities of air and water—then plummets. And, as it does so, everything suddenly changes.

Water is the crucial ingredient in this process. Not only does it lubricate the motion of the plates and help the subduction continue, but, even in very tiny amounts, its presence lowers the temperature at which the rocks of the mantle will begin to melt. And since the water also lowers the density of the wedge-shaped mélange that is being swept and smeared downward too, the molten rock that is being created below it finds that the rocks above it have suddenly become (thanks to the water) less dense, less rigid, less strong. They have become, in other words, a perfect exit route for the partly melted rock below, enabling it to rush upward, to melt even further because of the decompression mentioned earlier. Then, with the dissolved carbon dioxide and water vapor suddenly turning back into gas and frothing out of solution, the whole mass rushes up and out as a torrent of phenomenal explosivity into the unsuspecting open air: as a gigantic and classical subduction-zone volcano.

That is why Krakatoa exploded. As to why it exploded so powerfully and so very noisily—this entirely different debate continues apace today.

There are some clues. The geography of the islands indicates, for example, that there was once an ancient super-Krakatoa, and that it at some unspecified time in the past exploded and collapsed into itself, leaving behind a caldera; the parenthesis-islands of Lang and Verlaten were clearly the caldera walls, the cliffs at the edge of the old volcano. The volcano that succeeded this then had three distinct peaks—Rakata, Dana, and Perboewatan. Each was an exit passage for a gigantic magma chamber that clearly existed deep below the region.

So there can be little doubt from this simple evidence alone that the 1883 Krakatoa existed above a large chamber of magma, and that with the three exit pipes above it weakening its roof, it had a propensity, in times of violent stress, to collapse. The question that has occupied the minds of many specialists in recent years is this: Was it the fact that seawater managed to get into this chamber at the moment of collapse that was the primary cause of the deafening explosion and the tidal waves? Or was it simply one contributory factor, with some other process also at work to make things even more dramatic?

A series of experiments performed in pressure vessels in laboratories around the world has suggested that other factors were indeed at work—but that they were complex and subtle. They suggest that a pulse of fresh basalt from deeper in the earth may have been unexpectedly injected into the base of the magma chamber; that this new pulse heated the existing magma above it, causing a violent convection current and the sudden frothing of even more gas—and the sudden breach of the chamber roof. This idea of magma mixing has lately taken hold: Processes going on even deeper within the subduction zone may perhaps have contributed to the might of the Krakatoa event.

And then, finally, there is the overall siting of Krakatoa, halfway between Java and Sumatra. It lies directly above what might be called a *hinge point* around which the two islands are slowly swinging, the strait ever widening, the islands turning like the pages of a northward-closing book—Sumatra moving to the northeast, Java to the north, Krakatoa in the middle.

There is certainly a complex network of faults in the Sunda Strait. Their existence is one of the reasons why there is a strait there, an absence of island mass, in the first place. Slowly, very slowly, science is trying to make some sense out of the complexity of it all. Geophysicists in Troy, New York, have spent recent years placing a rash of Global Positioning System receivers on the nearby islands, and have found that all manner of tiny movements are taking place—that the main subduction continues to creep away, as it has done for millions of years, but that tiny little sideways jogs are taking place too, small weakenings, creations of a peppering of tiny faults that make the region into the most remarkable of geological

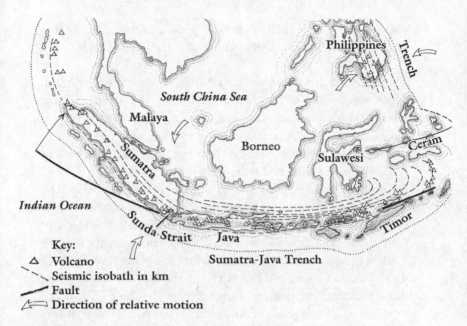

The basic tectonic structure of the region: the Australian Plate is moving north to collide with the Asian Plate, while all manner of stresses and faults build up along the collision zones by Java and Sumatra.

laboratories—a fascinating study, even if Krakatoa had never existed.

But it does exist and will play its tricks on the world once again, and before very much longer. The processes that led to the events of August 1883 are unstoppable. There is a subduction factory of monumental proportions to the south and east of Sumatra. It is uniquely sited around and beneath the small island that lies on the hinge point between Sumatra and Java. The island is surrounded by the volatile waters of the sea, water that causes mayhem if it gets within a mile of boiling magma. The island itself is surrounded by countless small faults and zones of weakness, as well as by a raft of components of basic rock—acidic rock, sedimentary rock—that is twisting and turning every which way under a barrage of stresses and strains that exist more notably here than anywhere else on earth. Small wonder, indeed, that there is only one Krakatoa. The place where it blew itself apart is so geologically dangerous that one can almost imagine there being room for a dozen more.

They counted their dead, and they buried them where they could, which was usually where they found them. The Dutch officials reacted with commendable speed, burying bodies at the rate of several hundred a day, drenching the swamps with carbolic acid, pulling down wreckage, setting cleansing fires. The king back in Holland opened a fund. Dutch mothers sent blankets, tents, food. A flotilla of ships traveled east to see what could be done. The Great World Circus staged benefit performances in Batavia, before packing up and going home, taking their distressed little elephant with them. The process of rebuilding began, with the Fourth Point lighthouse at Anjer, remade with iron plates, hurried back into operation to ensure the safety of the commercial shipping lanes, a symbolic beginning of the rebirth. The cable lines to and from west Java and south Sumatra were repaired. The aid workers came into town. The charities set up shop. The scientists fanned out to investigate, to report, to recommend.

But in time they all went home again, to deal with other problems and to answer fresher questions. They left the coastal people of Java and Sumatra, and those islanders of shore dwellers known as the Bantenese, among their patched-up ruins, and in time they forgot all about them. They did not stop to wonder where these people might eventually look for sustenance and succor.

Perhaps they should have. For it turned out that not a few of these unhappy, dispossessed, and traumatized people eventually looked to the west, to Mecca, and to the benevolent power of their Islamic religion to answer their needs. This was a political and religious consequence of disaster—a consequence entirely unanticipated by the ruling Dutch colonists—that was to have the most profound and longest-lasting fallout, for the Indies, for Europe, and beyond.

9

Rebellion of a Ruined People

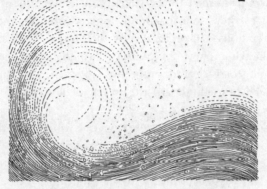

The leaders formed an élite group, which developed and transmitted the time-honoured prophecies or vision of history concerning the coming of the righteous king—the Mahdi.

—from *The Peasants' Revolt of Banten in 1888*
by Sartono Kartodirdjo, 1966

T he central market in Serang, an unremarkable crossroads
town a few miles inland from the little pepper port of Banten, sells everything and attracts everyone. On Tuesday,
October 2, 1883—five weeks after the eruption, while the dead of
Java and Sumatra were still being buried and the ruins of the
towns and villages on the coast were still being cleared—a Dutch
fusilier stopped by to purchase his weekly supply of tobacco.

In normal circumstances it would count for nothing that this
man was a part of the ruling foreign enginework of the Indies: The
merchants of Serang were ecumenical in their pursuit of trade, and
the tobacco seller, like everyone else, would happily accept the custom of anyone—brown, yellow, or, as in this case, a white European from the curiously privileged group who ruled over them.

But these were not normal circumstances. All of a sudden there seemed to be a curious feeling of tension in the air. The Dutch had been aware of it for some days now—a kind of muted resentment, a vague hostility that made people look away, or mutter among themselves in hushed tones whenever a Dutch official happened by. Perhaps they were imagining things. After all, the Krakatoa relief operations—all organized by the Dutch—were now in full swing, Dutch money was pouring into the area, shelters were being built, roads cleared, businesses reopened. The Dutch governor-general himself, aboard the steam yacht *King of the Netherlands,* had paid an official visit.

The local people had every reason, one would have thought, to be grateful for the colonists' help in speeding along their program of rescue and rebuilding. It wasn't the fault of the Dutch that the tragedy had happened: but, without a murmur of complaint, the Dutch were quite generously helping the people who were worst affected. True, the relief efforts organized from Rotterdam and Amsterdam were more concerned with ensuring that Dutch-owned businesses were able to stagger back on to their feet; but if the native people derived benefit as a result—well, that was in part what colonialism was all about, surely? And they thus had ample reason to look kindly on their faraway benefactors, *n'est-ce pas?*

But then here in Serang, the unimaginable happened. It was while the young soldier was in the process of handing over a fistful of guilders for a package of fine-cured and locally grown tobacco with which to pack his meerschaum that a bearded man, dressed entirely in white and armed with a curved dagger, suddenly threw himself on the soldier and began to stab him repeatedly in the back.

The astonished fusilier, gravely wounded, managed to stagger into a nearby Chinese shop. The attacker, his task completed to his satisfaction, ran away and was soon lost in the throng. The market was thrown into chaos. Nothing like this had happened before. Troops from the garrison—one of the five in Java's First Military Region, and one of the smaller—immediately flooded into the area, and promptly arrested a number of known malcontents. But none could be proved to have been the attacker: The would-be assassin seemed to have gotten away with it.

Then much the same thing happened again six weeks later. Another young man, also dressed ostentatiously in white robes, somehow managed to infiltrate the garrison headquarters itself, and when he was challenged, brandished a long knife and slightly wounded a locally raised sentry named Umar Djaman. The attacker was this time captured and interrogated: Military investigators, bewildered by the man's confusing answers to their questions, suggested in their report that the motive for his assault was an inexplicable case of "extreme religious zeal."

To the local military commander the attacks were unprecedented and sinister. True, there had only been two of them, and they might well have been perpetrated by the same, deranged man (though the assailant held in custody in November refused to confirm that he had also carried out the October incident). But two attacks or not, it seemed to the puzzled soldiery, and to those already aware of the sullen mood, that there had been a mysterious outbreak of a weirdly fanatical anger, and that it was for some reason directed specifically against the Dutch masters. It would be prudent, the soldiers suggested to their political superiors, that the authorities from henceforth be permanently on their guard.

They were right to be careful. The events in Serang that autumn turned out to be the beginning of a long and exceptionally violent period in western Java—a period that culminated in a memorably dangerous and politically ominous rebellion that erupted five years later, in 1888. The Banten Peasants' Revolt, as the episode is now generally known, is thought by many today to be a turning point in the region's colonial history—one of the milestones on the road leading to the eventual expulsion of the Europeans in 1949, and the creation of modern independent Indonesia. There had been many small rebellions in Java over the years; but what happened in Banten had a significance that transcended many of the other outbreaks of violence and hostility.

For what took place among the population of the northwestern tip of Java between 1883 and 1888 was not, however, of political significance alone. The time also had great religious moment, since it marked a period when Islam had become fully entwined with the local political developments of the day and—so many scholars now accept—had also begun actually to dictate at least some of the major political developments of the period. It was not the first time, of course. In India, the Mogul rulers had become amply caught up in the politics of the subcontinent; and in Spain, ever since the Arab governor of Tangier had invaded via Gibraltar in A.D. 711, Arabs were to run large tracts of the peninsula for much of the next four centuries. Islam had wielded enormous power in southern Europe too.

In the East Indies, matters proceeded rather differently. A close look at these five years, particularly among the unusually pious people of Banten, shows clearly the beginning of militant and anti-Western Islamic movements—movements spawned in Java but that have since become an important feature of the *realpolitik* of the modern world. An examination of the events that began with this pair of attacks on the colonial military in the autumn of 1883 suggests that the driving force behind most of the subsequent violent events in west Java—prefigured by the bearded men in white, acting out of their "extreme religious zeal"—was without a doubt fundamentalist, militant, anticolonial, anti-infidel Muhammadanism.

Many reasons have been put forward to account for the late nineteenth century's upsurge in anti-Dutch feeling in the region— and for the parallel upsurge in Islamic zealotry in the East Indies. Poverty, alleged colonial tyranny, corruption, the unbearable heaviness of the imperial yoke—all these factors, here in the Indies as in a score of other territories similarly burdened by the authority of outsiders—bore down on the local people and, as their education and awareness of the world increased, began to make them restless. Sporadic revolts, irritatingly persistent outbreaks in other regions of Dutch East India—in Aceh and Macassar and the Moluccas particularly—all appear to have been driven by what his-

torians would later accept was a growing mood of restlessness—
just as anticolonial movements were developing at much the same
time in India* and Malaya, and elsewhere.

In Java and Sumatra, however, there was one additional factor
that seemed to have played a role in fomenting a mood of general
popular unease—and that, surprisingly, was the devastating eruption of the huge volcano. The geological processes that destroyed
Krakatoa, in other words, appear to have played no small part in
creating the political mood of the moment. That the volcano and
the economic and social dislocation it caused had an effect of some
sort seems now undeniable; but whether its impact was limited
and peripheral, or whether it can be linked to the development of
movements that in turn led to the eventual seeing-off of the
Dutch, is a matter of some debate.

The immense geophysical turbulence of the East Indies—with
twenty-one active volcanoes and ten active solfataras on the island of
Java alone—has long played an important part in mystical belief.
Each volcano has a god—Krakatoa's being the widely feared Orang
Alijeh—and he readily displays his anger with earthly conditions by
spewing forth fire and gas and lava. But more: The Javanese in particular take a global view of their vulcanicity, believing that their
island is where the earth and the heavens have been arranged closest to one another, and where transmissions between one sphere
and the other are more common and more intimate than anywhere
else on the planet.

So, on Java, volcanic eruptions are much more than simple
expressions of dismay by distempered deities. They are astral messages sent directly down to the earth, and of an importance that
would be ignored only at man's peril. Given such a system of
beliefs, it might perhaps not be wholly unreasonable to suppose that

---

*The Sepoy Mutiny—or the First War of Indian Independence, as it is called retrospectively
in the subcontinent—took place in 1857 and, though eventually quelled, was a sure indication that the writing was on the wall for the British in India, as well, in time, as for the rule
of foreign settlers elsewhere around the world.

Krakatoa's almighty act of self-immolation in August 1883 was seen locally as possessing the most profound of inner meanings.

So did the eruption somehow act as a political catalyst? Did it, for reasons rooted deep in this Javanese mysticism, drive a wedge between the terrified and dispossessed people and the paternalistic Dutch authorities? Did it then nudge them toward the comforting stability of Islam? And did Islam's subsequent defiant stance against colonialism then somehow offer such succor and comfort to those who were dispossessed and terrified that they wholeheartedly accepted the offer to follow its precepts and demands, however extreme they might be?

And still further: Did Islam come to act as a banner under which these people might turn against the Dutchmen whom they could now, all of a sudden and with the clarity of a new perspective, see not as their benevolent leaders and well-intentioned mentors but, as so many imperial agents are eventually viewed, as their oppressors? And if Krakatoa somehow played a part in this chain of events, then can the eruption of Krakatoa come to be seen in a sense as an unwitting, readily adopted *political* event in and of itself—an event with effects that would resonate in the East Indies for many years to come?

At the time of the eruption, the Dutch in the East Indies were showing signs of momentarily losing their grip. The imperial purpose, quite coincidentally, seemed to be faltering. The old self-confidence of the Hollanders had taken a beating, and a mood of reform and change was in the air.

The cause for their discomfiture, though they might be loath to admit it, had been the publication in Amsterdam twenty years earlier of the book that had sent a shudder through the conscience of an entire Dutch people, and forced them to wonder just why they were running, in so questionable a way, a colony so far from home. The short novel that had had this extraordinary effect, Netherlands-wide, was written by a man who briefly hid behind the pseudonym Multatuli, and it was called *Max Havelaar.*

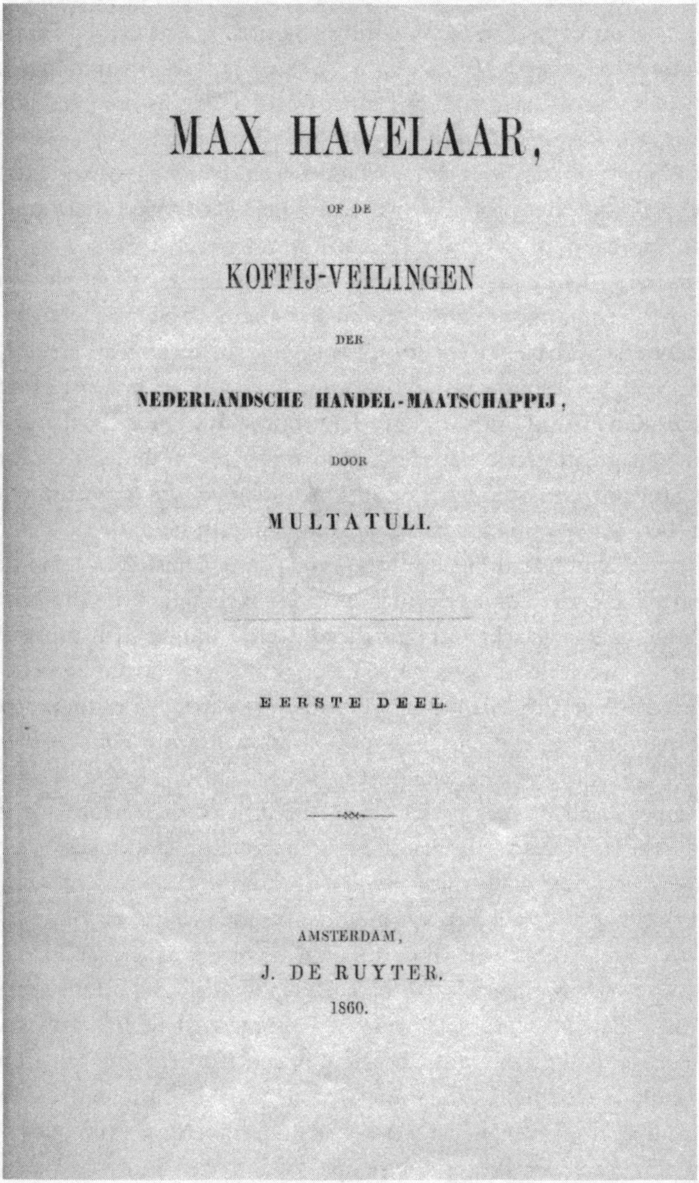

MAX HAVELAAR,

OF DE

KOFFIJ-VEILINGEN

DER

NEDERLANDSCHE HANDEL-MAATSCHAPPIJ,

DOOR

MULTATULI.

EERSTE DEEL.

AMSTERDAM,
J. DE RUYTER.
1860.

The title page of *Max Havelaar*, published in 1860.

The author was in fact a young colonial official named Eduard Douwes Dekker (*Multatuli,* a somewhat self-pitying alias, is Javanese for "I have endured much"). In 1855, as the protégé of the governor-general of the day, he was appointed assistant resident of a small west Javan regency called Lebak, which is coincidentally not far from the garrison town of Serang. He arrived there under the impression that he was on a secret mission to correct a slew of injustices that he knew had been visited on the local people—but over the three months he remained in the post he managed, by a relentless campaign of whistle-blowing, to ruffle the feathers of the entire Dutch administration. After uncovering tales not just of mismanagement and inadvertent cruelty, but of murder and corruption on a far greater scale than he had imagined, he resigned, returned to Holland, and eventually wrote the book that was to become one of the great landmarks of recent Dutch literature.

It was published in 1860, to the shock, astonishment, and dismay of an entire country, which learned for the first time the details of the manner in which their officials were running their most distant and wealthiest possession. The book was a savage indictment of the colonial attitudes of the Dutch—and in particular of that astonishingly exploitative Dutch invention known as the *Kultuurstelsel,* the cultivation system, which was introduced in 1830 and which compelled all villages to set aside one-fifth of their crops for the government in order to pay the cripplingly high land taxes. All villagers were held collectively responsible for the tax payment, and to ensure that responsibility was met no one could travel beyond his or her village without official permission—which was seldom given.

The system made the Dutch rulers fabulously rich; but upon its exposure in *Max Havelaar* one critic wrote that he had now seen demonstrated "that almost nothing of the great revenues from the island was devoted to the education or benefit of the natives; that no mission or evangelical work was undertaken, or even allowed, by this foremost Protestant people of Europe; and that next to nothing in the way of public works or permanent improvements resulted to the advantage of those who toiled for the alien, absen-

tee landlord, the country being drained of its wealth for the benefits of a distant monarch."

The Dutch—government and planter alike, for *Max Havelaar*
focused much of its attention on corruption within the coffee plantations—were condemned as "synonyms for all of rapacity, tyranny,
extortion and cruelty." Dekker cheekily dedicated his book to King
Willem III—"as Emperor of the glorious realm . . . that coils yon-
der round the Equator like a girdle of emerald, and where millions
of subjects are being maltreated and exploited in your name." For
doing so, and for daring to write so intemperate an exposé and to
lay it before a smug, unconcerned Dutch public, he was vilified,
attacked, and forced into the same kind of exile (he died in Germany) that was suffered by the similarly evangelizing Dutchman
Vincent van Gogh.

But reform of the kind that Dekker was demanding did eventually come about. *Max Havelaar* was debated in the Dutch parliament. The iniquities of the *Kultuurstelsel,* so vividly described in
the novel, were slowly recognized, and through the years following the sensation it was gradually abolished.* Pepper was freed
from its strictures in 1862, two years after the book's publication;
clove and nutmeg were taken off the list in 1863; tea, cinnamon,
cochineal, and indigo in 1865; tobacco in 1866. And eventually a
wholly new approach to the governance of the colony took root.
By the end of the century the East Indies were ruled under the
principles of a brand-new and so-called Ethical Policy. The Dutch
now started to take sedulous care of their subject peoples. Under
the new scheme they employed officials not simply to repress and
squeeze profits from the territory, as in the past, but to take charge
of public health, to improve education, and to offer agricultural
help, the better to advance the condition of the people.

This reform—too little, too late, and not enough to still the
nationalist mood—was perhaps Eduard Dekker's greatest legacy.
But it was not to be in place until the beginning of the twentieth

---

*A mere coincidence, the government insisted.

century. At the time of the Krakatoa eruption—and at the time of
the events that led to the Banten rebellion—most of the old colonial attitudes and most of the old colonial establishment still held
sway. Matters were beginning to change and to improve, but they
had not yet done so, and the unreformed state of the colony left
ample room for those who were determined to agitate against the
Dutch and their unrequested mastery to make such mayhem as
they could. Among those most eager to lead the agitation, and to
make the most mayhem, especially in Banten and the ultrareligious
west of Java, were the more conservative-minded Muslims.

One hundred seventy million Indonesians are currently members,
notionally or devotedly, of the Islamic faith. It is the most populous Muslim country in the world, and the greatest of all success
stories—if numbers are the best indicator of success—in fourteen
centuries of Muhammadan proselytizing. All of its people are
either converts or descendants of converts: It is easy to forget that
the world's greatest Muslim populations—in Iran, Malaysia, India,
Pakistan, Bangladesh, and Indonesia—all belong to a faith that is
quintessentially and inescapably Arab. Indonesia was converted by
Arabs, and it looks to Arabia and Arabians still for spiritual guid-
ance and direction.

Islam, it should not be forgotten, is at its heart an imperial religion, and Arabism is perhaps the greatest of all contemporary imperial movements—one of the many reasons for its current collision
with the West, which of course has its own competing, profit-driven
imperial agenda. The collisions between Arab-inspired Islam and its
agents, and the money-driven, trade-driven West have been many
and various: Those that occurred in the East Indies in the latter part
of the nineteenth century are now, when seen from today's perspective, classics of the kind.

Islam first came to the East Indies in the thirteenth century, and
ironically (considering the present schism between East and West,
between spirituality and materialism, between God and Mammon)

it came with Arab traders who were in search of business there. There is a grave of a sultan in north Sumatra that dates from 1211. There is another, at Gresik in east Java, which sports designs indicating that it had been carved by masons from India in 1419. A late-fifteenth-century mosque in Demak, on the north coast of Java about three hundred miles east of Batavia, was clearly the result of architectural compromises between Javanese and Arab builders. It had a holiness to it that the local mullahs regarded as both profound and ineffable, and they declared that to visit it seven times—though there is no spiritual explanation for the number—had the spiritual worth of the single *hajj* pilgrimage to Mecca.

In Banten itself Islam became properly established at the beginning of the sixteenth century, a little later than in much of the rest of Java, and considerably later than in north Sumatra. It caught hold immediately, spread with extraordinary speed, and before long became a model to which Arabs and hajjis alike could point with pride. Among the Bantenese and their coastal compatriots the Sundanese, the religion achieved a degree of penetration that was almost unrivaled in the archipelago. The west Javanese soon had a reputation for being more assiduous, more spiritual, and more fundamentalist than almost any others. It was difficult for a traveler to pass between Batavia and the coast without seeing scores of mosques, and without hearing the five-times-daily cry of the muezzin, calling the willing faithful, here in their millions, to prayer.

Yet, it is important to remember that East Indian Islam was always of a much milder stripe than that practiced in the Middle East and in Africa. The lingering influence of Hinduism, in particular, led to a significant local dilution of Islam's rigors. Regional superstitions, pockets of animist beliefs, and a whole host of religious oddities washed over the mullahs' teachings—with the result that the Islam that developed in Java, especially, turned out to be highly syncretic, a maze of compromises that drew influences not merely from Mecca but from a patchwork of other beliefs as well.

In spite of this rather off-center aspect of Javanese Islam, a cen-

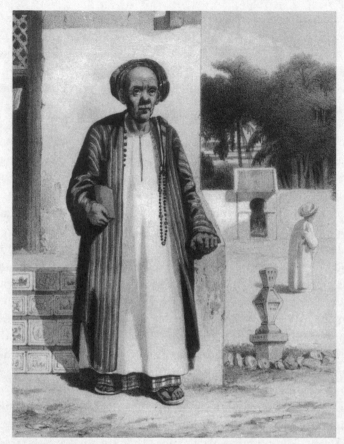

A Javanese Muslim imam who had performed the same
pilgrimage to Mecca as the rebel leader Hajji Abdul Karim.

tral feature of life on the island, and in particular of the lives of
most Banten Muslims, remained the pilgrimage to Mecca.
Although all good Muslims everywhere were required to take part
in the *hajj*, figures compiled by the Dutch government suggest
that more Bantenese and Sundanese went to perform the obliga-
tions of orthodoxy than any other group in Java. And the figures
were all the while rising steadily: In 1850 just 1,600 undertook the

hajj; by 1870 it was 2,600; and in the 1880s an average of 4,600 took off on boats to Arabia.*

The Dutch authorities were understandably wary of the practice, and they could have used the harsher regulations of the cultivation system—to curb the travel of suspect individuals, for instance—to prevent it. But they soon realized that to forbid a custom of pilgrimage that was centuries old would have been fatally imprudent: all they could do was to try to make sure the pilgrims were persuaded to keep their potentially corrosive sojourn in Arabia as short as possible, and to monitor the behavior of them all just as soon as they came home.

What bothered the Dutch was that the longer the pilgrims stayed away, the more "Arabized" they were on return, the more they held the Dutch infidels in contempt, the more they tended to take part in violent acts against the colonial power. "Mecca was nothing but a hotbed of religious fanaticism," wrote Snouck Hurgronje, the leading Dutch scholar on Islamic matters of the time, "where people were inculcated with hostile feelings against Christian overlords in their homeland."

(The history of militant Islam is long and extremely complex, and well beyond the scope of this chapter's account of the political effects of Krakatoa's eruption. Yet it is perhaps worth mentioning that the rise in extreme anti-Western Islamic feeling in some corners of the world—like the Dutch East Indies—came about when it did and as it did because elsewhere in the Muslim world, in the late nineteenth century, Islam was coming under an increasing threat from Western imperialism. In North Africa and the Middle East, for example, European powers—the French and the British most notably—were seizing territory or assuming influence on all sides, to the dismay of the mullahs and the mosque.

---

*Three times during the 1880s there was what was called a *Hajj Akbar,* a "Great *Hajj,*" when the ceremonies on Mecca's Plain of Arafat happened to take place on a Friday. Such a *hajj* occurred in 1880: no fewer than 9,544 Javan and Sumatran pilgrims took part that year, with Banten supplying by far the greatest proportion.

The mullahs in Mecca, not unreasonably, saw those pilgrims who were coming from the East—from lands that were already under the unwelcome control of Dutch infidels themselves—as having a use. They could perhaps be messengers, men who could return to their homes to spread the Word, try to reassert Islamic purity and authority, and somehow eventually—as their supreme goal—wrest the archipelago from the menace of the unbelievers' control. The fact that the Krakatoa tragedy took place just when these developments were beginning to unfold is one of those historical coincidences too attractive to ignore.)

It is usually the case with the upsurge of any political or religious movement that one figure, a charismatic leader, or a demagogue, or both, becomes the identifiable personality of the movement. Such was very much the case in west Java, with the steady rise to prominence of a Java-born mystic named Hajji Abdul Karim. Abdul Karim, whose teachings played no small role in the Banten rebellion, was from the middle of the 1870s the leader of a powerful local Sufi movement, which he and a corps of acolytes managed from his headquarters in Banten town.

Abdul Karim had started his Islamic education early. He had been to Mecca when he was a mere child; by his teenage years he spoke and read fluent Arabic, he had a scholarly knowledge of Islamic theology. He returned to his birthplace in his late twenties, and, as his mentor back in Mecca suggested, set himself up as a seer and messenger of Allah—a role that endeared him mightily to the Javanese masses. He had been in trouble with the Dutch authorities for violating passport regulations, and so was already regarded as a scourge of the infidels, a thorn in the imperial flank.

By the late 1870s his home had become a place of pilgrimage: Tens of thousands of fanatic Bantenese and Sundanese would come each day for a laying-on of hands or a few words from this remarkable man. He was showered with alms. The Dutch resident—suspicious no doubt, but well aware of the power of the man—paid an official visit. At first the message that this *wali Allah* offered to his growing army of followers was simply one of the need for piety, orthodoxy, and asceticism.

But as the number of his disciples grew, so his message dramatically changed—though whether it did so at the behest of Mecca or on Abdul Karim's own initiative is not clear. His revised version was considerably more alarmist, and for the ruling Dutch it was deeply ominous. For Abdul Karim began to predict what other devout Islamic seers were busily preaching, much to the worry of authorities in other regions of the world—that the Mahdi, the messianic figure who would appear to save the world from godlessness in its last days, was shortly about to appear.*

And in this pronouncement appears at last the key, the single link of chain that connects two apparently unconnectable features: on the one hand a volcanic eruption and on the other a movement for political change. The prediction made by this charismatic Islamic mystic and ardently accepted by tens of thousands of his followers, *that the Mahdi was about to come,* turns out to be intimately connected to the eruption of 1883. And it is so connected because the version of Islamic teaching that deals with the Mahdi and his holy war against the infidel holds that the arrival of the Mahdi is always accompanied by a series of definite signs. *There would be diseases of cattle. There would be floods. There would be blood-colored rain. And volcanoes would erupt, and people would die.*

And it so happened that each and every one of these predictions had occurred in Banten *in precisely the manner that Hajji Abdul Karim had forecast.* Cattle were dying on all sides because of an uncontrollable outbreak of murrain, the foulest of all bovine plagues. The west Javan coastal villages from Merak south to Labuan had been devastated by tsunamis. The rain was still tinted

---

*A similar phenomenon was noted in the Sudan, where between 1880 and 1885 a number of insurrectionary leaders described themselves as Mahdis, and caused enormous political and social unrest.

brown with the ash that swirled ceaselessly in the skies over Java. The island of Krakatoa had blown itself to pieces. And thirty-six thousand people had died in the tidal waves from the ash flows and the gas flows that had resulted.

What clearer signs could any devout believer possibly demand as an indication that the Day of Judgment was at hand, the Mahdi was on the way, and the holy war against the infidel was about to begin? Small wonder, some might say, that two of the foot soldiers in the coming war, dressed in martyrs' white, pressed their attack against the unbelievers, just weeks after the eruption was done. The fact that Abdul Karim had himself long since returned to Mecca* to assume a senior post in the Sufi hierarchy there made no difference: His teachings had been heard, his disciplines were in place, and a network of his so-called *tarekat,* the brotherhood he had established to carry on his work, was functioning like a well-oiled piece of machinery.

One of those who would be bold enough to link the two events, to put Krakatoa at the head of a long chain of happenings that culminated in the 1888 rebellion, was by chance one of the eruption's eyewitnesses. He was called R. A. van Sandick. A technical-school teacher from Deventer in central Holland, he had been hired by the colonial government for his knowledge of hydraulics. He just happened to be aboard the official vessel the *Gouverneur-Generaal Loudon* when Krakatoa exploded.

He watched with horror as the events unfolded; he plunged promptly into the relief effort, learning as he did so as much as he could about the social conditions of west Java, and how they had been changed by the tragedy. In 1892 he wrote a short, seven-chapter book, *Leed en Lief in Banten* (*Sorrow and Love in Banten*), which was the first to offer details of Abdul Karim's predictions. The following translation may be a little shaky, but the message in the relevant chapter is abundantly clear:

---

*He left in 1876, given the kind of sendoff one might expect for a pope or a saint.

The *mullahs* and teachers of religion in the *pesantren*,* who were stirring up the people in Banten, took the opportunity given by the enormous and deep-felt impression left by the Krakatoa eruption, to expand their influence. Was it not, they said, the revenge of Allah, not only against the unbelieving dogs, but also against those Bantenese people who were serving these *kafirs*, these infidels? There was no doubt: the disaster of Krakatoa was a sign of God, the great omen of which the holy Abdul Karim had spoken. Had he not predicted heavy earthquakes, and the end of the world? And see, the sun was darkened for hours, and now after the eruption the sun shone as a red or sometimes as a grey or blue ball on a grey firmament. Was this not strange, these nameless colours shining these days at twilight?

Did not God create the tidal waves that rose 30 metres above normal sea-level? And did he not speak in a thunder, as a result of which the whole of Banten shook in deepest darkness? And, ask the fishermen of the Sunda Strait—has not the bottom of the sea been raised by a God? Has not three-quarters of the island of Krakatoa disappeared? Are you blind to all these deeds brought about by God? Be humble for the Almighty! Pay for your sins! Can you still doubt, said the *mullahs*, now that you know that Abdul Karim has predicted all this?

A somewhat more unsettling development was the discovery that autumn of a number of documents and letters, written in Arabic, which were circulating in the towns of Banten. The Dutch colonial police, who managed to intercept specimens, said immediately that they were attempts by foreigners to foment trouble in

---

*The *pesantren* were Islamic boarding schools or seminaries that still exist in large numbers across modern Indonesia. They exerted a powerful influence on the attempt to spread Muslim orthodoxy and dogma in the nineteenth-century East Indies, and have considerable social force today. That Indonesian Islam is so mild remains irksome to the *pesantren* leaders: their eventual hope is to bring the stray sheep of Java and Sumatra fully back into the Muslim fold.

the wake of the eruption. And though most of the letters in fact seem to deal with the immense social problems caused not by Krakatoa but by the cattle plague, suggesting it had been the work of a wrathful Allah demonstrating his displeasure with the Dutch infidels and their local stooges, the letters' interest today turns more on their origin than on their content.

Most of the letters appear to have come, via messengers, from Arabia itself. And though the entire peninsula was then under the rule of the Ottoman Turks, it was still the fountainhead of Islamic orthodoxy. Fundamentalism prevailed most especially on that particularly devout desert southern tract of the peninsula known as the Hadhramaut, now part of the eastern side of Yemen.

The fact that nineteenth-century Java was being whipped into a fury in part by Yemeni fundamentalists—acting directly (it is said, though not proven, that Arab mullahs were in Banten soon after the eruption), or indirectly (the letters and distributed propaganda documents), or through their proxies (of whom Abdul Karim was the most prominent of scores of Arab-educated *hajjis*)—is oddly mimicked by similar events that appear to be taking place around the world today. The spiritual vigor behind today's Islamic militancy comes in large measure from the mosques of the Hadhramaut; and there are figures abroad today, Saudis and Yemenis both, who are every bit as defiantly anti-Western as was the East Indies' Hajji Abdul Karim more than a century ago. The historical parallels are intriguing, and with implications that will keep scholars busy, no doubt, for many years.

The first two strikes on the soldiery of the Serang garrison turned out to be part of what would become a much larger plot. The planning was meticulous and took years—though the first suggestions of rebellion were made just after the eruption in 1883, the outbreak itself did not occur until five years later, July 1888.

In all cases the leaders and instigators were *hajjis*—the rebellion was Islamic-inspired, Islamic-led. Teams of forty were selected. Oaths were demanded and offered, with all the participants solemnly agreeing, in writing, to perform the killing. Fighters were chosen, schooled in the techniques of *pentjak*—an East Indies version of fencing—and heavily armed with newly made swords and lances and the viciously sharp curved daggers called *goloks,* all of them made by sympathetic metalworkers in Batavia. White robes and white rag turbans were made and collected for the warriors. Lists of targets—all of them European, all of them *kafirs*—were selected. Rumors were circulated that a holy war was at hand, creating frightened apprehension among the Europeans, a mood of eager anticipation among the natives in their kampongs. Come the morning of Monday, July 9, 1888, and the stage was properly and fully set.

The first attack was made at a village called Sanedja, which today is a suburb of the industrial town of Cilegon.* Long before dawn teams of partisans cut the telegraph wires, blocked the escape roads, then at first light swarmed into the compounds of the various Europeans—the assistant resident, the salt-sales manager, the collector, the junior controller—and hacked them and their families down wherever they were found. There was nothing pretty about the assault, and little that was noble: the first family to fall victim was that of a clerk named Dumas: while he managed at first to escape through a window, his *amah*—whom the assailants mistook for his wife—was attacked with lances. The Dumas baby was sliced to pieces in her arms, and when the servant was later found she was alive but terribly lacerated, still cradling the dead infant. Dumas himself was later discovered sheltering with a Chinese: He was dragged outside and shot.

The prison was attacked and broken open, the inmates freed. The assistant resident, a man named Johan Hendrik Gubbels, was chased from one end of town to the other: His young daughter Elly was stoned to death, her head crushed with a boulder. Her sister Dora was hacked down and killed by men who, as the morning

---

*The site today of the enormous Krakatoa Iron & Steel Works.

wore on, were increasingly frenzied, seemingly enjoying what was fast becoming an orgy of bloodletting. Gubbels himself was eventually found and stabbed to death. His body was dragged out into the open air for the rebels to see—and they howled their approval.

There were occasional acts of unexpected mercy. Mrs. Dumas, for example, was found and—somewhat improbably—released after she had signed a paper saying she agreed to be converted to Islam. But otherwise the killing and sacking went on and on for hour after hour—until late in the day there arrived a full battalion of Dutch infantry and a squadron of cavalrymen. The infantrymen were armed with a new and quite terrifying weapon, just arrived from Holland: the repeating rifle. It was this weapon, above all else, that finished this fierce but, in the end, very brief rebellion.

The white-robed rebels believed, as holy warriors in any conflict are wont to do, that their piety would surely protect them against the Dutch bullets. But it did not. They died under the hail of Netherlands lead just as surely as would any godless infidel. And the Dutch military were quite evidently in the mood for killing that afternoon: When they opened fire with bullet after bullet, they did not intend to take prisoners. Thirty of the rebels were killed and thirteen more were wounded, out of a total of perhaps a couple of hundred. When the identities of the casualties were checked days later, almost all of them were found to be *hajjis,* men who had waged war in the name of Allah the Merciful, as the legacy of their long-absent spiritual leader, Hajji Abdul Karim.

The *hajjis'* twenty-four victims were Hollanders and their Javan employees—*kafirs* too, in the eyes of the dagger-wielding fighters. There were civil servants, merchants, prison-guards, wives, daughters. They were regarded by the Muslims as the initial victims of what would be an eventual *perang sabil,* a holy war, in which all trace of infidel behavior and attitude—and people—would be rubbed out.

But, of course, no such thing happened. The rebellion had been crushed; an inquiry was staged; the Dutch slowly instituted reforms; taxation was eased; strictures on travel were relaxed; a new mood of tolerance and ethical standards took root. In the *kampongs,* talk of the Mahdi's coming evaporated, as did the fanatical mood for war. An accommodation—uneasy at first but more comfortable as the years progressed—was eventually reached between the competing requirements of the Muslim and the Christian faiths. The Peasants' Revolt of Banten faded in popular memory.

From today's perspective the rebellion is regarded very much as a way station on the route to eventual Indonesian independence, the beginning of the end for this alien peculiarity of Dutch rule so very far from home. The Islam that had driven the Bantenese to fight so brutally in 1888 became, in time, more of an organizational structure for the coming revolution, and less the banner under which the revolutionaries might fight. But the Indonesia that was born in 1949 was a Muslim state, and it remains one today—with Islam in Java and Sumatra today in a much more aggressive mood than it has been for a long, long time.

When the Muslims first turned aggressive, with the attacks on the soldiers in the late autumn of 1883, Krakatoa had just erupted, and the ruin and devastation that was its legacy made a wretched contribution to the miserable lives of millions of people around the Sunda Strait. Their misery was swiftly exploited—cynically, some might say—by the calculating Islamic leaders of the day. The melancholy condition of the Javanese and Sumatra peasantry was exploited by a corps of mullahs and scholars who had come back from their pilgrimages to Mecca and the Hadhramaut, eager to recruit like-minded East Indians to help wage an essential first strike against the godless Western infidels and *kafirs* who were posing such a threat to the purity of Islam.

To that degree, the eruption of Krakatoa did indeed help to ignite a political and religious movement that flared briefly and violently in Java, and that left an indelible mark on the polity of the East Indies. Had anyone been in prescient mood, the Banten rebellion might also have struck a tocsin note—a warning of similar events that could well occur very many years later.

The bombings that took so many lives on the island of Bali in the autumn of 2002, for example, seem a haunting echo of those happenings in northwestern Java more than a hundred years before.

10

The Rising of the Son

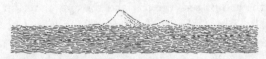

And then,
the most humble of plants,
a moss.

And then,
one morning,
the first sound of an insect,
so dry
you would think it
still mineral.

And then,
hope.

—Max Gérard, 1968

It would be a bold local fisherman indeed who, any time after the summer of 1883, would dare steer his craft between the cliffs and islets and shallows that form the outer shell of old Krakatoa, and drop nets into the waters directly above where the volcano had once stood.

The sea in these parts by now had a fearsome reputation. For a long while after the cataclysm the Sunda Strait was a place that terrified mariners, and those who had to pass through it did so as quickly as they might. The idea of lingering to catch fish was quite out of the question: Merely to peer down into the seas and wonder what might be going on deep below would give most passersby the shudders.

The danger of a further catastrophe remained for a long while uppermost in the minds of everyone who had survived the first. For months no one would venture near the charred ruins of the remaining islands; only those who went out on the scientific expeditions dared, and they were reckoned to be foolhardy in the extreme.

But economic necessity is a hard taskmaster; and by the time six months had passed, the fear and the apprehension dissipated, as they always do. The various Javanese gods had been placated (with flowers and sweetmeats and offerings) and their anger assuaged (in part by the attacks on the Dutch soldiery). Slowly, one by one at first, and later in companionable little fleets, the *prahus* and the smacks were out bobbing among the Krakatoa reefs again, looking for fish. This time they may have been dodging the occasional clumps of pumice and stranding on the odd unanticipated shoals, but their doughty and long-experienced captains soon took to behaving as though nothing much had ever really happened.

And such remained the case, the fishing improving, the waters placid and unremarkable and comfortingly deep, the volcanic dangers steadily receding in the public memory—until one day, after almost exactly forty-four years had passed. Then, on the evening of June 29, 1927, a group of fishermen who were hauling up their nets after a day spent innocently trawling for wrasse and sweetlips, and trolling by line for grouper and skipjack, witnessed something entirely unexpected and wholly extraordinary.

With a great roaring and rumbling sound, a clump of enormous gas bubbles suddenly broke the surface of the sea. The bubbles seemed to be all around, to be rising in strange and random combinations to port, to starboard, ahead, abaft.* It was very con-

---

*A concentration of escaping gas from the seabed can, in certain circumstances, create a zone of negative buoyancy, where passing ships are at risk of sinking. This happened in the 1970s when a seabed gas well in the North Sea fractured, and an inverted cone of small bubbles rose from it to form, at the surface, half a square mile that was more gas than water, and not at all buoyant. Moreover, the escaping gas was highly flammable, which nicely doubled the hazard.

fusing, and very frightening. Exactly where under the sea these bubbles—which exploded in clouds of spray and ash and foul-smelling sulfurous gas—were coming from, and whether they were coming from one point or from many, was difficult to say. The panicky sailors, caught in the midst of them, seemed to think they were concentrated a point more or less above where Danan, the middle of the three former Krakatoa peaks, had once been.

As a seamark, a point of reference, there had long been near the center of the caldera a curious clump of needles of light-colored rock that mariners (and the hastily printed new navigation charts) named Bootsmans Rots—Bo'sun's Rock. These guano-stained pillars, nearly vertical sheets of andesite that are, with the peak of Rakata, the only real relics of the old Krakatoa, rise fifteen feet out of the ocean directly above the old central peak of Danan—like "a gigantic club which Krakatoa lifts defiantly out of the sea," as one early visitor had it. Around the islets the waters, alive with scores of sharks, are six hundred feet deep at least: The pillars rise from some unfathomable depths, and most fishermen still regard them today as they did back in the twenties, as a warning to *keep back*, a reminder that the volcano, if not all fully there in person, was still there in spirit, at the very least.*

And as if to underline the warning on this evening in June, the eruption of bubbles appeared to be coming almost from directly beneath the pillars, perhaps just a little to the northwest of them, and thus a little to the west of the old crater. Moving through the bubbles the fishermen felt the water quickly grow warm†—and there came a point where steam could be seen rising from where

---

*It will be recalled that the 1883 eruption threw up a pair of low spits to the north of the ruins, which were called Steers and Calmeyer Islands. The breakers had eroded Calmeyer back down to sea level at the time of a visit in May 1884, and Steers lingered only until the end of the same year. At the time of the renewed volcanic activity in the twenties all that remained of them was a pair of oddly shaped and navigationally inconvenient shallows.

†The crater lake in the caldera of the notoriously destructive Mount Pinatubo in the Philippines has a similar hot spot above a vent; steam hangs above the lake at this point, and up close the water hisses with bubbles and is hot to the touch. I once paddled a plastic canoe into the hot zone, and it began to melt, sagging alarmingly under my weight.

the bubbling had become most fierce. As the men paddled and sailed away as fast as they could, and as night began to fall, so they saw a diffuse red glow settle on the water, as if the bubbles were somehow mingled with fire.

Dr. Verbeek, who had been the first to step on to the remains of the ruined mountain back in October 1883, and who had compiled his masterly study of the eruption two years later, would have known well what was happening. He had died in 1926, but in 1885 he had written, with nicely measured prescience:

> . . . in any renewed activity of the volcano it is to be expected that islands will arise in the middle of the sea basin that is surrounded by Rakata Peak, Sertung and Panjang, just as the Kaimeni arose in the Santorini Group, and just as formerly the craters Danan and Perboewatan themselves formed in the sea within the ancient crater walls.

The bubbles were the first indication at the surface that a new volcano, lurking somewhere deep on the bed of this bathymetrically uncharted corner of the sea, was trying to build itself up. It came as something of a surprise, despite Verbeek's prognosis: when a survey was made in 1919 a shoe-shaped ridge was found to have developed to the northwest of Bo'sun's Rock, but there was no evidence that it was part of a developing volcano. But swiftly, in the aftermath of the random eruptions of bubbles and clouds of steam of June 1927, a distinct line of froth and bubbles and plumes of steam started to develop in the water—such that by the end of the year scientists were able to map a quarter-mile course through the water that appeared to mirror a rent in the seabed's surface, a thousand feet below.

The tenor of the activity then changed. The bubbles became fiercer, large fountains began to play from the surface of the sea, black froth, steam, spurts of ash, and bombs of pumice began to surge from between the waves. Cones of water sixty feet high shot into the air, with rays of black magmatic material, like needles of

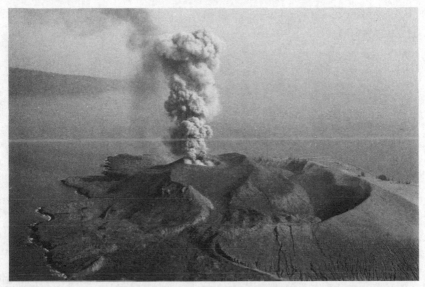

Anak Krakatoa—the highly eruptive "son of Krakatoa"—shown in 1979, nearly half a century after its explosive birth from below the sea.

jet, rising fifty feet further above the water cones. As the eruptions became ever stronger, so domes of water, half a mile across, rose out of the sea, and the mixture of volcanic material within them gave a curious flecked appearance, with mottled layers of black, white, and gray, so very different from the vivid blue seas all around.

And then, most bizarre of all, flames started guttering on the surface of the water, and then shooting out in huge yellow jets and sheets of fire: Observers had the impression that the water itself was now ablaze, or had been covered with flaming oil, as if the scene were the aftermath of some terrible maritime tragedy.

Finally, on January 26, 1928, the volume of bubbles and flame transmuted into ash and solid rock, and broke surface: A thin curve of brand-new land appeared for the first time above the sea. This new land grew, black and sicklelike, for several days, until it formed a humped, scimitar-shaped island. It looked like a sand

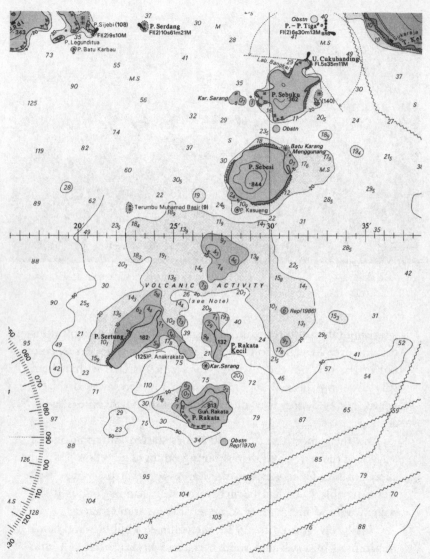

A current official depiction of what the hydrographer of the Royal Navy styles *Anakrakata*—the island of Anak Krakatoa, growing steadily at a rate of five inches every week.

dune five hundred feet long and ten feet high, with a concave steeper side on its southwestern edge and a rage of smoke and explosions discharging from its base. A Russian geophysicist named W. A. Petroeschevsky was on hand to see* the birth of this new piece of the world's real estate, and he gave it the name its successor still has today: *Anak Krakatoa*—the "son of Krakatoa."

The mortality rate for new marine volcanoes is very high, and this one son did not survive for long: After a week the relentless power of the surf wore it away, and all that was visible in mid-February was a patch of muddy-colored water, occasionally pierced by a funnel cloud of smoke, steam, ash, and, every so often, by small ragged pellets of hot and plastic lava. Then, some months later, a new island appeared once again. This time eruptions from two separate points created a pair of cones that rose seventy feet above the sea and were joined to each other by a slender spit of land: But then the volcanic activity died away, the waves attacked and chewed away, and this new confection, after all too brief a life, slipped back beneath the surface.

Corrosive encounters between this matched pair of seemingly equally relentless and powerful forces—the volcano and the ocean—went on for the better part of the next three years. On some occasions the process of volcanic creation won the day, and land was produced that survived for a while. Sometimes the equally awesome power of the rainy-season seas and tides and currents made maritime mincemeat of it all, reducing it to a pile of submarine grit.

The number of explosions was prodigious: during the twenty-four hours from noon on February 3, 1928, no fewer than 11,791 separate detonations were counted; on June 25, an even more remarkable 14,269—ten eruptions every minute of the day and night. The second of the islands stayed where it was long enough

---

*He later built a concrete and corrugated iron bunker on Panjang Island, to the northeast, from which to view the new activity.

for scientists who were in Batavia attending the Fourth Pacific Science Congress in May 1928 to arrange an excursion there and do some real-time fieldwork: they were vexed when it slid back beneath the sea again, destroyed by the relentless power of the waves.

Sometimes it was not just the waves that did the destroying: the newly appearing volcanoes occasionally self-destructed, blowing themselves to pieces, as happened with the third of the Anak Krakatoas in early August 1930. But gradually, and as a tacit reminder of the gigantic and unopposable forces of the subduction zone that was working away beneath it, the volcanic forces began to gain ground, and the island they were trying to create began to achieve a certain permanence.

On August 11, 1930 the submarine vents made their fourth concerted attempt to sculpt a lasting memorial on the surface. They put up a ring-shaped island that had the appearance of a large black doughnut—and it stayed put for two days. On its second day of existence a monumentally large phreatomagmatic eruption—the kind that results from the mixing of hot lava and hot gas with cold seawater—tore upward into the sky, reaching a height of almost a mile, and then dumped an enormous quantity of volcanic detritus back on top of the fragile wisps of island below.

Those that witnessed the explosions noticed a feature that would soon be regarded as typical of this kind of eruption—the so-called cock's tail jets. The upward jets of the explosions are black with the material they carry, but they have outer edges that are rich with condensing steam, and so are starkly white. The whole phenomenon looks much like the tail of the more dramatic kind of male chicken.

And the dumping of an enormous volume of new material, all at once, seems to have done the trick. It allowed the island a degree of permanence that enabled it to stabilize and consolidate itself. Ever since then, the rate at which fresh ash and rock accumulate on the surface of the new island has managed to exceed the rate at which the island's edges are eroded away by the ceaselessly ruinous action of the sea.

The moment that Krakatoa's son was fully born it could be assured of the continued existence that it enjoys today. On the charts of the region the hydrographers of the various navies, recognizing the new status, steadily changed the color of the island's outline from stippled blue, signifying *new, temporary, and uncertain,* to the unbroken black that means *established, permanent, and fixed.* It has been designated so ever since: Anak Krakatoa (or Anakrakata, as the Royal Navy's latest chart of the region now calls it) has been from August 1930 onward as permanent a feature of the East Indies as Java and Sumatra have long been or, to be more realistic, has been as permanent as the islands of the Krakatoa complex that came before it.

It has, however, been an extraordinarily active volcano, growing rapidly and unstoppably ever since its birth. The observatory set up on Panjang Island by Petroeschevsky proved invaluable for the steady monitoring of its progress—as it grew from the twenty-foot-high weakling, half a mile long, that started life in 1930, to the five-hundred-feet-tall peak, a mile long and half a mile wide, that it had become in 1950, to the fifteen-hundred-foot-high double-cratered monster of an island that it is today.

Its growth in size—from nothing—has been matched precisely by the growth—also from nothing—of its population of plants and animals.

For when Anak Krakatoa rose from the sea it was in all probability totally empty,* devoid of life and, in essence, quite sterile. Both its surface and its interior were, it was thought, far too hot to permit the existence of almost any kind of living thing; and the island was far too new to have any history, to possess any former biology or botany that might have the potential to generate any kind of resurgence of life. The mountain was, to the fascination of biologists from around the world, a *tabula rasa,* ready to have the

---

*The question of whether *all* life was totally destroyed, and whether the remnants of Krakatoa were totally sterile, remains open. Certainly all initial science undertaken on the islands assumed that life had been wiped out. But in later years this view was challenged, and severely.

polychrome marvels of life painted up it, layer upon layer, year upon year. It was a Garden of Eden, yet one without plants, without animals, and without mankind—and with a whole world of scientists waiting to see what might grow there and what might not.

But of course what remained of old Krakatoa was a clean slate too, the island remnants most probably all burned and sterilized by the flames of 1883. These two locations—the ruined remnant of old Krakatoa, and the newborn innocent of Anak Krakatoa—have thus become sites of huge international interest, where answers are still being sought to two fascinating questions. On the ruins of old Krakatoa—*how did and how does life recover*? And on the virgin island that was later created in the midst of those same ruins—*how did and how does life start*? What were the differences, if any, between *life returning* and *life beginning*?

Of course these two questions could not at first be either asked or answered at the same time: Biologists had forty-four years with only one site in front of them that they were able to study. In those early days there could thus be no kind of comparison. Ever since 1930 the three-island remnants of greater Krakatoa (if the scorched masses of Verlaten and Lang Islands—Sertung and Panjang respectively—are to be counted, in ecological terms, as relics of the old volcano) and the brand-new single island of Anak Krakatoa have been standing there, beside one another, inviting inspection and comparison. Biologists from then onward have been able to examine each, and try to work out if the mechanics of life-recovering-from-ruin (on the old islands) and life-starting-from-scratch (on their new neighbor) operated in the same way, in similar ways, or in altogether different ones.

It is a study that continues today. More than a century after the creation in the Sunda Strait of what has turned out to be the extraordinarily useful biological laboratory of the new Krakatoa community, not all the answers are in, by a long shot.

In considering the recovery of Krakatoa itself—which was all that could initially be considered—nineteenth-century scientists had one significant problem: Because they could not be entirely certain what kinds of life had existed on the island before the eruption, they could not say in what direction the island was likely to attempt to recover—what state it might be under pressure to return *to*.

Clearly it had once been richly endowed: John Webber's famous sketch,* made when Captain Cook's expedition made its melancholy[†] pause there in 1780, shows palms and ferns in gloriously feral abundance. Present-day botanists have pored over the picture to try to ascertain the various species Webber depicted: There is a type of grass that returned to the island in 1920, a palm-tree called *Licuala spinosa* that came back in 1982, and a fern that was first seen again in 1987.

Following the *Resolution*'s sojourn, there had been some cultivation on the northern flank of the island, with a scattering of people (once even a small prison), and with goats, a vegetable garden, and trees grown for firewood. The very few scientists who had landed on Krakatoa during the seventeenth, eighteenth, and early nineteenth centuries had collected specimens of grasses, pepper plants, orchids, and mahogany trees, as well as an unusual kind of parasitic mistletoe. One Dutch biologist went there looking for snails and found five types.

But no one had ever tried to produce a definitive systematic list of the entire populations that existed before the eruption: All that could be said with certainty was that Krakatoa had been covered with a snail-rich, orchid-rich, pepper-infested, and grass-floored tropical rain forest, more or less similar to that found in today's Sumatra.

Despite this irksome lack of knowledge about what had gone before, the biologists from around the world who headed off in

---

*Less famed for its artistic quality than for the fact that it is the only known drawing depicting Krakatoa's preeruption botany.

[†]The expedition had been in a cheerless mood ever since Cook himself had been killed, in Hawaii, some months before.

the direction of Sunda Strait knew that what awaited them in the aftermath of the eruption could well be fascinating. It would, one of those heading to the island wrote, "be very interesting to follow step-by-step the progress of the development of new life on this land now dead but which, in a few years, thanks to the intense heat of the sun and the abundance of equatorial rains, will surely have been recovered in its green mantle."

The scientists fanned out with dispatch—and with care. Just like investigators who take elaborate precautions not to contaminate the scene of a disaster or a crime, so most of the biologists who came to Krakatoa did their best not to sully, with such usual contaminants as bacteria or seeds or rats or measles, the rare posteruptive purity of the islands. It was a purity that remained untouched for some considerable while.

Rogier Verbeek was the first on the island, in October, a scant six weeks after the eruption. And he was too early: When he first clambered on to the dusty shore, at a time when the islands' floors were still almost too hot to touch and mud flows were still pouring from the lava cliffs, he could see no living thing in sight, nor any evidence that anything alive might be lurking nearby. From a personal point of view it was a shame: Given his heroic dedication to the story of the eruption, it might have seemed appropriate for him to be the first to find clear evidence of new life.

In fact it was a Belgian biologist named Edmond Cotteau who spotted the first stirrings, when he visited with a French-government-sponsored expedition six months later, in May 1884. The leader of the expedition had reported finding much the same lifeless devastation as seen by Verbeek: "the magnificent vegetation which had been so often admired there remains nothing but a chaos of enormous trunks, whitened and desiccated, among the surrounding desolation." It was dangerous too: Rakata's main cliff was eroding fast, and rocks were rolling down its sides in a ceaseless tumble, making an unending noise "like the rattling of distant musketry."

It took the men a frustratingly long time to find a sheltered beach out of range of the fusillades of bouncing boulders. But then Cotteau, who had walked southward from the eventual first landing site on the northwest corner of Rakata, and who had managed to get all the way around the point to what is now called Handl's Bay, suddenly spotted something.

It was nestled between two rocks, on an otherwise arid and seemingly death-filled spit of gray and dusty land. This was the very first living thing that could with certainty be said to have appeared after the catastrophe: and it was, Cotteau wrote with measured excitement, *a microscopic spider.* He looked hard for another, but could find only this single specimen. Yet significantly, and with a nice symbolism, "This strange pioneer of the renovation was busy spinning its web!" The lonely little spider was hoping, in other words, that it would eventually be lucky enough to catch a fly.

The arachnid's optimism was admirable, considering the trauma that must have otherwise unsettled it. Yet as it happened (though whether it happened for this one animal we cannot say), it was an optimism that was not at all misplaced. For within months an abundance of life—and that almost certainly included at least the modest sufficiency of flies required to satisfy a small army of spiders—began to return to the islands, in earnest.

This first spider's appearance on the island—and then the appearance of many of the life forms that subsequently resettled Rakata—in time triggered a question that has continued to vex the biological community for the many years since. Was there a chance that some of the animals and plants that first appeared did already in fact still exist on Krakatoa, as survivors? Or had life on what remained of Krakatoa been utterly extirpated—had it really been blasted and bombed and withered by the searing heat, and then buried under a hundred feet of ash?

If this last was the more likely explanation, then all the new life that went into the repopulating of the islands must have arrived on their shores and crags from across the intervening stretches of sea. And if this was the case—if all the newly established forms of life were newcomers, in other words—how, exactly, had they come?

Had this one spider been lurking in a crevice and, though perhaps badly singed, been sufficiently *compos mentis* to spin his web for M. Cotteau? Or had a passing seabird, the spider lodged in one of his claws, landed on the otherwise uninhabited island, and left the insect behind when it flapped off into the air again? Had the spider floated in on a coconut husk, clinging grimly to the mesh of hairs until all the bobbing stopped? Or had it been borne in on the air, wafted on the breezes?

Today it seems most likely that wafting was the means by which this one baby spider reached the seared shoreline of Rakata. It is well known now that some spiders and other wingless creatures have a propensity for "ballooning"—extending a strand of thread of silk from their bodies and allowing the wind currents to catch it and bear it and each of the creatures off to an unknown destination. Ballooning spiders can travel effortlessly for scores of miles: The sea crossing to Rakata from either Java or Sumatra would take no more than a trivial few hours. And in drifting and wafting on the breezes, this creature becomes a member of what has recently—and delightfully—been named

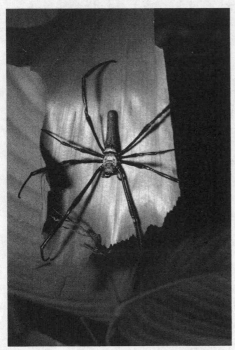

Nephila maculata, a ballooning spider.

the *aeolian plankton,* windborne kin to the microscopic drifters of the sea.*

Once properly under way, the rush of life returning to the islands was impressive in both its speed and abundance. Crews of passing ships first thought they saw patches of green about a year after the eruption. In June 1886 an expedition spent four days on Rakata, and found they were right: They counted no fewer than fifteen flowering plants and shrubs—mostly beach plants, suggesting that these at least had arrived by sea—two mosses and eleven ferns. The visitors also found a gelatinous layer of blue-green algae in a moist film on top of the volcanic ash: This layer—wherever it might have come from—probably helped spores to germinate, it was later thought, and provided some kind of support for young and feeble rootlets.

Clearly something had helped—because by the time a new set of visitors arrived the following year there had been a positive orgy of rampant growth. There were now dense fields of grasses, so tall that a man could hide himself. The variety of grasses included Java's well-known *alang-alang,* which is always the first to grow (conveniently) after a forest fire and (rather less so) after a farmer has cleared a field for planting. There were tufts of wild sugarcane poking out of the ash piles, many more ferns than the eleven that had been counted the previous year, scores of types of sand-binding creepers, hibiscus plants, self-fertilizing orchids, a very recognizable red-leaved coastal tree known as the Indian almond, three varieties of fig trees and a pinelike tree that the Australians named (for its resemblance to the lush plumage of their native cassowary bird) the casuarina.

By 1906 the forests were thickening, the trees maturing, the

---

*Southern Californians became briefly alarmed in the early autumn of 2002 when clouds of ballooning spiders wafted onshore from the Pacific. After the terrorist attacks of the year before, puzzling phenomena like this became suddenly invested with sinister overtones.

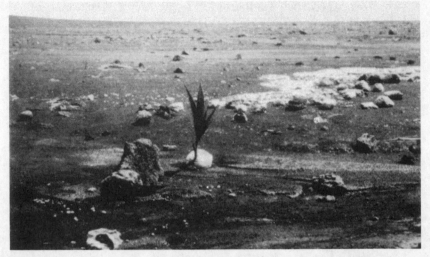

Life returns to the Krakatoa beaches: a coconut husk, a morsel of random flotsam, sprouts a shoot, ready to become island vegetation.

climb to the two-thousand-foot summit becoming every more difficult as proper forests began to cloak the mountainside. Parties of visitors now had to hack their way in, getting attacked en route by red and black ants and carpenter bees, being surrounded by swarms of colorful butterflies, slipping on large earthworms, taking pleasure in seeing kingfishers, nightjars, green pigeons, wood swallows, bulbuls, and orioles. Some visitors had reported seeing a large monitor lizard, and said that coconut palms had returned and were lining Rakata's southern shore.

Ever since then the island has become entirely overgrown. Casuarina trees on the coast stand well over a hundred feet tall. The mixed forests inland are alive with birds (among which are, prettily, the zebra dove, the pied imperial pigeon, the greater coucal, the white-bellied fish-eagle, the brahminy kite, the orange-bellied flower-pecker, the pied triller, the mangrove whistler, and the magpie robin), together with beetles and centipedes, geckos, dragonflies, grasshoppers, and tree-dwelling snails, a pompilid

wasp that is designed solely to hunt and kill spiders (to the con-
sternation of the descendants of M. Cotteau's first find) and a large
snake that was initially identified as a boa constrictor but that later
turned out to be the species of the somewhat more congenial
reticulated python.*

And civilian mankind's arrival, which was not always as carefully
choreographed as the visits of those early biologists, resulted in the
introduction of less comely animals. In 1917 a German named
Johann Handl, to take one of the stranger examples, arrived on the
southern end of Rakata with his family and servants, and
announced that he was settling on the old volcano as a pumice collector. He built himself a small house on Rakata during (and presumably to get away from) World War I; he planted a garden and
lived comfortably in this somewhat unusual hideaway for the following four years. But it turned out that he had, presumably inad-
vertently, brought along with him in his boat a breeding pair of
most unwelcome guests. A substantial population of *Rattus rattus,*
the black rat, is now happily established on the island, and makes
mayhem among the nests of most of Rakata's coastal birds.

And as the years wore on, so the numbers and types of plants
and animals on the Krakatoan remnants waxed and waned, and in
time created for themselves (by the way of what outsiders call the
law of the jungle) some kind of biological equilibrium—a state
that is never perfectly reached, a kind of biological nirvana, endlessly sought in every kind of complex community, yet rarely
entirely attained.

Herr Handl's garden, once abandoned, filled quickly with ten
kinds of weeds. In the early 1920s a number of newly discovered
large bees, velvet ants,† fungus gnats, scoliid wasps, mosquitoes,
crane flies, swallowtail butterflies, fruit bats, and woodpeckers
were found. In fact Rakata was shown, forty years after the erup-

---

*More than twenty-four feet long, and a very powerful swimmer, easily able to reach the
island from the mainland.

†These are in fact a kind of wasp.

tion, to be home to fully 621 species of animals. And by 1931 there were fully one hundred species of spider, according to a British spider expert called Bristowe—the number and variety of everything having swelled almost exponentially.

Assuming for the moment that most of what exists today on the three islands of Rakata, Panjang, and Sertung was brought in from across the sea (and was not born out of hidden survivors of the islands' original population)—how did it all arrive?

That first spider, as mentioned, very probably was of the ballooning variety. Other creatures and plants invaded by quite different means—all of them, when taken together, offering a powerful demonstration of the insistence of biology on bursting forth wherever it can, all showing the unquenchable nature of the fire in what the more romantic of today's biologists like to call "the crucible of life."

The first samples of colonizing greenery, for instance, the creeping beach plants and shrubs and the small coastal forests, came by sea, rafted across on driftwood, pumice, and other debris,* or hitchhiking in with birds that feed on fruit and then excrete the seeds. The pioneering plants found in the inland areas were almost certainly created from windborne or birdborne spores or seeds—and they had to be tough enough to survive with very little water and in the full glare of the sun. (Both they and their sea-borne coastal cousins would have an additional advantage if they were hermaphroditic, and could get along in an environment somewhat wanting for sexual partners.) The presence of the thin veneer of blue-green algae helped—and it may well have been the key to getting the inland plant populations properly under way. The figs, especially, were typical of what could be brought in by this method—their presence in abundance being evidence of the way the fig tree, once settled, then manages to propagate itself.

---

*There is an early sepia photograph of a coconut's half-shell lying on an ash beach, with a stalk sprouting a leaf and rising from its upper side, a pale root feeling its way into what passes for earth. (See illustration on page 354.)

As the spread of those first plants increased, and as new ones were introduced and started to compete and then struggle for existence and room—as the botanical ecology of the island began to change, in other words—so the zoological ecology changed too. As the coastal grassland began to give way to casuarina forests, as the woods on the mountain's upper slopes began to get darker and wetter and danker, so the populations of animals altered. The butterflies, beetles, and open-country birds and reptiles that were the first to colonize the island, and that positively liked its dryness, started to be replaced by forest animals, geckos and skinks, bats and birds—the hobby, the barn owl, and the eagle-hawk—that favor shade and like to live in a warm world dripping with moisture.

The overall result, a century and more after the eruption, is the existence today of a group of islands that have a markedly different biological and botanical makeup from the two great landmasses that lie fifteen miles to the north (Sumatra) and the east (Java).

One example will suffice: On the Javanese and Sumatran mainland there are twenty-four species of termite. Six of these live in hardwood trees, seven live in the dead trunks of living trees, six in wet soft wood, and five make their nests in the soil. On Rakata Island, though, there are a total of only eight types of termite. Not one of them lives in the soil; only one likes to inhabit the dead parts of living trees, just two like softwood, and by far the greatest number—five species—prefer to live out their days in hardwood trees alone. Just why this marked difference might be is still a topic for much reflection. Perhaps Rakata's soil is still too newly volcanic, or the dead wood is perhaps just not dead enough. Perhaps the environment is fresher, more raw—there must be some good reason why termites do not much care for the island that once blew itself to pieces, and why they like the islands that have gone untouched for thousands of years.

Because of unanswered questions just like this, Krakatoa has remained for many decades the focus of intense studies worldwide, a fascination for a score or more of the world's leading botanists, and the target of expeditions and field trips by the dozen. And yet, though the mechanics of the repopulation of the islands are now reasonably well established, the principal question that we set aside earlier remains sturdily only half answered: Was the first population entirely fresh and new? Did it arrive on the devastated islands by sea and by air, coming from the volcanically wounded, but far from biologically sterile outside? Or was it not really new at all? Did it start off with seeds or with living animals that somehow managed to survive the original inferno?

The controversy became so established a part of botanical lore that it long had a name: *the Krakatoa problem.* Its central question—survivors or outsiders?—dogged entire generations of scientists, and was the trigger over the years for some bitter, angry, recriminatory, and downright insulting invective. Most of the ugly words swirled around the person of a forthright and feisty Dutchman named Cornelis Andries Backer, who held the delightfully titled post of "Botanist with Special Responsibility for the Flora of Java" at the Buitenzorg Botanical Gardens; he first visited the islands in 1906.

Backer looked skeptically at the notion that all Krakatoan life had been destroyed, a theory agreed upon by almost the entire botanical establishment of the day, and proclaimed it to be near-total nonsense. He looked in particular, for example, at sweeping statements like that written by his superior, the botanical garden's director, Melchior Treub, who had led the major posteruption expeditions to Krakatoa:

> . . . at the time of the eruption the trees felled or smashed by violent outbursts must have been half-carbonized, in view of the extremely high temperatures that certainly prevailed over the whole island. After that, Krakatoa had been covered, from the summit right down to sea-level, in a layer of burning ash and pumice. This layer had a thickness varying between one and sixty meters. In those conditions it is clear than no vestige of the flora would have been able to exist after the cataclysm.

The most persistent seed and the most protected rhizome must have perished.

Backer weighed into them with the kind of rhetoric hitherto unheard of in the genteel world of Edwardian botany. In 1929 he wrote and self-published a three-hundred-page monograph that denounced the "hasty" and "careless" botanical work that had gone before, which, he insisted, resulted in the lazy, easy conclusions that all Krakatoan life had been wiped out.*

He had a venomous disregard for those who had carried out the early studies. Treub, he declared, was a prime example, a man who "was no florist, and whose knowledge of tropical plants was very limited." Treub's expedition to Krakatoa had been "much too short," the data acquired "very incomplete," and his trip so shoddy an exercise that it was not worth calling it a scientific exploration at all but rather "a mere excursion of persons interested in the problem but not seriously devoted to trying to solve it."

Other botanists went later to catalog the growth of new plants. Backer accused them of "misspending their money," acting under "childish delusions," indulging in "vague speculations," and making a series of expeditions that was "a complete failure."

The tone of his remarks set off a firestorm of criticism from scientists around the world. But the underlying points he made were all seen to be valid: The principal one being that no one, he said, had ever spent enough time or carried out a systematic enough survey to be totally sure that nothing had survived the original cataclysm. It was easy to take a look at the island and surmise that nothing could have lived through the fire; but it had since become well known that some kinds of seed not only survive fire but actu-

---

*Backer had himself been at first lazily and easily convinced of the likelihood of total annihilation of all life. But then came evidence to the contrary—most notably, twenty-two years after the eruption, the discovery of a cycad, a plant that looks like a cross between a palm and a fern, which seemed much too large for something that had grown entirely from scratch. The idea that this might have been a survivor of the blasts set Backer thinking—and it ended up completely changing his mind.

ally need high temperatures of this magnitude to germinate properly. Was it not in fact highly likely that such seeds had existed, and survived, on Krakatoa?

The ash layers on the upper slopes of Rakata were thin, and would have been washed away by heavy rains. What if some roots and rhizomes that had been lurking beneath the earth's surface had lived—surely they would begin to sprout once tropic warmth and moisture and sunshine flooded their now ash-free setting?

Then again—the first expeditions had all noted (but had never properly investigated) the presence of greenery in the deeply scored valleys that lay between Rakata's summit ridges. Why had this greenery, whatever it was, never colonized the two lower islands, Sertung and Panjang? Why had these islands been given their vegetative resurrection much later than the upper slopes of Rakata? If airborne and seaborne immigrants were the source of this life, then all three islands should be gifted with new life at the same time. But they weren't. The upper slopes of Rakata got it first.

The answer to this and a number of similar unconsidered conundrums, Backer suggested, was that Rakata's colonization probably came *from within*—that surviving plants on the island's upper slopes had created the new life there; this was an upper-slope phenomenon that was entirely lacking on the islands Sertung and Panjang, which were lower in altitude.

It all made excellent sense—but it was a theory that was now, thanks to the sloppy fieldwork and hastily written conclusions of the first years after the eruption, quite impossible to prove. And that was the shame of it: No one could henceforth *ever* be sure what had happened on Krakatoa's relic islets. Bad science, in short, had left mankind puzzled, a raft of questions unanswered, and the Krakatoa problem essentially unsolved and, very likely, forever insoluble.*

---

*Although modern ecologists are still fascinated by and so still debate the question, a consensus appears lately to have emerged: that while perhaps some species *did* survive the cataclysm, rendering the Krakatoa relics less than totally sterile, there probably was not enough relic life to mess up nature's experiment—conducted in full view of a fascinated world—in what was effectively true primary colonization nor to inhibit or alter the steady succession of the plants and animals that did that colonizing.

*　　*　　*

The same was not the case, however, over on Anak Krakatoa. For this was an island that was wholly new, and one whose newness precluded any possibility of there being survivors at all. Survivors *of what,* one might well ask.* And science was not about to make the same mistake that Backer accused it of making with Krakatoa. A later director of the botanical gardens, Karel Dammerman,[†] whose 1948 study of the islands became revered as the biological equal to Verbeek's famous geological study, said that in Anak Krakatoa

> . . . we have an island originally entirely devoid of animal and vegetable life, with even a completely sterile soil. It is therefore of the utmost importance that the flora and fauna of this island should be constantly examined, and at regular intervals, and it is greatly to be hoped that this unique opportunity *will not be neglected, as it was in the case of Krakatoa itself* [Emphasis added].

The first creatures to appear and to be noted on any of the four Anak Krakatoas (the first three islands were eroded to nothing by the waves: It was the second of these that was visited by scientists) were insects— first a very desiccated black cricket and next a female brown ant, both of which were already known to exist on Rakata, two miles distant. But their existence was cut short by the vanishing

---

*It is perhaps not too far-fetched to suppose that one primitive life form could have survived the birth of Anak Krakatoa, and that is the kind of heat-loving bacteria that clusters around the hydrothermal deep-ocean sulfurous vents known as black smokers. These creatures, which are magnificently called chemolithoautotrophic hyperthermophilic archaebacteria, are thought to be similar to the first life forms that were ever on this planet, which lived in the ancient seas of three and four billion years ago. None has so far been found on Anak Krakatoa; and this part of the Sunda Strait remains too dangerous and unstable for the kind of submarine investigation that would be necessary to prove their existence clustered around old Krakatoa's still-active vents, if there are any.

[†]Dammerman's modesty and reticence are memorialized in a mollusk that he discovered, which is named *Thiara carolitaciturni*—one of the rare examples of a nomenclature honoring a persona rather than a person.

of their home; and the third Krakatoa did not survive long enough or peacefully enough for anyone to venture out to survey it.

But once the fourth edition of the island appeared in August 1930, and once it seemed to be enjoying some permanency, the scientists came out in their droves. One of the first was William Syer Bristowe, the English arachnologist who had enumerated the wild profusion of spider species on Rakata.* He was sailed over to the ashy shores of Anak, and in short order discovered a beetle, a mosquito, some ants—and three species of spider.

Plants came next: Fifteen months after the emergence of the fourth Anak Krakatoa, the northern coast of the island was littered with floating tree stumps, bamboo stems, roots, and decaying fruit—eighteen seeds were discovered, ten of which had already taken root. Another visit added four more plant species to the list, all of them already well bedded in, together with moths, fungi, and a number of migratory birds, mainly sand-plovers. All seemed set fair for an explosion of new life—when the volcano at the edge of the new island unexpectedly and catastrophically erupted once more, and all life appeared to have been snuffed out.

This was to happen three more times. Not until after a spasm of truly devastating eruptions in 1953 did some kind of volcanic stability return to the troubled young island—and yet then, and very unfortunately, it turned out to be another quarter of a century before scientists returned to the region in any significant numbers.

For some curious reason, scientists stayed away from the new island. Perhaps biologists were just too fearful of the region's unpredictable vulcanicity; or perhaps they were afraid of the political troubles of the Indonesia of the time; or perhaps there were more mundane reasons, such as cuts in departmental budgets or the reordering of academic priorities. Dammerman's plea of 1948, that the world keep a very close watch on the island, was widely

---

*Bristowe is perhaps the most famous figure in the spider world, known for calculating that the weight of insects devoured by British spiders in an average year exceeds the total weight of all British people combined.

ignored—with the result that considerably less is known today about the resettlement of the island than should be known.

Krakatoa had itself suffered from bad and lazy science at the beginning of its posteruptive phase. Now Anak Krakatoa was suffering from twenty-five further years of unconcern and benign neglect. It was left until the 1980s for research work on both mountains to begin again in earnest. It continues to this day—with every botanist and zoologist suddenly realizing that in order to learn as much as possible about these uniquely interesting and remarkably contrasting biological situations, there is much catching up to do.

The geological makeup of the island has changed in recent years, and in the 1960s lava flows—hitherto unknown in the Krakatoa region—began to cascade from the crater and flow down to the sea. More than half of today's island is currently covered by congealed black lava—meaning that large areas of Anak Krakatoa that were previously vegetated are now barren once again. Students of island biology find this irksome, to say the least.

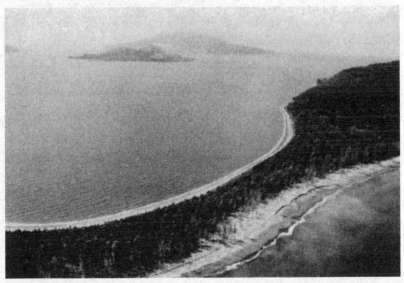

A thick coastal casuarina forest on Sertung island, and with Anak Krakatoa and Rakata in the background.

Ian Thornton, an Australian biologist with a long-standing fascination for Anak Krakatoa, devised a simple experiment to see
how rapidly insect life flooded into one of these seemingly barren
lava flows. He placed plastic containers filled with seawater on top
of the flows—and waited to see how many airborne arthropods*
they might catch. In ten days they found seventy-two species,
including wasps, earwigs, moths, and beetles: the Aeolian tide of
life appears to be quite as unstoppable as King Canute once found
the sea to be.

About 150 species of plant now inhabit the island: There are
casuarina forests on the low and soft-soiled northeastern shore; so-
called wild sugarcane, the aptly named *Saccharum spontaneum*,
bursts out of the hot and ashy slopes; delicate ferns grow in the
shaded parts of some of the jagged lava cliffs.

Once the plants were there, providing seeds, fruit, shade, and
moisture, and once the insect colonies that were first to be established had grown more varied and abundant, so the birds arrived
and settled. The first to do so had to be ground nesters (there
being in the early years an inconvenient absence of trees)—birds
like the savannah nightjar, the collared kingfisher (which can build
its nest in the sides of gullies), and the white-breasted waterhen,
which can settle in the mesh of salt grasses at the margin of the sea.

Then, when the casuarinas were tall enough to allow nesting, in
came the fruit-eating and insectivorous birds like the bulbul and the
crow. And as the forests thickened, some amphibians that had
somehow found their ways across the sea began to slink in and
make their nests—monitor lizards, paradise tree snakes. And,
inevitably, rats.

Next a generation of fig trees started to flourish—and fruit-
eating and fig-eating pigeons and doves arrived to feed on them.
They began to jostle for space and, in doing so, began to drive out
some of the lesser first arrivals, among which were the ground-
nesting birds whom the jungle law dictated had now probably out-

---

*The *Arthropoda*, a vast division of the animal kingdom, includes both spiders and insects.

stayed their welcome. Raptors then took over from the doves. In time the peregrine falcon arrived, established itself as Anak Krakatoa's avian monarch and, together with the barn owls, began to feed sturdily on the rats.

Thus, steadily and slowly, the population strengthened, evolving in appearance and flavor from being just the result of casual happenstance, becoming the ever more determined and rugged collection of island species that it remains today. And yet: Nature proposes but fate disposes. All this happy development has been interrupted again and again—in the seventies and the nineties—by the mountain blowing its top.

Today the population that has consolidated itself is distinctly different from that on the relics of old Krakatoa, as well as from that on Java or Sumatra. But the degree to which the animals and plants are actually going to be permitted to sustain themselves is far from clear: The ash and the lava that fall nearly continuously now are wreaking havoc with what could otherwise be a vastly theatrical biological experiment, arranged for the world to see. What could be grand theater turns out, all too often, to be just a tease.

The "great chain of being," an idea that has been in existence since the time of Aristotle, connects every living thing into one vast and seamless hierarchy—a system in which the Almighty, with man just below, stands at the apex. If one wishes to look at illustrations of the processes that lie at the lower end of this grand and somewhat fanciful scale, one need only look at the creation of life that can be seen on Anak Krakatoa and at the re-creation of past life that can be seen on the relics of old Krakatoa. In both places the processes are in fact remarkably similar, though varied in detail.

Both processes seem to have begun with the making of films of bacteria, which clung to the beds of volcanic ash and coated the ragged, burned ruins of once-hot rock. Matters got properly under way with, on the one hand, the arrival of fungi and simple grasses, and with, on the other, the addition of spiders, crickets, and beetles. Things continued with the establishment of ferns and orchids and more general plants, a mix of botanies into which were eased the herbivorous birds and a scattering of small animals; then came fruit and fruit-eating insects and the perching, nesting birds—and finally the larger animals with teeth and snapping jaws that would, in time, prey on all of these.

The list is seamless, the great chain set down for all to see: Bacteria, plant, insect, fruit, herbivore, carnivore. It is the classic development of life—a development a little more complicated on old Krakatoa perhaps, because of what lived there before and may have survived—but in all other senses, and assuredly so on Anak Krakatoa, entirely classic.

To the outside world the eruption of 1883 may have spelled death and devastation. To the world of biology and botany, however, the subsequent energetic happenings on islands in the Sunda Strait represent nothing more nor less than a freeze-frame picture of the future of life itself—a demonstration of the utterly confident way that the world, however badly it has been wounded, picks itself up, continues to unfold its magic and its marvels, and sets itself back on its endless trail of evolutionary progress yet again. The crucible of life turns out to be the most difficult of vessels to break: Not even the world's most dangerous volcano could do it truly irreparable damage.

Epilogue:
The Place the World Exploded

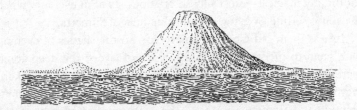

T he cars that sweep swiftly southward along what passes for west Java's corniche have to be prepared for all manner of delay and interruption—lumbering bullock carts, stalled trucks, lurching cyclists drunk on palm wine, sprawling street fairs, impromptu political demonstrations, undisciplined scatterings of chickens and goats and cattle, and on every corner of the roadway small children, children everywhere. It is frustrating driving; by the time the cars reach the outskirts of the small town of Carita, few on board will be in any mood to notice a small, undistinguished, and yellowing wooden structure that is set well back from the road at the north end of town, on the brow of a low hill above a tapioca plantation.

Most probably those passing, especially if they are from afar—

and Carita is these days a seaside holiday destination, so the coast
road is frequently crowded with prosperous Indonesian families
trying to escape the oppressive heat and crush of Jakarta—will be
looking to their right, and not to where the building stands on
their left. For on the right-hand side, the west, the view is quite
soothingly magical—especially so at about six each evening, when
the sun is setting over the distant blue hills of Sumatra.

Just as in an old Chinese watercolor, so the shades of evening
blue seem to merge in an infinite series of layers: the deep aqua-
marine of the sea in the foreground, the bold azure of the darkening twilight sky, the pale powder blue of the Sumatran mountains
behind, and in between them the scattering of islands of the
Krakatoa archipelago. Edged in pastel blues that are dark or light
according to their distance, the islands change with the shadows
they cast upon one another, or as the smoke that is usually rising
from the peak in the center of the group drifts and curls above and
around about them.

And yes, you find yourself saying to yourself in an almost incred-
ulous whisper, This is where it all happened. This is Krakatoa, the
place with a name now firmly annealed into the language, welded
into the world's public consciousness, a name that has become a
byword for nature's most fearsome potential for destruction.

The steep, sharp, instantly recognizable peak on the left-hand
side of the little cluster of islands is Rakata, the dead relic, the
ruined husk of the great eruption. The low and less distinguished
islands of Panjang and Sertung hug the horizon, the wrecked parentheses of the ancient caldera. Sometimes, in certain lights, they can
be tricky to distinguish against the pale backcloth of Sumatra: Their
distances seem compressed, so that from the Java shore they look as
though they are a single uninterrupted island, even though they are
in fact two, and one is a good deal closer to shore than the other.

And rising from the midpoint of the pair, in fact almost exactly
in the middle of the neat little archipelago, is a peak that, though
lower than Rakata, has at this distance and from this angle the
shape of a perfectly formed cone—with a plume of smoke some-

times rising slowly above it and, on the line where mountaintop and smoke cloud meet, a sinister, beautiful, ominous-looking orange glow of fire. This is the centerpiece of the tableau, both in fact and in fable: This is the dangerously fast-growing adolescent child of the cataclysm, Anak Krakatoa.

The beauty and strange menace of the scene is memorable: Small wonder that passing southbound motorists are enraptured, and look steadily across the sea to their right. The yellow-painted buildings above the tapioca trees on the left hold no attraction at all, and they are generally ignored, passed by, and, if glimpsed at all, instantly forgotten.

Perhaps they should not be. For the buildings are the field station of the Krakatoa Volcanic Observatory, and in one of the simply furnished rooms within the small cluster of structures there is a device—its electronics a little long in the tooth these days, but its metal casing, dials, and instrumentation cleaned and oiled and cared for—that measures as exactly as it is possible to measure what is going on beneath the earth below Krakatoa. This is the machine that will warn the region, the country, and the world, one hopes, in the unlikely event that things begin to go awry.

The device, which many years ago was gifted by the Americans, is designed to alert all those in Indonesia whose task it is to watch out for signs of trouble—the civil defense agencies, the army, local hospitals, managers of food depots and blanket stores, and everyone who lives in those low-lying coastal areas that might be inundated by tides—if another catastrophic eruption is imminent.

The technology is simple enough. Some years ago a group of geologists from the Geological Survey of Indonesia and the U.S. Geological Survey installed an array of seismic sensors on Anak Krakatoa, scattering some into clefts in the lava surface and settling others into holes drilled some few feet into the slopes of the mountain. The sensors were connected to a coder and to a radio transmitter built into a tough aluminum box. This box was then buried in a trench dug on the eastward-facing, Java-facing slope of the island, a couple of hundred feet above the sea and in line-of-sight with the observatory that was at the same time being built north of Carita.

A solar panel on a mast above the buried box assured an uninterrupted source of power, and therefore a seamless flow of signals—such that night or day, every day of every year, what was happening seismically on this most potentially violent center of the Krakatoa complex would be sent instantly across the strait, to be picked up by the cluster of radio aerials that rises above the yellow-painted, rarely noticed, and insignificant little observatory.

The rules laid down by the Volcanological Survey of Indonesia require that there always be at least one observer on duty inside the building. When I stopped by, the point-man was a thin, tired-looking and rather nervous forty-year-old named *mas* Sikin,* who lived in a nearby village. When we met he was halfway through his week-long shift, keeping the world's most notorious volcano under close invigilation—a duty for which he was paid the equivalent of fifty dollars a month, together with a substantial ration of rice.

He occupied two of the observatory's rooms—the others were assigned to the survey's chief, whose office was eighty miles away in Bandung and who stopped by only in the event of an emergency. In one of Sikin's pair of rooms there was just a cot and a handbasin. In the other, under a formal portrait of the current Indonesian president, Megawati Sukarnoputri, on a trestle table beneath which were four large twelve-volt car batteries, was the monitoring machine itself, a black-and-silver chrome box with a metal plate identifying it as a radioseismograph, made by a venerable American earthquake-measuring company known as Kinematics, Inc.

There is a needle, a large revolving drum, and a sheet of paper. On this paper, which is changed every twelve hours, is a series of traces in purple ink, recording second by second the activity across,

---

*It is often the case, as here, that Indonesians have just one name. The prefix *mas* is simply an honorific, akin to "brother," and is a way of describing and addressing one who is younger than oneself. Had Sikin been much older, he would have been deserving of the title *bapak*, or its more widely used diminutive *pak*, meaning "father."

upon, and deep beneath the island. Today's trace indicated only the merest of trembles; the night before the needle had waved unsteadily for a few hours in the middle of the night; and the week before—Sikin took the trace down from a cardboard folder on a shelf above the recorder—the lines were crazed, as though some-one had shaken the inked needle in a sudden fit of anger. The lines were blurred and entwined together, because the needle had been vibrating and waving back and forth both at a great frequency and with considerable amplitude: That week Anak Krakatoa had clearly undergone a spasm of some sort.

Not enough of one, however, for Sikin to have called out his superiors in Survey headquarters in Bandung. A routine eruption, of which last week's had been a classic of the kind, was nothing to get excited about: Anak Krakatoa is like a great safety valve, blow-ing off steam and a good deal else besides on a regular basis, never holding itself back and distorting itself dangerously, with an enor-mous relieving eruption as the culminating climax. Generally speaking it is the active volcanoes that do not erupt that are the dangerous ones: Inside them energy is being stored up, little by lit-tle, until the stresses become too great and there is a catastrophe. On Anak Krakatoa there is continual release—which may look dra-matic and may on occasion make for trouble and cause casualties, but which suggests that, in the short term at least, the danger is predictable and any crisis manageable. So long as the volcano is watched, current belief has it, the neighborhood is safe.

A routine eruption is invariably visible from shore some minutes before it is detected by the seismometers and is then written on to the seismograph drum. That recent bad day's eruption—it was a Sunday afternoon, Sikin remembered, and he was standing at the observatory door puffing on his clove-flavored *kretek* cigarette— began with a sudden whoosh of dust and smoke that could be seen pouring from a vent just behind the obvious main summit of Anak.

He said that he watched for a moment only, counting the seconds by snapping his fingers. He walked back to the Kine-matics machine and, sure enough, ten seconds after the first sight of the smoke, after ten snaps of his fingers, the needle began to move.

There was at first a great swing, several inches to one side, after which it slammed back to the other side so fiercely it might look to a stranger as though the needle would break. Then it moved back again, and again and again, the trace writing on top of itself, but with movements that were now diminishing somewhat as the drum unrolled, so that the record began to look like an arrowhead, tapering toward a point. It was a familiar pattern, remembered distantly from old newspaper pictures or images on museum walls at places that had suffered badly from seismic shocks, places like Skopje and Anchorage and Istanbul, or the volcanoes of Mount St. Helens or Unzen.

Sikin was still clicking his fingers while he watched the machine's passionless unrolling—until another five seconds had passed, when he went over to the open doorway and cocked his ear. Sure enough, right on cue, there came from across the strait a rumble, like distant thunder, or the shaking of a theatrical thunder sheet. And then silence.

Below him, behind the tapioca plants, he could see that the masts of the *prahus* in the river had begun to swing back and forth as the craft shook at their moorings. The sea surface had the sudden look of hammered pewter—until the swell took over, the breeze ruffled away any pattern that might have been briefly imposed upon it, and everything returned to normal. Across on the mountain the smoke had now lifted away from the summit, which was now quite clear; the billow of black had risen well into the sky and was being borne away southward in the streaming winds. Otherwise the sky was cloudless.

Soon there was just a single ragged blot floating in the air, together with the traces written on the paper drum, to stand as a record of how Anak Krakatoa had once again reminded the surrounding world that it was still very much alive. And only the paper would survive. In six months' time a car would arrive to take it and all the rest away, to be stored in a damp basement in Bandung.

*   *   *

It is not entirely quite clear whether one is formally allowed to visit Krakatoa. The archipelago is a part of a national park,* is a protected treasure, can be dangerous on occasion and in theory is officially off-limits to those without government sanction. The bureaucrats safely in their offices in Jakarta fret over the scrapes that visitors might get into. People have been killed or injured on the island, hit by flying lava bombs. The Sunda Strait suffers from notoriously fickle winds and seas, is crowded with fast-moving cargo ships and has deep waters that are alive with hungry sharks. The local vessels commonly used for the crossing break down with depressing regularity: Few Americans will forget the fate of the twenty-seven-year-old Californian women Rickey Berkowitz and Judy Schwartz, who, trying to cross to Krakatoa in 1985, found themselves drifting in a leaky open boat for three weeks, surviving on a diet of peanuts, rainwater, and Crest toothpaste.

In Bandung I had been handed a piece of paper dripping with official permit stamps and signatures, and told that *just this once,* an exception to the normal rule, I might be allowed to go to Krakatoa on my own. But in Carita it was perfectly evident that no one gave a thought about ever getting permission, and that whether it was allowed or not, going across to Krakatoa was simply, like so much in the East, a matter of nothing more than supply and demand. If you wanted to go, then, for a price, go you most assuredly could and would.

And so, early one morning on a Carita beach that was already swarming with boys selling shells, sarongs, fried squid, coconuts, and kites, and with cheerful young women offering very un-Islamic-sounding full-body massages and broad winks that promised even more delights, I found myself among groups of young

---

*The Krakatoa archipelago is a detached part of the Ujung Kulon National Park, most of which includes the peninsula in Java to the south of the islands and known as Java Head. The one-horned Javan rhinoceros is plentiful in the park. One of the long-term benefits of the 1883 eruption has been the reluctance of a superstitious people to live and settle in large numbers anywhere near the volcano—a reluctance that has protected a very rare species from what would otherwise have been, in the face of population pressure, almost certain extinction.

men who pestered me slyly, as though in a Saharan souk, hissing theatrically, "You like go Krakatau?" Consequently, within an hour or so, after I had looked critically at the small armada of boats on offer, I had selected a sleek yellow wooden *pinisi*, a fishing craft with a reliable-looking seventy horsepower Evinrude engine, had found a guide with the improbable name of Boing, and had waded out into the surf and clambered aboard.

The skipper first made a drama of unfurling his national flag and tying it with twine to the jackstaff. It turned out that just by chance, we were sailing on August 17, which was the anniversary of Sukarno's famous proclamation of independence in 1945.* With a mischievous grin, the captain then pulled up a case of Bintang beer from the cool waters down in the bilges, tossed a couple of cans each to his two shipmates and to Boing and to me, waved a salute to something that sounded like the glories of Indonesian freedom and the likely continuation of good postcolonial living, burped, hiccupped, gunned his engine noisily, bounced his boat through the green heaps of surf and finally sped off unsteadily into the open waters of the strait and toward the hazy western horizon.

Early on this hazy morning, it wasn't possible to see the archipelago from sea level—until we had been cruising steadily west for about half an hour. I spent those first minutes gazing up at a frigate bird high above us, flicking its scissorlike tails to change direction. There were dolphins too, playing under the bow; and flying fish launching themselves like small missiles, darting across the troughs between adjacent swells.

And then the skipper nudged me out of my reverie. "Rakata!" he cried—and sure enough, rising through the haze directly ahead of us, was the perfect pyramid-shape of the last relic piece of old Krakatoa. To its right, as I squinted into the distance, I could soon make out the smaller cone of Anak Krakatoa, a puff of white smoke rising above it, and a slash of white running across the face

---

*The formal end of Dutch rule came four years later, just after Christmas 1949. The August date is nonetheless the preferred time for celebration.

of the mountain, a lava flow that had so far escaped weathering by the elements. The driver indicated that it was as well that the smoke rising above the volcano was white. If it turned gray—if it was filled with debris, in other words—then it would be time to think about leaving. Fast.

Between our position now and the islands ahead were the main north- and southbound shipping lanes of the Sunda Strait, and for a half hour negotiating them seemed to occupy the skipper's mind more keenly than did the color of Krakatoa's smoke. For the lanes were very busy, and when you are in a small craft, down at sea level, the immense speed of the onrushing cargo ships is almost impossible to calculate.

There was, for example, the northbound container vessel that, when we first saw her, was a mere speck on the port horizon. Within five minutes, however, this speck had become a massive ship whose hullside proclaimed that she belonged to COSCO, the China Overseas Shipping Company. She was bearing down on us with an awful speed. We cut our engine and let her pass us, and when she did so she must have been no more than five hundred yards ahead. The nub of her bulbous bow elbowed the waves aside, setting us rocking alarmingly as we sat, hove to, and watched her immense bulk slide cinematically in front of us.

And then she was past, safely. Her enormous screw thrashed wildly in front of the huge rudder; and as she drew away to our starboard we could see her name written in Chinese, and her port of registry: Shanghai. She was on her way home, probably back from a voyage to ports in Africa and the east coast of South America. There was no indication that she ever saw us; such crew as were on the bridge were most likely standing on her own port side, gazing from their vantage point across at the volcanoes.

Once she was safely gone, we sliced and bumped across her mile-wide wake, dodged another couple of smaller southbound ships and edged in toward the lumpy, confused-looking waters of the volcanic caldera itself. The boat began to lurch alarmingly again, as though we were passing through a line of rapids. The skipper grinned. He knew how strange a phenomenon this was. The submerged lip of the old caldera, it seemed, was somehow playing havoc with the tides and currents, tripping up the streams, creating a ring of intersecting overfalls that broke all around us in ragged rings of white and green.

We edged closer, coming to within half a mile of Rakata's huge vertical cliff, which was blindingly lit by the midday sun. Seen from this distance and angle, it displayed a textbook cross-section of a wrecked volcano: Before us was the dark central vertical spire of the old core, and its sheared-off surface pierced by dozens of veins of long-frozen pipes of lava, the whole cone capped high up in the sky by a ragged crown of trees.

And then we were through the broken waters and into the belly of the thing, the omphalos of the Oriental world, a place where the waters were unimaginably deep once more; and down in their depths, unimaginable things were happening. Within the caldera, Rakata was large enough to shut out some of the northbound swells and to divert the currents, and so once we were inside the submerged walls and in Rakata's lee the sea suddenly became calmer, more congenial. The boat stopped its corkscrewing and resumed a more steadily upright putter to shore.

We set a course first toward the strange tilted aggregation of guano-covered pillars called Bo'sun's Rock, around which the waves gnawed hungrily. We turned slightly to port just as we seemed ready to touch it, and left it to our starboard; then we headed to the south side of Anak Krakatoa itself. The plan was to circumnavigate the island before landing on the beach on the east side. The skipper was showing off now, and liked to hug the shore for my benefit: The island, the child of Krakatoa, now an enormous black bulk a few feet off to our right, smoked contentedly and calmly; and the smoke, we all noted, remained steadily white.

Most of the western and northern shores of the island consisted of dark gray lava flows—and the one paler flow, newer than the rest, that I had seen from afar. They were all horrible gnarly things, mostly quite unweathered, with rocky tentacles that probed into the

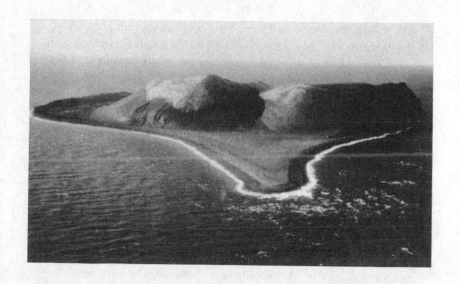

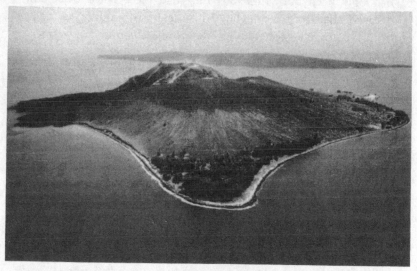

Aerial photographs show the uncanny similarity between Anak Krakatoa, below, and its sub-Arctic cousin Surtsey, off Iceland. Geologically and tectonically, however, the islands and the volcanoes that created them could not be more different.

sea, where the waves washed around and sucked hungrily underneath them. The land above the flows was spiky and crammed with razor-sharp cliffs of frozen basalt. Once in a while I could see a single red flower, sunning itself from within a cleft; and seabirds would perch on the highest points, surveying the world around them. But otherwise the flows were lifeless and harsh, the raw front edge of the volcanic process, all solidified into hundreds of yards of dead, unyielding, and ugly black rock. The crater itself was invisible from here; the smoke column, though, wafted ever upward.

Once we were around the eastern side, however, the island took on a completely different aspect. The slope down from the now-visible summit became much less pronounced, and instead of lava there was ash and, in places closer to the beach, ashy soil. It had been transmuted into soil because, fringing the eastern and northern flanks of the island, there was now a long copse of trees—and over time (though not very much time, since the island had only been born in 1930) they had died, fallen over, and decayed, their corpses mingling with the ash to produce, to judge from the gorgeous fecundity of the growth all around, a wonderfully humus-rich substrate.

From this vantage point, indeed, Anak Krakatoa looked like a perfectly normal tropical island, with no suggestion that it was one of the youngest islands in the world.* And it was toward its fringing woodlands that the skipper directed his boat. We dropped anchor fifty feet from shore and splashed through the shallows, then up on to the scalding beach. There was a path through the casuarina trees: We walked in as quickly as we could—the sand was far too hot (whether from the pitiless sun or the volcanic origins of

---

*In size and shape Anak Krakatoa looks much like one of its immediate successors, the island of Surtsey, off the south coast of Iceland. It was created in 1963, now has grazing animals, and perhaps in time will have a human population too. There are even newer islands, however, all of them volcanoes: An island appeared near Iwo Jima in 1986, and the Japanese christened it Fukoto Kuokanaba; an islet near the Metis Shoal in Tonga's Vava'u Group appeared in 1995 and was named in honor of the then much favored rugby international Jonah Lomu; and in 2000 an undersea volcano named Kavachi erupted in the Solomons—this may yet consolidate as the world's youngest island.

the sand, I could not be sure) to allow any dawdling—and set up camp in a grove, where we would build a fire and take our lunch.

It looked for all the world as though we were in a tropical forest. The vegetation was thick and dripping with moisture. Butterflies swirled in the thermals. Birds called from high up in the canopy; and one of the boat's crew showed me the nest of what he said was an owl. There were the footprints of rats and geckos; the fine pumice sand was alive with ghost crabs, which skittered delicately along on tiptoe, as if keeping away from the heat; and along one patch of sand close to where we were eating there were the tracks of some much larger animal, about which the crewmen only grinned, ominously.

After eating grilled fish and rice and drinking the remains of the Bintang, Boing and I set off uphill. The crewmen had in any case fallen asleep in the shade—making us promise beforehand that we would return two hours before nightfall or risk being run down by a freighter if we tried to cross the Sunda Strait after dark. It should in any case take no more than an hour to get to the summit: Anak Krakatoa was only fifteen hundred feet high.

But of course climbing a third of a mile up a high hill made only of newly fallen ash is a far from easy occupation. First, there was the simple matter of getting out of the coastal forest—a hundred feet of thick, lush vegetation through which we had to hack our way until it ended abruptly, with a patch of low grasses and a sudden blaze of brilliant sunlight as we emerged right on to the naked flank of the volcano. Not only was it unbelievably hot here—and this was the kind of heat that drills painfully up through the soles of even the best-insulated shoes—but the ash that rose up steadily ahead slipped and slid, and opened up into scores of small chasms and runnels and rills, such that for every five steps upward, I fell backward four.

We passed the aluminum box with its radio aerial, pointing directly at the faraway and, at this time, invisible coast of Java: Sikin would be there, I fancied, watching the paper roll steadily out from the Kinematics drum with nary a vibration coming from his charge.

The island was quiet this day and, judging from the column of pure white smoke that coiled up from the summit ahead, was likely to remain so for a while.

But it had evidently not been so quiet very recently. As we climbed higher, we saw hundreds of rocks on all sides that had clearly whistled down from on high after a furious eruption and crashed into the sides of the hill. Craters had formed in the ash around where the rocks had fallen, just as though bombs or artillery shells had exploded: The craters gave the surface the curious appearance of the landscape of Ypres or Passchendaele, only here it was dry, not muddy, and sloping, not flat.

Some of the lava bombs had been huge, some as large as a bus—and whenever we edged passed one of those, with Boing laughing in the way that only Javanese can laugh, with their fatalistic attitude to all happenings good and bad, I confess that I looked up in the sky, moderately apprehensive. If Anak Krakatoa decided on a tectonic whim to hurl one of those into the sky, then gravity would ensure that life for those below would become very short and concentrated indeed.

Up we trudged; until we reached a point where the last grasses—tall clumps of the so-called wild sugarcane, *Saccharum spontaneum*—petered out, and the landscape became nothing more than ash, bomb crater, smoke, and the ever widening panorama of the sky. The skyline itself was quite near, and when after ten minutes more or so I breasted the ridge—panting, perspiring, glad it was all over—I was actually surprised at how easy it had all been.

I stood tall on top of a truck-sized lava bomb, and took in the immense view. I gazed down at our island, at the curved cusp of the grove of trees by the beach below, at the black-sand shore and at our yellow *pinisi*, tiny and fragile, swinging on her anchor. I gazed down at the lava flows, great black frozen rivers of coiled and frozen magma that had not so long ago been pouring down into the boiling ocean. I looked over at the jagged outcrop of Bo'sun's Rock, welling out of its private froth of surf, and over to the low island of Panjang to the east, and the immense half-conical wall of Rakata directly ahead.

And then I tried to look across at where the archipelago's other island of Sertung should lie—only at this point I was suddenly confronted by an immense wall of sloping gray ash that rose high, high above me. What foolishness! I suddenly realized. What a mistake to have made! The relief I might have briefly felt at supposedly reaching the summit of Krakatoa's child vanished in an instant: It turned out that I was not even halfway up the mountain. Where I stood was no more than a false crater, the mere relict rim of a much earlier eruption. Anak Krakatoa's widely feared active crater, the summit of the volcano and the site from behind which the eruptive smoke was even now billowing, was yet another long, long slog away.

There was nothing for it. Boing and I manfully slogged away upward once more, as we were duty bound to do, slipping and sliding ourselves generally and painfully up and up, in the fierce heat and dazzling sunshine, ever wary of what mischief the mountain might be planning. The sky widened farther and farther; the sea below became a dazzling, gleaming sheet of steel; the temperature rose in an almost terrifying way that seemed to have no relation to the tropical sun. After thirty wearying minutes more, we stepped over a rock edged with a crust of yellow-stained sulfur crystals, and across the teeth of a ridge—below, spread ahead like some infernal dish of hell, was the crater itself.

The white smoke had by now enveloped us: It was hot water vapor, mixed with the curiously attractive smell of sulfur dioxide and dust. The surface I could see ahead was a fragile-looking crust, newly baked and broken in places, with plopping gobbets of hot mud spurting into the air and hissing, machinelike jets of gas roaring and whistling up into the cloud. From afar, the volcano had appeared quiet; but from up here, on the very lip of its mouth, at the working end of the heir apparent to the greatest volcano the modern world has experienced, it seemed anything but.

The mechanics of the making of the world were all in evidence, just a few feet ahead of where I stood. All this talk of subduction zones, of the collisions between two of the world's immense tectonic plates, of the unfolding of the ring of fire—it all came down to this. Here, in this hot, crystalline, yellow-gray, wheezing, whistling, mud-boiling cauldron, was where the consequences of subduction were being played out.

The power of the process was all too apparent here also, in the strangely compelling symphony of grindings and snortings and sulfurous snarlings, in all the rushings of yellow and green gases, and in the snapping and straining of the rocks and crystals and crusts. This was a place that was filled with nameless and unfathomable activities, and it had a terrible, fascinating menace. It was a place that was all too evidently primed, ready at any instant to explode again—and, in exploding, to do goodness knows how much harm to goodness knows how many souls waiting unwittingly down below.

After a while the sulfur began to catch in my throat, and Boing started to become anxious that we might in any case be staying too long. And so we trudged downhill for the last time, soon glissading through clouds of ash, passing at a run the observatory radio transmitter and the clumps of sugarcane, before meeting the fringe of the woods and diving thankfully through the final few hundred feet into the comparative cool of the shade of the seaside casuarina trees.

The crew had gone out to ready their boat for the journey back to Java, and Boing swam over to talk to them. I was hungry, and pulled from my knapsack a squashed chicken sandwich, made for me in the hotel before we left. I sat on the trunk of a fallen tree in the quiet, reflecting on the afternoon, and on the awesome, epicentral, deeply symbolic place above that I had been fortunate enough to see.

And then I heard a rustling, crackling sound in the woods, a strangely unearthly noise that made the hairs rise suddenly all along the nape of my neck.

I stopped eating, and looked around. There was no other sound—not even from the boatmen, who were by now probably well out of earshot. The rustling continued, and it got louder—until, suddenly breaking out of the cover of the jungle, emerged the head and then the body of a six-foot-long lizard, waddling slowly and steadily toward me, its jaws wide open.

It was a horrible, strange, weirdly shaped creature—a long fat brown body with what looked like a thick seam of flesh running the length of its midsection. As it walked its tail thrashed from side to side, and from its small flat-topped head came a tongue, a foot or more long, that flickered in and out menacingly.

The beast as a whole did indeed look menacing, and very dangerous indeed. Deep down I knew that it was probably quite harmless, and that it was in all likelihood simply a specimen of the great five-banded monitor, the wonderful swimming lizard known to Javanese as the *biawak* and to science as *Varanus salvator*. But

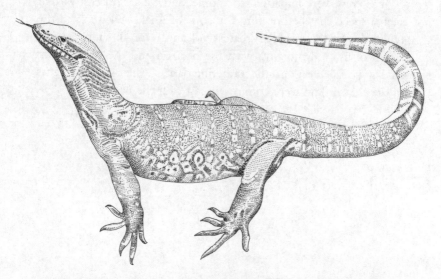

Varanus salvator, the five-banded swimming monitor lizard, now an all too common inhabitant of the ruined islands.

that deeper realization came only later in the day; at the precise moment on that August afternoon when it emerged from the trees, when I was sitting alone in the jungle on the side of a hot and very active volcano, the animal looked like nothing so much as a fully fledged dragon, and I was more than a little alarmed by his arrival on the scene.

So I threw him my sandwich. He took one disdainful look at it, gazed up briefly at me—and then grabbed the Carita Beach Hotel kitchen's doubtless carefully prepared confection of chicken and white bread between his wicked-looking teeth, and skittered off back into the darkness of the jungle, his armor-plated tail thrashing its valedictory.

I got up cautiously from my tree and, with as much dispatch as was consonant with the dignity of the occasion, walked down to the shore and on to the sand. I hurried through the cooling waves, out to where the yellow fishing boat was waiting. I decided not to say what I had seen.

The crew were impatient for home, and we set off eastward at such a speed that before long the cone of Anak Krakatoa and the relic ruins of the world's greatest volcano were all sliding down astern of us, and merging with the horizon and the setting sun. And as we sailed on into the gathering dark, so the twinkling lights of the west of Java were coming up fast over the bow.

Recommendations for
[and in One Case, Against]
Further Reading and Viewing

"Wake up, wake up; you've got to get in the shade!"

I shook my head and opened my eyes again. There was a man kneeling over me. He wasn't a native, and didn't suggest an explorer or a traveller. He was wearing a correctly tailored white morning suit, with pinstripe pants, white ascot tie, and a white cork bowler.

"Am I dead?" I asked. "Is this heaven?"

"No, my good man," he answered. "This isn't heaven. This is the Pacific island of Krakatoa."

—from *The Twenty-one Balloons*,
by William Pène du Bois, 1947

Just after 8:32 on the crystal-clear Sunday morning of May 18, 1980, the long-awaited, universally expected eruption of Mount St. Helens, in the southwestern corner of Washington State, blew away the entire northern face of what was then America's most notorious volcano. The event turned out to be a classic of the volcanic art, camera-ready for the textbook: an ash cloud rising sixteen miles into the sky, and visible two hundred miles away; the mountain's summit suddenly reduced in height by thirteen hundred feet; scores of square miles of countryside burned and devastated; twenty-two thousand further square miles blanketed with debris; billions of trees swept flat; and fifty-seven people killed, most of them suffocated by clouds of boiling grit.

And yet, though the eruption of Mount St. Helens—which was televised, filmed, photographed, and chronicled in more loving detail than any other eruption in history—was to become briefly so very famous, it never even came close to dislodging Krakatoa from its position as the most notorious volcano of all time. For some curious reason—and part of that reason quite probably no more than the euphonious nature of the volcano's given name—the saga of Krakatoa has remained firmly and immovably welded into the popular mind.

The principal elements of the story of its great eruption of August 27, 1883—the immense sound of the detonation, the unprecedented tidal waves, the death-rafts of drifting pumice, the livid sunsets—all still play their part in the world's collective consciousness. They remain welded into the popular mind in a way that the spectacular eruptions of the planet's other truly great volcanoes, like Etna, Santorini, Tambora, and St. Pierre—and even the Vesuvius of Pliny and Pompeii—have still never quite managed to match.

Krakatoa—the name. That may well account for it. But there are other reasons too—among them quite probably the timely appearance of two items of popular culture relating to the event. One is a slim volume of a children's book, published to near-universal praise in 1947; the other, a Hollywood film released twenty years later to near-universal scorn. More than any other external factor, these two creations quite probably account in large measure for the extraordinary durability of the Krakatoa story.

The children's story was *The Twenty-One Balloons* by William Pène du Bois: it won America's renowned Newbery Medal in 1948, and has never been out of print since. It tells the story of a math teacher from San Francisco named William Waterman Sherman, who flies in a balloon westward across the Pacific, crash-landing (after seagulls pecked holes in the silk fabric) on what turns out to be "the Pacific island of Krakatoa." Here the impeccably dressed locals are all fabulously rich, since the volcano in the island's center sits directly on top of an immense diamond mine.

The resulting story is all about the professor's adventures among the remarkable people of a utopia, which, because of the eruption of 1883, swiftly becomes a dangerous dystopia. All have to flee in a specially built balloon-lifted platform. The book—180 pages, endearing, illustrated by its thirty-year-old author—is enchanting; most intelligent children will have read it, and they will in consequence know Krakatoa as, at the very least, a place both dangerous and beautiful, and wondrously exotic.

Children who were born in time to read the first editions of *The Twenty-One Balloons* would have been in their early thirties in the year 1969. They would thus have been a precise demographic target for one of Hollywood's archetypal B-movie directors, the otherwise little-known Bernard Kowalski, who in that year made the universally known, much derided, utterly improbable, irredeemably mediocre, and magisterially mistitled epic *Krakatoa, East of Java.**

The ensemble cast—Maximilian Schell, Diane Baker, Rossano Brazzi, Brian Keith, and Sal Mineo among them—might possibly have saved a stronger script or storyline. But the sheer lunacy of the plot, which involved sunken treasure, wayward hot-air balloons, long-legged and half-naked female Japanese pearl divers, escaped prisoners, and a series of very obvious polystyrene models of a volcano, inevitably forced whatever grand vision Kowalski might have had to disintegrate into farce.

Despite the lavish technological promise offered by both Cin-

---

*Kowalski directed some fourteen films and countless episodes of popular American television shows. Prior to *Krakatoa, East of Java*, he made *A Night of the Blood Beast, The Hot Car Girl*, and *Attack of the Giant Leeches*; four years after *Krakatoa* he directed an epic, which presumably involved the activities of dangerous snakes, called *Sssssss*. The director is, however, not entirely responsible for siting the volcano at the wrong end of Java: His film was based on a book of the same name by an even more obscure writer named M. Avallone: It is to others we should look for matters of geographical exactitude.

erama and Technicolor, the film performed very badly at the time, remains generally a cinematic joke today, and is thought of as merely a less costly precursor to such titanic disasters as *Ishtar, Waterworld,* and *Heaven's Gate. Krakatoa, East of Java* can be seen very late at night on some obscure American television channels; by contrast in Britain, where for some reason the film still enjoys the status of a minor cult classic—a fondness for kitsch, some say—it was part of the expensive and much touted television schedules as recently as Christmas 2001.

In the late 1980s Lorne and Lawrence Blair, two seemingly indefatigable and irrepressibly enthusiastic British explorers,* produced a series of extraordinary television documentaries about the island of Indonesia called *Ring of Fire.* In the way of such things, the television company then produced a book (*Ring of Fire: An Indonesian Odyssey,* London: Inner Traditions International, 1991), which is copiously illustrated and informative. One of the films, cheekily titled *East of Krakatoa,* has two minutes of memorable footage of the early eruptions of Anak Krakatoa in the thirties.

In 1999 Channel Four showed an ambitious two-part television series based on David Keys's remarkable book *Catastrophe: An Investigation into the Origins of the Modern World* (London: Century, 1999), which, as chapter 4 indicates, speculates that an early eruption of Krakatoa may have thrown the entire known world of the time into profound disarray. The idea has supporters and detractors in equal measure: It ought to be read, skeptically, for a good analysis of the possible early history of the volcano.

There have been surprisingly few books about the volcano's 1883 eruption in recent years, other than an immense number of specialist and technical volumes. One of the few is *Krakatoa* by Rupert Furneaux (London: Secker & Warburg)—but it was published in 1965, an unfortunate two years before the establishment of the theory of plate tectonics arrived to answer all questions about why volcanoes erupt, and so the book has a necessarily lim-

---

*One of the pair wore a monocle, which made them highly caricaturable.

ited value. It is, however, a stirring tale, and exceedingly well told, and I made liberal use of some of the eyewitness descriptions that Furneaux so assiduously dug out of various Dutch and maritime archives of the day. Ian Thornton's *Krakatau: The Destruction and Reassembly of an Island Ecosystem* (Cambridge, Mass.: Harvard University Press, 1996) is thoroughly up to date and much more readable than its title suggests; but, on the other hand, it concentrates heavily on the biogeography of the island, which those hoping for the more general story may lament.

The enormous and well-nigh definitive *Krakatau 1883: The Volcanic Eruption and Its Effects* by the distinguished vulcanologists Tom Simkin and Richard S. Fiske (Washington, D.C.: Smithsonian Institution Press, 1983) is required reading for anyone with a serious interest in the event and its aftermath: My own copy is thumbed to the point of near destruction. It has scores of illustrations, diagrams, tables, and a vast bibliography, all of immense use to someone like me. But it is at heart a scientific book, and its appeal will tend to be limited to the specialist: The fact that no one ever answered the authors' appeal for yet more eyewitness descriptions suggests either that there are no more to be had (which is not true: At least two entirely fresh accounts appeared while I was doing my own research) or the audience for the book was limited to scientists and somehow missed the kind of people who hoard old letters and journals from long-dead relatives who once traveled Out East.

The Royal Society's famous report, *The Eruption of Krakatoa and Subsequent Phenomena* (London: Trübner & Co, 1888), can still be found, expensively, in antiquarian bookshops; as can the heroic *Krakatau* by R. D. M. Verbeek (Batavia: Government Printing Office, 1886), with copies available—at a price—in either Dutch or French. Simkin and Fiske very obligingly translated much of Verbeek's work into English (for the first time) in their own 1983 volume. Serious students of the volcano should make all possible attempts to read at least some of this marvelously enthusiastic work, in whatever language available.

Finally, in the must-read category: Anyone with the available funds should buy for their shelves the massive, astonishingly detailed and beautifully written *Encyclopedia of Volcanoes* (San Diego: Academic Press, 2000), not least because it is edited by the Icelandic vulcanologist and world-renowned Krakatoa enthusiast, Haraldur Sigurdsson, presently a professor at the University of Rhode Island.

Other books I found useful and interesting include:

Abeyasekere, Susan. *Jakarta: A History*. Singapore: Oxford University Press, 1987.

Angelino, A. D. A. de Kat. *Colonial Policy. Vol. 2: The Dutch East Indies*. The Hague: Martinus Nijhoff, 1931.

Armstrong, Karen. *Islam: A Short History*. London: Weidenfeld & Nicolson, 2000.

———. *Muhammad*. London: Gollancz, 1991.

Bangs, Richard, and Christian Kallen. *Islands of Fire, Islands of Spice*. San Francisco, Sierra Club Books, 1988.

Barty-King, Hugh. *Girdle Round the Earth: The Story of Cable & Wireless and Its Predecessors to Mark the Group's Jubilee*. London: Heinemann, 1979.

Berger, Meyer. *The Story of the* New York Times *1851–1951*. New York: Simon & Schuster, 1951.

Bertuchi, A. J. *The Island of Rodriguez, a British Colony in the Mascarenhas Group*. London: John Murray, 1923.

Blue, Gregory. *Colonialism and the Modern World: Selected Studies*. Armonk, N.Y.: M. E. Sharpe, Inc., 2002.

Blussé, Leonard. *Strange Company: Chinese Settlers, Mestizo Women and the Dutch in VOC Batavia*. Dordrecht: Foris Publications, 1988.

Bruce, Victoria. *No Apparent Danger: The True Story of Volcanic Disaster at Galeras and Nevado del Ruiz*. New York: HarperCollins, 2001.

Cardini, Franco. *Europe and Islam*. Oxford: Blackwell, 2001.

Carson, Rob. *Mount St. Helens: The Eruption and Recovery of a Volcano*. Seattle: Sasquatch Books, 1990.

Clarke, Arthur C. *Voice across the Sea*. London: Frederick Muller Ltd., 1958.

Colijn, H. *Neerlands Indie Land en Volk*. Amsterdam: Elsevier, 1912.

Conrad, Joseph. *An Outcast of the Islands*. London: T. Fisher Unwin, 1896.

Couperus, Louis. *The Hidden Force*. Amherst, Mass.: University of Massachusetts Press, 1985.

Cribb, Robert. *Historical Atlas of Indonesia*. London: Curzon Press, 2000.

Daum, P. A. *Ups and Downs of Life in the Indies*. Singapore: Periplus, 1999.

Daws, Gavin, and Marty Fujita. *Archipelago: The Islands of Indonesia*. Berkeley: University of California Press, 1999.

Decker, Robert, and Barbara Decker. *Volcanoes*. New York: W. H. Freeman, 1979.

De Vries, H. M., ed. *The Importance of Java Seen from the Air*. Batavia: H. M. De Vries, 1928.

Fairchild, David. *Garden Islands of the Great East*. New York: Scribner, 1943.

Forster, Harold. *Flowering Lotus: A View of Java in the 1950s*. London: Longman, Green & Co., 1958.

Friederich, Walter L. *Fire in the Sea. The Santorini Volcano: Natural History and the Legend of Atlantis*. Cambridge: Cambridge University Press, 2000.

Geertz, Clifford. *The Religion of Java*. New York: Free Press, 1960.

Gilbert J. S., and R. S. J. Sparks. *The Physics of Explosive Volcanic Eruptions*. London: Geological Society of London 1998.

Haigh, K. R. *Cableships and Submarine Cables*. London: Adlard Coles, 1968.

Hall, R., and D. J. Blundell, eds. *Tectonic Evolution of Southeast Asia*. London: Geological Society of London, 1996.

Hamilton, Warren. *Tectonics of the Indonesian Region* Washington, D.C.: U.S. Geological Survey, 1979.

Helsdingen, W. H. van, and Hoogenberk, Dr. H., *Mission Interrupted: The Dutch in the East Indies and Their Work in the Twentieth Century.* Amsterdam: Elsevier, 1945.

Heuken, Adolf, S. J. *Historical Sites of Jakarta.* Jakarta: Cipta Loka Caraka, 2000.

Hicks, Geoff, and Hamish Campbell. *Awesome Forces: The Natural Hazards that Threaten New Zealand.* Wellington, NZ: Te Papa Press, 1998.

Hillen, Ernest. *The Way of a Boy: A Memoir of Java.* London: Viking, 1993.

Hobhouse, Henry. *Seeds of Change: Five Plants That Transformed Mankind.* London: Sidgwick & Jackson, 1985.

Huntington, Samuel P. *The Clash of Civilizations and the Remaking of the World Order.* London: Simon & Schuster, 1997.

Johnson, George, ed. *The All Red Line: The Annals and Aims of the Pacific Cable Project.* Ottawa: James Hope & Sons, 1903.

Kartodirdjo, Sartono. *The Peasants' Revolt of Banten in 1888.* The Hague: Martinus Nijhoff, 1966.

Keay, John. *Empire's End: A History of the Far East from High Colonialism to Hong Kong.* New York, Scribner, 1997.

Kemp, P. H. vander. *De Administratie der Geldmiddelen van Nederland-Indie (The Financial Administration of the Dutch East Indies).* Amsterdam: J. H. de Bussy, 1881.

Keys, David. *Catastrophe: An Investigation into the Origins of the Modern World.* London: Century, 1999.

Kuitenbrouwer, Maarten. *The Netherlands and the Rise of Modern Imperialism.* Providence, R.I.: Berg Publishers, 1991.

Kumar, Ann. *The Diary of a Javanese Muslim: Religion, Politics and the Pesantren 1883–1886.* Canberra: Australian National University, 1985.

Krafft, Maurice. *Volcanoes.* New York: Harry N. Abrams, 1993.

Legge, J. D. *Indonesia.* New York: Prentice Hall, 1964.

Levelink, Jose. *Four Guided Walks through the Bogor Botanic Garden.* Bogor: Bogorindo Botanicus, 1996.

Lewis, Bernard. *The Middle East: 2,000 Years of History from the*

Rise of Christianity to the Present Day. London, Weidenfeld & Nicolson, 1995.

———. *What Went Wrong? The Clash between Islam and Modernity in the Middle East.* London: Weidenfeld & Nicolson, 2002.

Lucas, E. V. *A Wanderer in Holland.* London: Methuen & Co., 1905.

Merrillees, Scott. *Batavia in Nineteenth-Century Photographs.* London: Curzon Press, 2000.

Milton, Giles. *Nathaniel's Nutmeg: How One Man's Courage Changed the Course of History.* London: Hodder & Stoughton, 1999.

Money, J. W. B. *Java, or, How to Manage a Colony.* London: Hurst & Blackett, 1861.

Multatuli [Dekker, Eduard Douwes]. *Max Havelaar, or, The Coffee Auctions of a Dutch Trading Company.* London: Heinemann, 1967.

Naipaul, V. S. *Beyond Belief: Islamic Excursions among the Converted Peoples.* London: Little, Brown, 1998.

Netherlands Royal Mail Line. *Java the Wonderland.* Arnhem: [n.d.].

Nieuwenhuys, Rob. *Mirror of the Indies: A History of Dutch Colonial Literature.* Hong Kong: Periplus, 1999.

——— *Faded Portraits: E. Breton de Nijs* Amherst, MA, University of Massachusetts Press, 1982.

Oey, Eric, ed. *Java* Singapore: Periplus, 1997.

Oosterzee, Penny van, *When Worlds Collide: The Wallace Line.* Ithaca, N.Y.: Cornell University Press, 1997.

Oreskes, Naomi, ed. *Plate Tectonics.* Boulder, Colo.: Westview Press, 2001.

Ponder, H. W. *Javanese Panorama.* London: Seeley, Service & Co. [1942].

Poortenaar, Jan. *An Artist in the Tropics.* London: Sampson Low [1927].

Pope-Hennessy, James. *Verandah: Some Episodes in the Crown Colonies 1867–1889.* London: George Allen & Unwin, 1964.

Preger, W. *Dutch Administration in the Netherlands Indies*. Melbourne: F. W. Cheshire, 1944.

Quammen, David. *The Song of the Dodo: Island Biogeography in an Age of Extinctions*. New York: Scribner, 1996.

Raby, Peter. *Alfred Russel Wallace: A Life*. Princeton, N.J.: Princeton University Press, 2001.

Raffles, Sir Thomas Stamford Bingley. *The History of Java*. London: Black, Parbury & Allen, 1817.

Read, Donald, *The Power of News: The History of Reuters 1849–1989*. Oxford: Oxford University Press, 1992.

Reitsma, S. A. *Van Stockum's Travellers' Handbook for the Dutch East Indies*. The Hague: W. P. van Stockum & Son, 1930.

Ross, Robert, and George Winius. *All of One Company: The VOC in Historical Perspective*. Utrecht: HES Uitgivers, 1986.

SarDesai, D. R. *Southeast Asia, Past and Present*. Boulder, Colo.: Westview Press, 1989.

Scarth, Alwyn. *La Catastrophe: The Eruption of Mount Pelée*. Oxford: Oxford University Press, 2002.

——— *Vulcan's Fury: Man against the Volcano*. New Haven, Conn.: Yale University Press, 1999.

Schama, Simon. *Patriots and Liberators: Revolution in the Netherlands 1780–1813*. London: Collins, 1977.

Schoch, Robert M. *Voices of the Rocks: A Scientist Looks at Catastrophes and Ancient Civilizations*. New York: Harmony Books, 1999.

Scidmore, E. R. *Java, the Garden of the East*. New York: Century Co., 1897.

Severin, Timothy. *The Spice Islands Voyage: The Quest for Alfred Wallace, the Man Who Shared Darwin's Discovery of Evolution*. New York: Carroll & Graf, 1997.

Shepard, Jim. *Batting Against Castro*. New York: Alfred A. Knopf, 1996.

Shermer, Michael. *In Darwin's Shadow: The Life and Science of Alfred Russel Wallace*. New York: Oxford University Press, 2002.

Sigurdsson, Haraldur. *Melting the Earth: The History of Ideas on Volcanic Eruptions.* New York: Oxford University Press, 1999.

Sitwell, Sacheverell. *The Netherlands: A Study of Some Aspects of Art, Costume and Social Life.* London: B. T. Batsford, 1948.

Soebadio, Dr. Haryati, *et al.,* eds. *Indonesian Heritage Encyclopedia.* 10 vols. Singapore: Editions Didier Millet, 1996 et seq.

Standage, Tom. *The Victorian Internet: The Remarkable Story of the Telegraph and the Nineteenth Century's Online Pioneers.* London, Weidenfeld & Nicolson, 1998.

Stephens, Mitchell. *A History of News: From the Drum to the Satellite.* New York: Viking, 1988.

Suárez, Thomas. *Early Mapping of Southeast Asia.* Hong Kong: Periplus, 1999.

Taylor, Jean Gelman. *The Social World of Batavia.* Madison: University of Wisconsin Press, 1983.

Thornton, Ian. *Krakatau: The Destruction and Reassembly of an Island Ecosystem.* Cambridge, Mass.: Harvard University Press, 1996.

Turner, Peter, ed., *Java.* Melbourne: Lonely Planet, 1995.

Umbgrove, J. H. F. *Structural History of the East Indies.* Cambridge: Cambridge University Press, 1949.

Vlekke, Bernard M. *Nusantara: A History of the East Indian Archipelago.* Cambridge, Mass.: Harvard University Press, 1945.

———. *The Story of the Dutch East Indies.* Cambridge, Mass.: Harvard University Press, 1945.

Vandenbosch, Amry. *The Dutch East Indies, Its Government, Problems and Politics.* Berkeley: University of California Press, 1941.

Vissering, G. *Geweldige Natuurkrachten (Nature's Power).* Batavia: G. Kolff & Co., 1910.

Wallace, Alfred Russel. *The Malay Archipelago: The Land of the Orang-utan and the Bird of Paradise. Narrative Travel, with Studies of Man and Nature.* London: Macmillan, 1869.

Weyer, Robert van de. *Islam and the West: A New Political and Religious Order Post September 11.* Alvesford, Hants.: O Books, 2001.

Wilkinson-Latham, Robert J. *From Our Special Correspondent.* London: Hodder & Stoughton, 1979.

Williams, Stanley, and Fen Montaigne. *Surviving Galeras.* Boston: Houghton Mifflin, 2001.

Woodcock, George. *The British in the Far East.* New York: Atheneum, 1969.

Zebrowski, Ernest Jr. *The Last Days of St. Pierre.* Piscataway, N.J.: Rutgers University Press, 2002.

Zeilinger de Boer, Jelle, and Donald Sanders. *Volcanoes in Human History.* Princeton, N.J.: Princeton University Press, 2002.

Acknowledgments, *Erkenningen,*
Terima Kasih

To try to write an account of a long-ago explosion in a faraway place—far away, that is, when the volcano is to the west of Java and I live 9,944 miles from it, in western Massachusetts—requires a mule train of helpers to bridge the distance, and a congress of libraries and librarians to compress the years. To meet both requirements I was exceptionally fortunate in finding the most congenial and knowledgeable legion of able and interested helpmeets—some of them old friends, many more of them brand new—without whose enthusiasm, skills, and wisdom this story would have been much more difficult to relate. As I list them here, with gratitude and pleasure, I must nonetheless reassure each and every one that though all did their very best to guide, advise, and warn me, such errors and misjudgments that may appear in the book—and I would like to think there are few—are my fault alone, and should not, to any degree, be laid at their door.

In the early stages of my research I was lucky enough to be directed, by courtesy of Atlas, my Dutch publishers, to the delightful Alicia Schrikker, a graduate student of Dutch colonial history at the University of Leiden. Alicia, whose professional area of interest encompasses the Dutch possessions in the East Indies, Ceylon, and Japan, leaped at the chance of working on the story of Krakatoa, and in short order she led me to sheaves of long-forgotten papers and archives and introduced me to battalions of long-overlooked and little-known players in the drama, fast making herself my most indispensable ally in the entire endeavor. While I owe much to many, my debt to Alicia is singular and incalculable; I wish also to thank her partner, Job Weststrate, for his forbearance during the many days she worked with me, and for his later practical help when Alicia took off for her own studies in Colombo, Galle, and southern Sri Lanka. Dr. Wim van den Doel, a distinguished historian at Leiden, also gave readily of his advice. And I must record my personal gratitude to Hessel Stamhuis, a wise and kindly academic at the same university who came to stay with me in America: His death some months later was a sad shock.

In Indonesia, where I traveled several times, my old friend Toni Tack, whom I had met some years before when we were exploring little-known Hindu temples in central Java, was enormously kind and helpful, and accompanied me to Jakarta on several occasions, as well as to those coastal communities of west Java that were worst affected by the calamity of 1883. Toni is a game and courageous lady—and, unhappily, she needs to be even more so nowadays, since the island of Bali, her chosen home for the past many years, is no longer so peaceful a place as once we all supposed—but she wisely elected not to come with me to climb the scorchingly hot slopes of Anak Krakatoa on those occasions when I felt so moved. Instead it was left to the redoubtable and engagingly named Boing, a guide from Anjer, to haul me up the ever-growing mountain—a duty he performed with agility and eternally good spirits.

Also in Indonesia I was given much help and hospitality by the British ambassador, Richard Gozney; by Professor Sartono Kartodirdjo, the renowned historian at Jogjakarta's Gajah Mada University (who, in addition to offering his advice on the nature and meaning of the Banten Rebellion of 1888, also kindly gave me a pot of his wife's homemade Balinese nutmeg jam in return for the jar of TeaTogether's incomparable Lemon with Earl Grey Tea marmalade, which I sometimes use—as when turning up at teatime at Professor Kartodirdjo's house—as my calling card); by

Father Adolf Heuken, Jakarta's kindly (and Jesuit) urban historian; by George Benney and the staff of the Mandarin Oriental Hotel Jakarta; by Trina Ebert and the staff at the Amanjiwo resort, close by the temple of Borobudur; by the Jakarta-based writers Scott Merrillees and Mark Hanusz; by the Indonesian bookstore owner and Krakatoa enthusiast Richard Oh; by my longtime friend Hannah Postgate, whose fondness and enthusiasm for all things Javanese are inextinguishable and wonderfully infectious; and by Robert Hall of the University of London's South-East Asia Research Group, whose expertise in the arena of the tectonic evolution of the region that includes Java and Sumatra—and thus Krakatoa—is well-nigh unrivaled.

Others who gave generously of their particular knowledge of the geology and geophysics of the region include Professor John Dewey, now at the University of California, Davis, and formerly Professor of Geology at Oxford; Rob McCaffrey and David Wark at the Rensselaer Polytechnic Institute in Troy, New York; Charles Mandeville; Stephen Self; Vicky Bruce. John Rucklidge, who led the 1965 Oxford University Expedition to East Greenland, of which I was the most junior member, and who is now at the University of Toronto, was most helpful in reminding me of the scientific purpose and value of that remarkable adventure. Richard Fiske, of Washington's Smithsonian Institution and of the Global Volcanism Program—and with Tom Simkin author of the incomparable and near definitive 1983 Smithsonian Press study of Krakatoa—was also generous with his time and insights.

My friend Andrea Hsu, in Washington, D.C., was as helpful with answering the more arcane research questions for this project as she has been in connection with similarly obscure queries for the last three of my books, and I cannot fully express in print my deepest gratitude to her. At rather longer distances, Penny van Oosterzee, Robert Cribb, and Nicholas Pounder (in Australia), Eloise Van Niel (in Hawaii), Rob Whittaker, and Professor Ceri Peach (in Oxford) all gave valuable advice on matters that ranged from the dietary preferences of nineteenth-century Batavians to the story of the botanical repopulation of the ruined islands around Krakatoa. I would like also to acknowledge the help of Adrian Beeby of Lloyd's, Gina Douglas of the Linnean Society, Stephen Gillies, Amanda Green of the Foreign and Commonwealth Office Library, Ann Kumar, Laila Miletic-Vejzovic, Zyg Nilski, Margaret O'Clair, Vanessa Raizberg, and Paula Szuchman. The newly formed Gieben-Wulf Cultural Research Centre in London answered a single question that had long been nagging me, and they did so accurately, efficiently and very, very quickly—a trinity of attributes that I think all would agree augurs well for their success: I do wish Emma and Andrea well. My son Rupert Winchester performed yeoman service on my behalf at the London Library (whose staff were unfailingly helpful) and the Public Record Office, and in his dealings with Cable & Wireless (where Mary Godwin, he reports, was particularly nice).

Norah O'Donnell, a wonderful old friend at the Humanities Division of the University of Chicago, helped me by borrowing on her staff card, and *for two full years,* a pair of extremely rare and crucial books from the University's Joseph Regenstein Library. I owe the very tolerant librarians there a great deal for not asking for the volumes to be returned (though both are now quite safely back in their carrels).

Sophie Purdy kindly read and made most useful comments on the early drafts of the book: She deserves the very warmest of thanks.

In London I wish once again to record my deep gratitude to my agent and friend Bill Hamilton. My new editor at Viking, Mary Mount, handled my rather less than elegant first telling of this complicated story with consummate brilliance, managing first to spot, with an uncanny and instinctive ability, all the inconsistencies and structural flaws and infelicities that littered the submitted script, and then to iron them out and so save me, at the very least, from the critical mauling I am certain I at first deserved. Mary would not be surprised to learn that I truly cannot find words adequate to describe my pleasure in working with and for her: I hope that we may team up for many more fascinating projects like this in the years to come.

Mary Mount's assistant Julie Duffy also made major contributions, most notably when it came time to find the pictures and other illustrations with which to leaven the text: I am most grateful to her. Soun Vannithone drew the line illustrations once again, and as with the last book, he did so perfectly and exactly on time: Natacha du Pont du Bois coordinated with an almost terrifying efficiency what was the complicated and linguistically challenging task (since Soun is from Laos, and many of the pictures are from sources in Holland) of seeing that all the illustrated material was prepared properly and to a very demanding schedule.

Once again Donna Poppy, surely the ablest of all the world's copy editors, took on with an unstinting good cheer the monumental business of keeping the devil away from the details in a book that ranged through unfamiliar complexities of science and history, religion, and sociology, where there were litanies of names in Dutch and Javanese that all needed checking, checking, checking. To Donna and to her colleague in New York, Sue Llewellyn: a thousand thanks and *stet!* to all you do.

And in New York I owe eternal thanks to my robustly splendid agent Peter Matson, and to his colleagues Saskia Cornes and Jim Rutman; and to Krista Stroever, a most wonderful editor.

Finally: If there was ever to be a *primus inter* all these distinguished *pares* above—and I realize there should not be, since all who helped should be equally blessed—then my vote would have to be for that most deservedly legendary of New York publishers, and a valued friend for the past six years, Larry Ashmead. Larry was the inspiration, the champion, and the encourager in chief: I only hope he finds that this slender book is worthy of all his many hopes he once entertained for it.

Index

Page numbers in *italics* refer to illustrations.

Aardenburgh, 132
Aaron, Eugene Murray, 260–61
Abell (Dutch official), 250–51
Actaea, 155, 159
Ade, Ebiet G., 37
"Admiralty Sailing Directions," 9
aeolian plankton, 352–53
Africa Plate, 110
Agamemnon, 185
air (shock) waves, 232–33, 247, 264–73
Alaric, king of the Visigoths, 11
Alaska, 306
Aleutian Islands, 306, 308
Alexander VI, Pope, 13*n*
Algemeen Dagblad, 199
Alpena (Mich.) *Evening Echo,* 177
Altheer, Mr., 164–65, 252
American Philosophical Society, 74
Anak Krakatoa, *363,* 368–84, *377*
  author's visit to, 373–84
  birth of, 3–4, 340–46, *343, 344*
  crater of, 381–82
  lava flows on, *363,* 376–78
  life on, *see* life, posteruption
  naming of, 345
  seismic monitoring of, 369–72
Anaxagoras, 301
Ancient Krakatoa, 116–18, *117,* 313
Andaman Islands, 262, 263
andesite, 118, 294, 305*n,* 309, 341

Andes Mountains, 305–6
andosols, 294
Anjer, Java, 166–67, 179–80
  in 1883 eruption, 213–14, 225–26, 233, 238, 248, 258
  population of, 210–11
  struck by tsunami, 256–57, 275
  swallows of, 153–54
Annerley, 231, 234
Arabism, 326
Archer, 155–56
Aristotle, 301, 365
A. R. Thomas, 155
Ascroft, William, 282, 287, 288
Asian Plate, 312, *315*
ASQ-3A fluxgate magnetometer, 91–92
asthenosphere, 107
Atlantic Monthly, 219*n*
Auden, W. H., vii
Australia, 53, 64, 111, 113, 143, 262
Australian Frozen Meat Company, 200
Australian Oceanic Plate, 312, *315*
Azores, 303

Backer, Cornelis Andries, 358–61
Bacon, Francis, 70*n*
Bahamas, 285
Baird, A. W., 273–74
Bali, 54, 65, 134, 338
ballooning spider, 352–53, *352*

Banda, 22
Banks, Joseph, 118–19
Banten, Java, 15, 16, 20–21, 28, 33,
    125, 317, 327
Banten, sultan of, 17, 30, 33–35, 40
Banten Peasants' Revolt (1888), 319,
    330–36
  Arabian influence on, 333–34
  attacks in, 335–36
  imperial system and, 322–26
  Karim's teachings and, 330–33
  militant Islam and, 319–21, 337
  planning of, 334–35
barographs, 265–67, 268, 272
basalt, 82, 294, 303, 310, 312–13,
    314, 378
Batavia, 231
Batavia, Dutch East Indies:
  "barring" practice of, 151
  Buitenzorg hill station of, 141–44,
    199, 223–24
  colonial life in, 133–34
  early building of, 38–40
  in 1883 eruption, 213–15, 232–37,
    247, 263, 274–75, 290
  establishment of, 32–36
  first settlers of, 38–39
  gasworks of, 215–18, *216,* 233, 247
  Great World Circus visit to, 197–99,
    202–4, 207
  naming of, 37–38
  population of, 41–45, 141
  in prelude to 1883 eruption,
    136–47, *145*
  Reuter office in, 193–94
  slave population of, 43–45
  technological innovations in, 144
  telegraph cable connected to, 187
  tide-meter of, *251,* 252, 274–75
  Town Hall of, 46
  VOC rule in, 46–47, 133
  Weltevreden suburb of, 139–40
  see also Jakarta, Indonesia
Batavia Plant and Animal Garden,
    199–200
Bates, Henry Walter, 57

Batting Against Castro (Shepard), 149
Bay of Naples, 231, 296–97
Beagle, HMS, 59
Behaim, Martin, 21–22
Belgium, 29*n*
bentonite, 294
Berbice, 222–24, 296
Berkowitz, Rickey, 373
Berouw, 219, 230, 249, 253–56, *254*
Beyerinck, Mrs., 163–64, 226–30, 249
Beyerinck, Willem, 154–55, 226–30,
    244–45, 246
Billiton Tin Company, 146
Bintaing, 156, 243
biogeography, 52–54, 55
biosphere, 299
biota, 53
bird's nest soup, 20–21
Bishop, Sereno, 282*n*, 288
Bishop's Rings, 282, 286
black smokers, 361*n*
Blair, Lawrence, 388
Blair, Lorne, 388
Blavatsky, Helena, 52*n*
Blundell, D. J., 51
Bode, 160, 253
Boing (guide), 374, 379–82
Bonney, Thomas, 270*n*
Book of Kings (Ranggawarsita),
    121–25, 128
Borjild, 231
Borneo, 21, 22, 65, 130, 134
Bo'sun's Rock (Bootsmans Rots), 341,
    342, 376, 380
Bothwell Castle, 296
Brandende Eiland, Het ("The Burning
    Island") (van Schley), 136, *137*
Brani, 282
Brazil, 13–14, 69–70, 146, 223
breeze block, 294
Brewer, W., 193–94
*Brewer's Dictionary of Phrase and
    Fable,* 197
Bristowe, William Syer, 356, 362
British-Australian Telegraph Company,
    186–87

British Empire, 282
British Ornithologists' Club, 52
Bromo, Mount, 47, 300
Brunhes, Jean, 94
Buffon, Comte de, 70*n*
Buijs, Thomas, 210*n*
Buitenzorg hill station, 141–44, 199
  Botanical Gardens of, 223–24
Bullard, Edward, 88
Burma, 33*n*
Burney, Mount, 306
Byron, George Gordon, Lord, 293

Cabral, Pedro Alvares, 13–14
Cailendra Dynasty, 121, 132
Calmeyer Island, 309, 341*n*
Cambodia, 33*n*
Camellia 1 (Ade), 37
Cameron, Alexander Patrick, 151,
  257, 258*n*, 269
  1883 eruption report of, 234–39,
  235
Canada, 29–30, 91, 287
Cape Colony, 33*n*
carbon dioxide, 242, 299
cartography, *see* maps
Catastrophe (Keys), 130–31, 388
Cavendish, Thomas, 34
Cayman Brac, 263
Celebes, 22, 64–65
Cerro Hudson, 306
Ceylon, 21, 22, 33*n*, 262, 285
  struck by tsunami, 276–77
Ceylon Observer, 276
Chacana, 306
Chamberlin, Thomas, 74
Champa Kingdom, 126
Charles Bal, 218–22, 245, 296
Chicken Market Bridge (Hoender-
  pasarbrug), 39–40
Chile, 13*n*, 286, 305
China, 128–30, 156*n*
Christmas Island, 111
Church, Frederic Edwin, 280–82
cinnamon, 18, 325
Ciparis, Louis Auguste, 242*n*

Civil War, U.S., 193
climate:
  1883 eruption and, 6–7, 291–95
  sixth-century eruption and, 130–31
Cloos, Hans, 73–74
Clothilde, 43
clove, 10, *10,* 18, 295*n*, 325
cock's tail jets, 346
Cocos (Keeling) Island, 111
Coen, Jan Pieterszoon, 32–36, *33,* 39,
  41–42, 133, 139
Cold War, 104, 264
Colombia, 242, 306
Columbia University, 105
communications technology, 5–6,
  142–44
  global village and, 6, 182, 195
  news agencies and, 190–93
  telegraph, *see* telegraph
Compagnie van Verre, 15
Concordia Military Club, 151, 170,
  200–202, *201*
Congo, 146
Conrad, 156, 166
Conrad, Joseph, 12–13
conservative plate boundary, 110
continental drift, 51–114, 168, 301*n*
  convection process in, 88–89, *89,*
  95–96, 106–7, 299
  gravity anomalies and, 86–88
  Hawaiian hot spot and, 99–102, 105
  and magnetism in Greenland basalt,
  82–86, 84*n*
  in mountain formation, 109–10
  plate tectonic theory and, *see* plate
  tectonics
  remanent magnetism and, 89–95, *93*
  sea floor spreading and, 86, *93,*
  94–95, 103–4, 113, 168
  stabilists vs. mobilists on, 105
  subduction zones and, *see* subduction zones
  transform faults and, 103–5, *104*
  Wallace Line and, 63–67, 113–14
  Wegener's concept of, 69–73,
  88–89, 95

Cook, James, 118–19, *120,* 349
corposants (St. Elmo's fire), 221
Cotopaxi, 306
Cotteau, Edmond, 350–51, 352, 355
country trading, 30–31
Cressonnier, M. Louis, 206
Cretaceous Quiet Zone, 94
Crete, 243
Cribb, Robert, 28*n*
cricket, 151
cultivation system (*Kultuurstelsel*),
    324–25, 329
Curie, Pierre, 90*n*
Curie point, 83, 90*n*

Daendels, Herman Willem, 138–39,
    140, 142
da Gama, Vasco, 13
Dalby, R. J., 173
Dammerman, Karel, 361, 362*n*
Danan, 118, 157*n*, 174, 309, 313,
    341
"Darkness" (Byron), 293
Darwin, Charles, 56, 57–62, *59,* 68,
    273
Darwin, George, 273, 302*n*
Day, Arthur Louis, 301–2
"Deep-Sea Cables, The" (Kipling),
    188–89
Defense Department, U.S., 105
de Houtman, Cornelis, 15–16, 17, 19
de Houtman, Frederik, 15
Dekker, Eduard Douwes (Multatuli),
    322–25
Descartes, René, 302
de Vries, Mr., 213
Dewey, John, 102–3, 104
Dickinson, William, 105*n*
Diego Garcia, 52, 261
Discovery, 119
Djaman, Umar, 319
Dowd, Charles, 218*n*
Drake, Francis, 34
Dumas, Mrs., 336
Dumas (cleric), 335
Durant, Will, 298

Dutch East Indies, 28–29, 319, 320
    Chinese in, 42–43
    Christian-Muslim relations in,
        31–32, 40–41
    colonial system in, 322–26
    cultivation system of, 324–25, 329
    Napoleonic Wars and, 138–39
    peasant revolt in, *see* Banten Peas-
        ants' Revolt
    regional headquarters of, 32–34
    slavery in, 42–45
    see also specific islands

earthquakes, 110–11, 153, 205, 241
East Africa, rift valleys of, 303
Easter Island, 305
Eastern Telegraph Company, 187
East Pacific Rise, 305
Ecuador, 306
Edinburgh, 187
Eerste Scheepvart (First Ship Sailing)
    expedition, 15–18
Egeron, 168–69
Elisabeth, 156, 181, 212
Emden, 188*n*
Encyclopedia of Volcanoes (Sigurdsson,
    ed.) 294, 390
English Channel, 273, 278
Equatorial Smoke Stream, 288
*Eruption of Krakatoa and Subsequent
    Phenomena* (Royal Society), 389
Ethical Policy, 325
ethogeological prediction, 205
Etna, Mount, 110, 297, 301
Eurasian Plate, 110
Evelina, 277
evolution, theory of, 56, 58–62
Ewing, Maurice, 88

Faircloth, Captain, 285
Farina, Johann Maria, 197
Ferzenaar, H. J. G., 174–76, *175,* 204
Fiado, 200
field geology, science of, 302
First Ship Sailing (*Eerste Scheepvart*)
    expedition, 15–18

Fiske, Richard S., 389
Flint, Earl, 285
Flores, 29
Foley, Mr., 263
Forbes, H. O., 172–73
Forbes (plant collector), 166
Formosa, 33*n*
For the Time Being (Auden), vii
Fort Speelwijk, 20
Fourth Pacific Science Congress
    (1928), 346
Fourth Point Lighthouse, 234,
    256–57, 258, 316
France, 138*n*, 329
    tsunami's effects on, 278
Frankenstein (Shelley), 293
Franklin, Benjamin, 291
Freuchen, Peter, 78
Fukoto Kuokanaba, 378*n*
fumaroles, 174
Furneaux, Rupert, 311*n*, 388–89

Galeras, 306
Galileo Galilei, 59, 76
Galton, Francis, 62
gasometers, 215–18, *216*
Geikie, Archibald, 270*n*
Genesis, Book of, 68
Gentlemen Seventeen, 30, 32
Geography (Ptolemy), 21
Geological Society of London, 301, 311
Geological Survey, U.S., 369
Geological Survey of Indonesia, 369
Gérard, Max, 339
Germany, 156*n*
Gibbon, Edward, 11
glaciers, 242
Gladstone, William, 236
global village, 6, 182, 195
Gogh, Vincent van, 39, 325
Gondwanaland, 70*n*–71*n*, 72–73,
    88–89
Gouverneur-Generaal Loudon,
    170–72, 212, 214, 218, 227–28,
    230–31, 233, 245, 249, 253, 296,
    332

Granville, Earl, *235,* 236
gravity, 86–87
Great Britain, 12, 34–35, 90, 261*n*,
    267–68, 287, 321*n*, 329
    in Napoleonic Wars, 138–39
    trading cooperatives of, 29–30
Great Northern Telegraph Company,
    193
Great Plague, 130
Great World Circus, 197–99, *198,*
    202–4, 207, 316
Greenland, 71, 75, 76, 77–86, *79,* 306*n*
    author's expedition to, 77–81
    geologic origin of, 81–82
    magnetic basalt of, 82–86
Greenwich Observatory, 272
Gubbels, Dora, 335–36
Gubbels, Elly, 335
Gubbels, Johan Hendrik, 335–36
Guild of Pepperers, 12
gutta-percha, 185–86

Haag, 156
Hades, entry to, 300–301
Hadhramaut, 334
Hall, R., 51
Hamburg, Mr., 172
Handl, Johann, 355
Han Dynasty, 10
Harmonie Club, 151, 170, 200
Hatfield, Oscar, 151, 233
Haughton, Mr., 285
Hawaiian Islands, 99–106, *104,* 303
Heims, Father, 157
Hekla (Lakagígar), 291–92
Helen, 57
Hess, Harry, 88–89, 91, 95, 96–97,
    98
Hesse, Elias, 49–50, 132
Hibernia, 186, 187
Himalayas, 110
Hinduism, 126, 327
History of Java (Raffles), 139*n*
History of Ocean Basins (Hess), 96*n*
Hoenderpasarbrug (Chicken Market
    Bridge), 39–40

Hollmann, Captain, 156–57
Holotan, 129
Holtum, John, 203–4
Homo erectus (Java Man), 113
Hong Kong, 275
Hooker, Joseph, 61, 62
Hope, 173
Hopkins, Gerard Manley, 285
Huaynaputina, 306
Hudson's Bay Company, 29–30
Hurgronje, Snouck, 40–41, 329
Hutton, James, 68
Huxley, Thomas, 62
hydrochloric acid, 242

Iceland, 94, 291–92, 303
Illustrated London News, 154*n*
Imperial Beacons & Coastal Lighting
    Service, 168
India, 13, 18, 33*n*, 40, 110, 277, 320,
    321
  struck by tsunamis, 273–74
India Rubber, Gutta Percha & Tele-
    graph Works Company, 185–86,
    194
Indo-Australian Plate, 113
Indonesia, 38, 113, 142, 152, 305, 319
  Islam in, 326–29, 337
  rain forests of, 134–36
  volcanoes of, 306–7
  see also Dutch East Indies; *specific
    islands*
Indonesia Pilot, 9
International Date Line, 218*n*
International Meridian Conference,
    218*n*
Io (moon of Jupiter), 299–300
Iran, 326
Irian, 54
Islam, 130, 316
  Arabism and, 326–27
  Dutch and, 31–32, 145, 316; *see also*
    Banten Peasants' Revolt
  Hajj pilgrimage in, 328–30
  in Java, 16–17, 40–41, 326–29, 337
  militant movement in, 319–21

Italy, 241
Itinerario (Linschoten), 23–26, 27

Jacob, Frederik s', *146,* 147, 149,
    150–51, 169, 199
Jacob, Leonie s', 150
Jakarta, Indonesia, 38, 134
  see also Batavia, Dutch East Indies
"Jakarta 1" (Ade), 37
James I, king of England, 13
Japan, 33*n*, 42, 193, 243, 305, 306
Java, 2, 20, 54, 86, 112, 113, 124,
    127, 152, 153, 238, 241, 248, 365
  active volcanoes of, 306–7, 321–22
  British and, 34–35
  Cailendra Dynasty of, 121
  Chinese in, 42–43
  Dutch regional headquarters in,
    32–34
  in early maps, 21–22, 24, 27
  first Dutch expedition to, 15–18
  geology of, 81–82, 314–15, *315*
  histories of, 125–26
  Islam in, 16–17, 40–41, 326–29,
    337
  mysticism in, 321–22
  spice trade in, 9–14
  technological innovations in, 142–43
  see also Dutch East Indies
Java Man (*Homo erectus*), 113
Javasche Courant, 151–52
Java Trench, 87, 109, 111–12
Jayabaya (Javanese monarch), 126
Jeffreys, Harold, 74–75, 301,
    301*n*–2*n*
John of Ephesus, 115
jökulhlaups, 242
Judd, John, 311
Julius II, Pope, 13*n*
Jupiter, 299–300

Kamchatka Peninsula, 306
Karim, Abdul, *328,* 330–33, 336
Kartodirdjo, Sartono, 317
Katmai (Novarupta), Mount, 4–5, 308
Kavachi, 378*n*

Kedirie, 296, 308

Keeling (Cocos) Island, 111

Kennedy, Henry George, 235–36, 269

Ketimbang, Sumatra, 163, 226, 230, 244–45, 257

Keys, David, 130, 388

King of the Netherlands, 318

Kipling, Rudyard, 188–89

Knossos, 243

Kosrae, 295

Kowalski, Bernard, 387

"Krakatau" (Shepard), 149, 209

Krakatau (Verbeek), 389

Krakatau 1883 (Simkin and Fiske), 389

Krakatoa:

Cook's visits to, 118–19

early settlers on, 118–20

in eruption of 1680, 47–50, 132–36

geologic evolution of, 116–18, *117*

jungles of, *120*

main island of, 118

maps of, 21, 24–28, *25,* 174–76, *175, 229*

name of, 21, 27–28, 123*n,* 183*n,* 236*n*

new island of, *see* Anak Krakatoa

remains of, 1–3

sixth-century eruption of, 130–31

summits of, 118–19, 157*n*

written history of, 121–23

see also Ancient Krakatoa

Krakatoa, East of Java (film), 2, 387–88

Krakatoa, 1883 eruption of:

aftermath of, 308–9, 316

air (shock) waves in, 232–33, 247, 264–72

ash and dust fall in, 157–59, 166, 171, 205, 221–22, 223, 233, 264, 282–83

atmospheric dust in, 280–83, 286–91

Beyerinck family escape from, 226–30

climate affected by, 6–7, 290–95

clouds and smoke of, 157–59, 160, 173–74, 204–5, 212–14, 231, 234

culminating explosions of, 233–39, 246–47

darkness in, 220–21, 225–26, 233, 237

death toll of, 240, 243

electric displays in, 214–15, 221, 222–23

film based on, 2, 387–88

first tremblings and vibrations of, 153–55, 160–64, 167

gasometer pressure readings in, 215–18, *216,* 247

geographic siting of, 314–15, *315*

geologic explanation of, 298–300, 311–15, *312, 315*

inspections and visits after, 154–55, 160–65, 167–71, 173–76

Karim's prediction of, 331–33, 334

Lloyd's report on, 166–67

map of, 174–76, *175*

new islands formed by, 309

in news, 177–78, 182–83, 189–93; *see also* communications technology

painting inspired by, 280–82

poetry inspired by, 284–86, 293

as political catalyst, 319–22

prelude to, 136–47, 209–10

public attitudes toward, 162

pumice fall in, 218–20, 228, 232, 234, 293–96, *296*

quiet period in, 167, 172–74, 199

religious explanation of, 300–301

renewed activity in, 204–5

reports of, 179–84, 186–87, 189–90

sea disturbances caused by, 213, 221, 226–28, 233–34, 237, 240*n,* 245; *see also* tsunamis

sounds of, 211–12, 220–22, 224, 233, 236–37, 244, 247, 259–64, *271*

sullen political mood after, 317–19, 337

sunsets and afterglows after, 280–82

VEI of, 308

Krakatoa (Furneaux), 311*n*, 388–89
Krakatoa Committee, 270, 270*n*, 273, 274, 284
Krakatoa problem, 357–60
Krakatoa Volcanic Observatory, 369
Kramat, 258
Kultuurstelsel (Dutch East Indies cultivation system), 324–25, 329
Kurile Islands, 306

lahars, 242
Lakagígar (Hekla), 291–92
Lamongan, 154
Lancaster, James, 34
Lang Island, 157*n*, 309, 313, 348
Laos, 33*n*
Lascar, 306
Laurasia, 72–73, *72*
lava flows, 241–42, 310
  on Anak Krakatoa, 363, 376–78
Leed en Lief in Banten (van Sandick), 332–33
Lemuria, 52
life, posteruption, 347–66
  biological equilibrium in, 355
  first stirrings of, 350–52, *354*, 361–62
  Krakatoa problem and, 357–60
  lava flows and, 363
  pest species in, 355
  preeruption life compared with, 348–49
  windborne arrival of, 351–52
Lincoln, Abraham, 193, 218*n*
Lindeman, T. H., 171–72, 214, 218, 230–31
Linnean Society, 51, 53, 61, 63, 64
Linschoten, Jan Huyghen van, 23–25
lithosphere, 107, 299
Little Prince, The (Saint-Exupéry), 1
Llaima, 306
Lloyd's, Society of, 178–80, 189–90, 194
  agency system of, 178–79
  Java agent of, 179–80
Lochart, Nanette, 205–6

Locomotive, 150
Lodewijcksz, Willem, 26
Logan, William, 222–24
Lombok, 65
Lord Jim (Conrad), 12–13
Loudon, see Gouverneur-Generaal Loudon
Lourenço de Brito, Dom, 18
Lyell, Charles, 61, 62, 68

Macao, 18–19
Macassar, 263, 320
McColl, Mr., 179, 257–58
mace (spice), 18
MacKenzie, Captain, 155, 160
McLuhan, Marshall, 182, 195
Madagascar, 52
Magellan, Ferdinand, 22
Magnet, Project, 91
magnetism, 89–95, *93*
magnetite, 82–83, *83*
Magpie, HMS, 262, 269
Malacca, 28, 33
Malay Archipelago, The (Wallace), 61
Malay Peninsula, 21, 22, 28, 30, 40, 52
Malay States, 143
Manchester Literary and Philosophical Society, 291
Manley, W. R., 285
maps, 21–28
  of Krakatoa, 21, 24–28, *25*, 174–76, *175, 229*
  by Linschoten, 23–25, *25*
  by Waghenaer, 26–27
Mardijkers, 43
Marie, 218, 230, 233–34, 245
Martinique, 242
Mason, Ron, 91–92
Massachusetts Bay Company, 29
Matuyama, Motonari, 94
Mauk, 258
Maurice of Nassau, 16*n*
Mauritius, 33*n*, 259, 261*n*, 277
Max Havelaar (Multatuli), 322–25, *323*

Mayan civilization, 130
Mayon, 264
Medea, 214–15, 231
Merak, 245, 248, 250–51, 257, 258, 331
Merapi, 47
Merbapu, 47
Merchant Adventurers, 29
Merchant Staplers, 29
meteorology, 287
Mexico, 13*n,* 241
Michael the Syrian, 115
Minoan civilization, 243
Minto, Lord, 139*n*
Moluccas, 22, 28, 34, 60, 320
monitor lizard, 383–84, *383*
Morse, Samuel, 142, *143,* 194
mountains, formation of, 110
Mountain Script, 125
Multatuli (Eduard Douwes Dekker), 322–25
Muscovy Company, 29

Nansen, Fridtjof, 78
Napoleon I, Emperor of France, 138, 191
Napoleonic Wars, 138, 259
Naraddha, 126
Natural History (Buffon), 70*n*
Natural History (Pliny), 11
Natural History and Antiquities of Sel bourne (White), 292*n*
natural selection, 57–58
Nature, 99, 102, 108, 285
Navy, U.S., 91, 104
Nazca Plate, 305
Neale, Philip, 170, 240, 274
Neck, Jacob van, 19
Neptunists, 302
Netherlands, 29*n*
  Batavi tribe of, 37–38
  early cartographers of, 21–28
  first Indies expedition of, 15–18
  Krakatoa eruption investigation of, 308–11
  in Napoleonic Wars, 138

in spice trade, 13*n,,* 14, 15–16, 19, 138
  trading cooperatives of, 29–30
  see also Dutch East Indies; VOC
Netherlands Indies Steamship Company, 170
Netherlands Royal Mail Line, 170*n*
"New Class of Faults and Their Bearing on Continental Drift, A" (Wilson), 102
New Guinea, 13*n,* 262, 305, 306
news agencies, 190–93
Newton, Isaac, 59, 301
New York Times, 288
New York *World,* 288
New Zealand, 64, 305
Niagara (Church), 280
Nicaragua, 305
Nieuwe Rotterdamsche Courant, 194
Nieuw-Middelburgh, 49–50
Norham Castle, 231, 234
North American Plate, 110
"Notes on a Series of Barometrical Disturbances which Passed over Europe between the 27th and 31st of August, 1883" (Scott and Strachey), 268–69
Novarupta (Katmai), Mount, 4–5, 308
Nova Scotian, SS, 193
nutmeg, 10, 18, 325
Nyiragongo, Mount, 303*n*

obsidian, 168
"On Coleopterous Insects Frequenting Damp Places" (Bates), 57
Onrust, 258
"On the General Geographical Distribution of the Members of the Class Aves" (Sclater), 53
"On the Law which has Regulated the Introduction of New Species" (Wallace), 60
On the Origin of Species (Darwin), 57, 61
Orang Alijeh, 47–48, 162, 321

Origin of the Continents and Oceans,
    The (Wegener), 69
Oxford University Geological Society,
    96–97

Pacific Ocean, subduction zones of,
    305–6
Pacific Plate, 110, 305
paleomagnetism, 85, 96
Pangaea, *72, 73,* 86
Panjang Island, 116, 119, 347, 348,
    356, 360, 368, 380–81
Papal Donation of 1493, 13
Paris, Treaty of (1872), 138
Parliament, British, 178
Peasants' Revolt of Banten in 1888, The
    (Kartodirdjo), 317
Pechler (tsunami survivor), 250
Pele, 100–101
Pelée, Mount, 242, 263
Pène du Bois, William, 385
pepper, 10–13, *12,* 19, 20, 325
Pepperers Guild, 12
Perboewatan, xi, 118, 157*n,* 165, 166,
    167, 168, 171, 172, 173, 309, 313
Perry, Matthew, 193
Peru, 306
Petroeschevsky, W. A., 345, 347
Philippines, 13*n,* 22, 42, 64, 241,
    305, 306
phreatomagmatic eruption, 310, 346
Pinatubo, 264, 297, 341*n*
Pioneer, USS, 91, 92, 95, 104
plague, 130
plate tectonics, 63, 67, 301*n,* 303–4,
    304
    as explanation for Krakatoa explo-
        sion, 311–13, *312*
    mountain formation and, 110
    plate collisions and, 109–11
    subduction zones and, 107–9, *108,*
        110, 111, 112–13
    theory of, 98–106
    volcanoes and, 299–300, 305–6
Plinian eruption, 12
Pliny the Elder, 11–12

Pluto, 300
Plutonists, 302
Plymouth Company, 29
Poeloe Temposa, 237
Polish Hat, 116*n,* 118, 309
Polo, Marco, 21
Portugal, 16
    in spice trade, 13–15, 18–19
Poughkeepsie (N.Y.) *Daily Eagle,*
    288–89
Poughkeepsie (N.Y.) *Sunday Courier,*
    290
Princess Clementine, 188
Prinses Marie, 174
Prins Frederik, 231
Ptolemy, 21
Public Record Office, London,
    234–35
Pulau Anakrakata, 9
Pulau Sangiang, 159*n*
pumice, 218–20, 228, 232, 234, 241,
    293–96, *296*
P'u-tei, 129
pyroclastic flows, 241–42

radioactivity, 106
radioseismograph, 370–71
Raffles, Lady, 139*n*
Raffles, Thomas Stamford, 139–40,
    142
Raja Basa, 163
Rakata, xi, 118, 157*n,* 171, 173–74,
    227, 341, 368, 374–76, 381
    after 1873 explosion, 308–9
    recovery of life on, 348–62, *354, 363*
Ranggawarsita, Raden Ngabahi,
    121–27, 131–32
Rasmussen, Knud, 78
Rees, W. A. van, 144
remanent magnetism, 89–95, *93*
    Mason's study of, 91–92
    symmetry of, 94–95
Resolution, 119, 349
Reuter, Julius, 191–93, 194
Reuter's (news agency), 191–93, *192,*
    194

rift valleys, 303
Ring of Fire (television documentary),
   388
Rochester (N.Y.) *Democrat and
   Chronicle*, 289
Rodriguez Island, 259–61, *271*
Roe, Thomas, 19
Roman Empire, 11, 130
Roorda van Eysinga, P. P., 141
Ross, Captain, 156, 183
Royal Dutch Oil, 146
Royal Mint, British, 90
Royal Navy, British, 34, *344*, 347
Royal Society, 62, 236, 248*n*, 389
   Krakatoa eruption investigated by,
      268–73, 284–85, 286, 291, 295
   tsunamis investigated by, 274–78
rubber, 223–24, *224*
Ruiz, 242, 306
Runcorn, Keith, 90–91, 96, 97–98
Russia, 305

St. Elmo's fire (corposants), 221
Saint-Exupéry, Antoine de, 1
St. Helens, Mount, 264, 297,
   385–86
St. Pierre, Martinique, 242
St Telemachus (Tennyson), 284
Salwatty Island, 262
Samarang, 156, 159, 243
Samoa, 295–96
Sampson, Captain, 234
San Andreas Fault, 110–11
Sandick, N. M. van, 254–55
Sandick, R. A. van, 249, 332
San Francisco Earthquake of 1906,
   111
Sangay, 306
Santorini, 243
Schley, Jan van, 136
Schouten, Wouter, 27
Schröder (philanthropist), 199–200
Schruit, Mr. (telegraph master),
   211–13, 225, 244–45
Schuit (Lloyd's agent), 166–67,
   179–84, 186–87, 189–90, 211–12

Schuurman, A. L., 169–70
Schwartz, Judy, 373
Schweitzer, Christopher, 46–47
Sclater, Philip Lutley, 51–54, 55, 64,
   72
Scoresby Sund, 78
Scotia, 283
Scotia Arc, 109
Scott, Robert, 268
sea floor spreading, 86, *93*, 94–95,
   103–4, 113, 168
Sebesi, 164, 165
Sebuku, 165
Seneca, Lucius, 301
Sepoy Mutiny, 321*n*
Sertung, 116, 348, 356, 360, *363*,
   368, 381
Seuss, Eduard, 70*n*–71*n*, 72
Shelley, Mary, 293
Shepard, Jim, 149, 209
shock (air) waves, 232–33, 247,
   264–73
Sigurdsson, Haraldur, 131, 390
Sikin, 370, 371–72, 379
Simkin, Tom, 389
Singapore, 143, 156, 166, 184, 187,
   232, 263, 275
Sir Robert Sale, 231
Si-tiao, 129
slavery, 42–45
Smith, William, 68
Smithsonian Institution, 284*n*, 307
Snider-Pellegrini, Antonio, 70*n*
Solferino, battle of, 191
Sorrow and Love in Banten (van
   Sandick), 332–33
South Africa, 277
South American Plate, 305
South Georgia, 277
Spaan, Mr. van, 210*n*
Spain, 13*n*, 14, 29*n*, 320
Speelwijk, Fort, 20
Speenhoff, Mr., 46
Spice Islands, 30, 295*n*
   see also Moluccas
Spice Route, 11, 13*n*

spice trade:
country trading practice in, 30–31
Netherlands in, 13*n*, 14, 15–16, 19, 138
Portugal in, 13–15, 18–19
reform of, 325
Tordesillas Line and, 13*n*, 14–15, *14*
spider, ballooning, 352–53, *352*
Standage, Tom, 177
Steers Island, 309, 341*n*
Sterling, Edward, 190*n*
Stok, J. P. van der, 160–62, 215
Stok, Mrs. van der, 160
Stokes, George, 270*n*
Strachey, Richard, 268, 270*n*
stratosphere, 283
Sturdy, E. W., 219*n*
subduction zones, 107–13, 168
in explanation of Krakatoa explosion, 311–14, *312, 315*
in Indonesia, 306–7
magma mixing in, 313–14
in Pacific Ocean, 305–6
plate tectonics and, 107–13, *108*
volcanoes and, 110, 112–13, 304–5
Sudan, 331*n*
Suez Canal, 141, 181
Sufi movement, 330
Suharto, 142*n*
Sukarno, 142, 374
Sukarnoputri, Megawati, 370
Sulawesi, 24, 54, 64–65, 134
sulfur dioxide gas, 242
Sumatra, 16, 20, 30, 53, 112, 113, 124, 127, 134, 153, 167, 205, 233, 238, 241, 243, 312, 349, 365
in early maps, 21–22, 24
geology of, 314–15, *315*
Islam in, 16–17, 40–41, 337
Sunda, 155, 166
Sunda Kelapa, 134
Sunda Strait, 3, 22, 50, 155, 180, 275, 361*n*, 373
in early maps, 24, 27
network of faults in, 314–15, *315*

posteruption state of, 339–40
speed of tsunamis in, 247, 248*n*
Sunset over the Ice on Chaumont Bay, Lake Ontario (Church), 281–82, *281*
Surapati (gang leader), 44
Surtsey, *377,* 378*n*
survival of the fittest, 60–61
swiftlets, 20–21
S. W. Silver & Company, 185
Sydney Morning Herald, 232
Symons, G. J., 270

Tachard, Guy, 27
Tambora, Mount, 4–5, 47, 243, 280*n*, 292, 306*n*, 307–8, 311
Taupo, Mount, 4–5, 308
Taylor, Frank, 71*n*
Tectonic Evolution of Southeast Asia (Hall and Blundell), 51
telegraph, 194, 269
and discovery of gutta-percha, 185–87
electric, 181–82, 184
first, 184–85
routes of, 186–89, *192*
submarine, 182, 184–85, 188–89
Telok Betong, 230, 234, 245–46, 248, 249, 251–54, 257, 275
Tenison-Woods, Julian, 232, 233*n*
Tennyson, Alfred, Lord, 188, 284
tephra, 241, 243
Tethyan ocean, *72, 73*
Thailand, 33*n*
Theodore the Studite, Saint, 10
theosophy, 52*n*
Thomson, Captain, 214–15, 231
Thomson, Ken, 30
Thor, Mr., 203
Thornton, Ian, 364
Thwart-the-Way Island, 159*n*, 237, 258*n*, 275
tide meter, *251,* 252
Times (London), 177–78, 182–83, 189–90, 194, 269–70, 296
time zones, 218*n*, 247

Timor, 13*n*, 19, 22, 29, 54, 166, 263
Tjiringin, 251, 258
Toba, Mount, 4–5, 307, 311
Tonga, 109
Tordesillas Line, 13*n*, 14–15, *14*
transform faults, 103–4, *104*
Treaty of Paris (1872), 138
Treub, Melchior, 358–59
Trobriand Islands, 53
troposphere, 283
tsunamis, 110, *219*, 220, 234,
    237–38, 240–53, 272–78, *279*
  in Anjer, 256–57, 275
  in Ceylon, 276–77
  destructive capacity of, 246
  in France, 278
  in India, 273–74
  long, 276–77
  Royal Society investigation of,
    274–78
  short, 276
  speed of, 247, 247*n*–48*n*
  worldwide effects of, 276–79
Turkey, 110
Turkey Company, 29
Turner, J. M. W., 280*n*
Twenty one Balloons, The (Pène du
    Bois), 385, 386–87
Twilight in the Wilderness (Church),
    280
Typhon, 300
Tyringin, 245, 248, 257

Ujung Kulon National Park, 373*n*
United States, 138*n*, 261*n*, 305
Unzen, Mount, 243, 264, 297

Vail, Alfred, 142
van den Broecke (storekeeper), 34–35
Vening Meinesz, Felix, 86–88
Verbeek, Rogier Diederik Marius,
    168–70, *169*, 174, 215, 248*n*
  Krakatoa report of, 264, 310–11,
    342, 361, 389
  Krakatoa visited by, 169–70, 308–9,
    342, 350

Vereenigde Oost-Indische Compagnie,
  see VOC
Vereker, Foley, 262, 269
Verlaten, 157*n*, 309, 348
Vesuvius, Mount, 12, 110, 297
Victorian Internet, The (Standage),
    177
Vietnam, 33*n*, 126
Villarica, 306
Villumsen, Rasmus, 75
Vlekke, Bernard, 133
VOC (Vereenigde Oost-Indische
    Compagnie), 29–30, 118
  bankruptcy of, 138, 140–41
  Batavia ruled by, 46–47
  country trading of, 30–31
  logo of, *37*, 38
  slavery practiced by, 42, 45
Vogel, Johan Vilhelm, 48–50, 132
Volcanic Explosivity Index (VEI),
    307–8
volcanoes, 4–5, 294, 298–306
  classical theories of, 301–3
  geologic signatures of, 126*n*–27*n*
  hazards of living near, 241–43
  organic life and, 298–99
  phreatomagmatic eruptions of, 310,
    346
  plate tectonics and, 299–300, 305–6
  Plinian eruptions of, 12
  in religion and superstition,
    300–301, 321
  subduction zones and, 110, 112–13,
    304–5
Volcanoes (Judd), 311
Volcanological Survey of Indonesia,
    370
Vulcan, 300

Wafula, Dieudonné, 303*n*
Wager, Lawrence, 78
Waghenaer, Lucas Janszoon, 26
Waldseemüller, Martin, 22
Walker, Captain, 155
Wallace, Alfred Russel, 53–61, *54*,
    64–67, 68, 72

Index

Wallace, Alfred Russel (*cont.*)
  in Amazonia, 57–58
  background and personality of,
    56–57
  collecting zeal of, 58
  Darwin and, 56, 59–62
  evolutionary insights of, 58–61
Wallace Line, 55–56, 63–67, 113–14,
    135*n*
  extent of, 64–65
  geologic processes and, 65–67
Wallis, James, 260
Waterloo, Battle of, 151*n*, 191
Watkins, Gino, 78
Watson, W. J., 218–22
weather forecasting, 265, 287–88
Webber, John, 119, *120,* 349
Wegener, Alfred Lothar, 68, 69–76,
    70, 77, 86, 91, 310
  academic hostility toward, 74–75
  continental drift concept of, 69–73,
    88–89, 95
  death of, 75–76
Wegener, Kurt, 71*n*

Weitzel, A. W. P., 140
Weltevreden (suburb of Batavia),
    139–40
Westphalia, Treaty of, 29*n*
W. H. Besse, 231
White, Gilbert, 292*n*
Willem II, king of the Netherlands,
    150
Willem III, king of the Netherlands,
    146, 147, 148–52, 325
William of Orange, 29*n*
Wilson, Anna, 197–98, *198,* 204
Wilson, John, 197, 204
Wilson, J. Tuzo, 99–102, 303
Witti, Francis, 262
Wonderful Century, The (Wallace), 61

Yemen, 334
Young American Hose Company,
    289

Zanzibar, 295
Zeeland, 155, 159–60
Zoological Society of London, 52

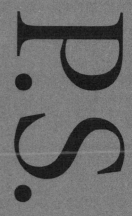

About the author

2 Meet Simon Winchester

About the book

5 The Haunting Legacy of Krakatoa and
   Other Natural Disasters:
   Simon Winchester Responds to the
   2004 Tsunami

Insights,
Interviews
& More...

Read on

11 A Reading Excerpt from *A Crack in the
    Edge of the World: America and the Great
    California Earthquake of 1906*

15 Have You Read?:
    More by Simon Winchester

Meet **Simon Winchester**

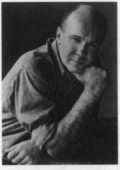

Marion Ettlinger

SIMON WINCHESTER, author, journalist, and broadcaster, has worked as a foreign correspondent for most of his career, although he graduated from Oxford in 1966 with a degree in geology and spent a year working as a geologist in the Ruwenzori Mountains in western Uganda, and on oil rigs in the North Sea, before joining his first newspaper in 1967.

His journalistic work, mainly for *The Guardian* and *The Sunday Times,* has based him in Belfast; Washington, D.C.; New Delhi; New York; London; and Hong Kong, where he covered such stories as the Ulster crisis, the creation of Bangladesh, the fall of President Marcos, the Watergate affair, the Jonestown Massacre, and the assassination of Egypt's President Anwar Sadat, the recent death and cremation of Pol Pot, and, in 1982, the Falklands War—during which time he was arrested and spent three months in prison in Ushuaia, Tierra del Fuego, on spying charges. He has been a freelance writer since 1987.

He now works principally as an author, although he contributes to a number of

About the author

American and British magazines and journals, including *Harper's, Smithsonian, National Geographic, The Spectator, Granta,* the *New York Times,* and *The Atlantic Monthly.* He was appointed Asia-Pacific editor of *Condé Nast Traveler* at its inception in 1987, and later became editor-at-large. His writings have won him several awards, including Britain's Journalist of the Year.

He writes and presents television films—including a series on the final colonial years of Hong Kong and on a variety of other historical topics—and is a frequent contributor to the BBC radio program *From Our Own Correspondent.* Winchester also lectures widely—most recently before London's Royal Geographical Society (of which he is a Fellow)—and to audiences aboard the cruise liners *QE2* and *Seabourn Pride.*

His books cover a wide range of subjects, including a study of the remaining British Empire, the colonial architecture of India, aristocracy, the American Midwest, his experience of the months spent in an Argentine prison on spying charges, his description of a six-month walk through the Korean peninsula, the Pacific Ocean and the future of China. Most recently he has written *The River at the Center of the World,* about China's Yangtze River; the bestselling *The Professor and the Madman,* which is to be made into a major motion picture by the distinguished French director Luc Besson; *The Fracture Zone: A Return to the Balkans,* which recounts his journey from Austria to Turkey during the 1999 Kosovo crisis; and the bestselling *The Map That Changed the World,* about the nineteenth-century geologist ▶

Meet Simon Winchester *(continued)*

William Smith. His latest book, *A Crack in the Edge of the World: America and the Great California Earthquake of 1906,* will be published by HarperCollins in Fall 2005.

Simon Winchester lives on a small farm in the Berkshires in Massachusetts. ∾

The Haunting Legacy of Krakatoa and Other Natural Disasters
Simon Winchester Responds to the 2004 Tsunami

The following article by Simon Winchester appeared in The Guardian *less than a week after devastating tsunamis struck the coasts of the Indian Ocean region on December 26, 2004. Winchester remarks on the haunting legacy of natural disasters, including the 1883 eruption of Krakatoa and the 1906 San Francisco earthquake.*

THE GUARDIAN (LONDON)—JANUARY 5, 2005

IT WAS AN EARLY summer's morning in nineteenth-century Zanzibar, and the boys from the mission school were out on their customary Sunday walk along the beach when one of them spotted something peculiar and sinister: a large, oddly shaped object, stranded at the tidemark. Their teacher, cautioning the children to stay well back, walked over to investigate: the object, she later wrote in a report to the Royal Society, turned out to be one of scores that had been washed up overnight from the Indian Ocean by the previous night's storm.

It was a great mass of pumice stone, some dozens of tons of floating rock. It had evidently been in the sea for a very long time: its bottom was crusted with barnacles and weed. But more bizarre still was what she found on looking more closely: welded thickly on to the rock's upper surface and giving it ▶

66 Welded thickly on to the rock's upper surface and giving it the most macabre appearance were dozens of human skeletons, together with the bones of monkeys and, distinguishable only later in the day when the school biology teacher had appeared, the skeletons of a pair of big cats, most probably Sumatran tigers. 99

the most macabre appearance were dozens of human skeletons, together with the bones of monkeys and, distinguishable only later in the day when the school biology teacher had appeared, the skeletons of a pair of big cats, most probably Sumatran tigers.

It took little detective work to ascertain that these rafts of pumice had floated to Zanzibar—and, it turned out, to dozens of other places along the east African coast—from where they had been created, 4,000 miles away to the east, by the devastating eruption of the Dutch East Indian island of Krakatoa. This entire island—11 cubic miles of it—had been blasted into oblivion in one of the planet's most titanic volcanic displays, on August 27, 1883. Now, nine months later, the floating pumice islands that had fallen from it, laden with their ghastly cargo of creatures killed in the blast, had found their way across an enormous ocean, a stark reminder of an event that the world was wanting to forget.

For most of the world back then knew about Krakatoa, just as almost everyone now knows of the recent Sumatran tragedy. There had been spectacular volcanic eruptions before—Toba and Tambora, both in the East Indies, as well as Santorini, Hekla and Mazama—but Krakatoa became an icon, remembered to this day, for the same reason that underpins the probably eternal notoriety of the Boxing Day earthquake: electricity.

Krakatoa was the first catastrophe of the age of communication. When President Lincoln was assassinated in 1865, it took twelve days for the news to get from Washington to London. When Krakatoa erupted, however, three new factors were in play: Morse code had been invented; Julius Reuter had set up a news agency; and the submarine telegraph cable had been developed. All were to play their part.

The first terse signal—"Strong volcanic eruption, Krakatowa Island" —from the Lloyd's agent who saw the flames spout from the volcano's summit, was carried over lines that were almost immediately broken by the tsunami that killed 36,000 people a few seconds later. But the message managed to get through to Batavia, the Dutch East Indian capital. From there it passed under the sea, to Singapore. It was then amplified, retransmitted on to Madras, passed to stations at Trincomalee, Colombo, and Bombay, then traveled via the newly built Suez Canal to Port Said, then sped by way of Malta and Gibraltar across Biscay to Porthcurno in Cornwall, and across the Atlantic to the Reuters receiving stations in Newfoundland and Boston.

The Boston Globe had the story on its front page not twelve days, but just four hours later. The next morning, horse-drawn commuters in Boston and New York would read of places such as Krakatoa and Sumatra and the Sunda Strait and the tragedy that had engulfed them, and would accord the names the same kind of familiarity as Baltimore and Cape Cod. The world suddenly became much smaller: the global village, one might say, became that August morning a distant reality.

But although the nineteenth-century world now knew what had happened, it still had no real understanding of why—and this highlights one very real contrast between the way that mankind once tended to react to the major, world-girdling calamities, those that occurred during what we might call early modern times, and the immensely more rational way that the world reacts today, in what we believe to be our properly modern scientific environment.

There is a generally accepted and consequently universal understanding of just why last Boxing Day's events occurred. The Indian Plate, subducting with immense force against the Burmese Plate, overcame a long, preexisting resistance, and snapped upward under the seabed off Sumatra; the sudden upthrust triggered tsunamis that caused an enormous loss of life around the Indian Ocean. The world's reaction has, as a consequence of the explanation, been measured and appropriate: there have been widespread appeals, generous outpourings of aid, promises of future prevention measures, and worldwide demands for ▶

new research. Seismology, not always the most favored of sciences, will probably gain new funding. And mankind dispassionately accepts—to a certain degree—that what took place was some kind of cruel cull.

At the time of Krakatoa, however, although the cable-connected world was fairly well informed about the event, it was also lacking any explanation and was terribly bewildered by what had taken place. For most people, something inexplicably strange had occurred on the other side of the world—something that caused the tides to rise as far away as Biarritz and Devonport, something that made the evening skies all over the world burn with a lurid fire (Edvard Munch's vividly colored painting, *The Scream,* recently stolen, was painted at the time of those Krakatoa-affected skies), something that caused an explosion so loud it could be heard 3,000 miles away. But as to why it had happened, no one—no scientist at least—could be found to offer any comforting nostrums. For all the world knew, the end of life itself could be at hand.

The world thus turned to its old standby: religion. It ascribed the event's ultimate cause to God. It ascribed its proximate cause to some act of man that had made God angry. And it acted to expunge the cause of that anger— and did so very quickly indeed.

In the case of Krakatoa, the Muslim prelates of Java first made this connection. The eruption that had killed so many and had ruined so much was clearly, they said, the work of Allah—a divine, who was—so the mullahs told their Javanese congregations of the day—supremely irritated that so many

> 66 At the time of Krakatoa, however, although the cable-connected world was fairly well informed about the event, it was also lacking any explanation and was terribly bewildered by what had taken place. For most people, something inexplicably strange had occurred on the other side of the world. 99

of their number were passively allowing themselves to be ruled by white infidel outsiders, the Dutch. To appease the sorely tried Allah, the mullahs said, the Dutch had to be killed and their influence expunged. Rise up, they advised.

And so they did—in a piecemeal fashion at first, in an organized rebellion five years later, and in a measured and defiant way in the decades that followed. The Dutch were eventually forced to leave; Indonesia, born out of the Hollanders' imperial fiefdom, remains today the world's most populous Islamic nation. Krakatoa was not the cause of the birth of Indonesia, far from it; but it was a sign, a trigger, and it remains a significant moment in Indonesian political history for that very reason.

Twenty years later, on the other side of the world, there was another seismic event that is now being seen as having had similar consequences. On April 18, 1906, San Francisco was leveled by an enormous earthquake caused by a rupture of the San Andreas Fault, where the North American and Pacific Plate press against each other. The physical consequences of the event were profound and immediate: death, destruction, appeals, aid, promises of better building codes, and more funds for science.

But the social and political consequences have only recently been realized—and religion played its part here also. For less than a week before the event, a meeting took place in a church in a Los Angeles slum, a meeting in which manic-seeming priests, all of them speaking in tongues and waving limbs, used any dramatic technique possible to win over the winnable—and promised that a sign from the Lord would come any day.

That was Sunday. The promised sign came, just before dawn on the following Wednesday, when San Francisco and most of the north of the state was ripped apart by the earthquake. The results were immediate. The following Sunday, the new-born church was filled to bursting—and the American evangelical organization, now broadly known as the Pentecostal movement, was, in essence, born.

And the effect? It may strain belief a little, but this same Pentecostal movement is the church that has, in recent times, given us such grotesqueries as Tammy Faye Bakker, Jimmy Swaggart, and Pat Robertson, all of them the spiritual keystones of the hitherto unrecognized religious structure that Karl Rove utilized last autumn to help win President Bush the election.

This, many will say, is going too far. Joining the dots can be a ▶

dangerous and foolhardy temptation to which to succumb. And yet to those who believe that Krakatoa led in part to Indonesian independence, the notion that the destruction of San Francisco may have played an unwitting historical role in the creation and sustenance of American conservative politics a century later is not as outlandish as it may seem.

When mankind was incapable of explaining major calamities, there was a tradition of turning to the skies for understanding and comfort. These days, believing that science answers all, we are less tempted to reach to religion and superstition to answer our needs.

Given the political consequences of our having done that in the past—some of these consequences good, some arguably less so, but all of them very long-lasting indeed—it is perhaps better for us all that we now claim to know the answers: to know why the plates collide, why the waves rush out so far, and why skeletons appeared on those beaches in Zanzibar on a peaceful Sunday morning more than a century ago. ∼

66 On April 18, 1906, San Francisco was leveled by an enormous earthquake caused by a rupture of the San Andreas Fault. . . . The physical consequences of the event were profound and immediate: death, destruction, appeals, aid, promises of better building codes, and more funds for science. 99

Reprinted with permission by *The Guardian* (London), January 5, 2005

A Reading Excerpt from
A Crack in the Edge of the World: America and the Great California Earthquake of 1906

In his newest book, Simon Winchester, the New York Times *bestselling author of* The Professor and the Madman *and* Krakatoa, *takes an adventurous and informative look at earthquakes, as seen through the devastating 1906 quake in San Francisco.*

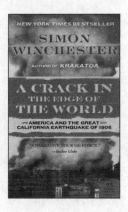

THE VIEW FROM THE MOON

HAD ANY GEOLOGIST at the time been able to look down at the planet in its entirety on April 18th, 1906, and had been able to see what then took place, he would at the very least have been utterly amazed by the physical context of the event, even if the event itself, when seen from on high, seemed less than overwhelming.

For as context, the planet would have been memorably beautiful. Had he been standing on the moon, say—had he been a 1906 version of Neil Armstrong, scanning with a hugely powerful telescope the surface of the blue and green and white ball that was hanging in his ink-black sky—he would have seen illuminated in front of him (assuming that the cloud cover was not too dense) a tract of the world that extended from what some of mankind called India to what others called the Rocky Mountains, and all of which was bathed in the brilliant white light of sunshine.

At the moment that we find interesting— five o'clock on that western American morning, give or take—he could see the terminator- ▶

line of western darkness pushing its way rapidly towards the Pacific. The earth would be moving relentlessly at a speed of some hundreds of miles an hour eastwards towards it, opening ever more populated parts of the landmasses to the light of the dawning day.

The line at this very moment seemed to begin in the north near Melville Island in the Canadian Arctic, pass on down through Banks Island and the unpopulated and ice-bound wilderness of the Northwest Territories and the Yukon, through Saskatchewan and Alberta, raggedly on down through the newly created state of Montana, through bison-and-Cherokee country of Wyoming and Colorado and New Mexico, across the Rio Grande towards Acapulco and to a point on the coast where it finally slid away off the North American landmass and on eventually to illuminate the still-inky emptiness of the Pacific Ocean.

To the east of the line, all was bright and daylight. To its west, an impenetrable dark. And on the line itself, an uncertain penumbra of a few hundred miles of a swathe of half-dark and half-light. On earth itself this penumbral vagueness would translate itself into the morning twilights that early risers were experiencing just now in cities and on farms and in small villages all the way from Vancouver island in the north down to Baja California in the south, and where the day designated at the 18th of April was just about to begin.

It is fanciful to suppose that anyone so far away, in distance or in time, would have had access to a large enough spy-glass. But assuming that such a device did exist, and that the person at this lunar viewing-point had its brass and glassware trained precisely on the

66 To the east of the line, all was bright and daylight. To its west, an impenetrable dark. And on the line itself, an uncertain penumbra of a few hundred miles of a swathe of half-dark and half-light. 99

northern coast of California at that very
particular moment in time, with the
terminator brightening his view inch by
faraway inch in his field of view—what,
precisely, might he have seen?

The answer is inevitably dismaying to all of
those who like to think that the earth and its
inhabitants and the events that occur upon it
have any importance at all, in a cosmic sense.
For from that distance he would have seen,
essentially, nothing.

Yet at a few minutes past five in the morning
of that day something did, indeed, happen.

The planet very briefly *shrugged*.

It flexed itself for a few seconds, perhaps a
little short of a minute. If our observer had
been acutely aware of his geography, and had
he been fortunate enough to have been staring
at a very precisely defined spot in the north of
California at exactly the right moment, then he
might have seen what appeared to him a tiny
ripple spurt in towards the coast from the sea.
He might, moreover, have seen that spreading
ripple as it moved slowly and steadily inshore,
and then watched as it moved, fan-like and
subtle, up and down the coastline as a tiny
shudder. It would have seemed to him a
momentary loss of focus, something that
would have made his vision suddenly blur very
slightly, and then just as quickly clear again.

He might have noticed this; or more likely
he might have blinked and missed it. If he
noticed at all, he would probably have assumed
it was more of a problem with his lens and his
telescope than with the surface of the planet
below. And even if he had concluded that the
ripple and the shudder had in fact occurred on
the green-and-blue-and-white planet that ▶

A Reading Excerpt *(continued)*

floated serene in the lunar sky, he would also have been quick to conclude that whatever it was had been momentary, trivial, and utterly forgettable. No more, for the earth entire, than a gentle and momentary shrug.

It was all so very different down on the surface of the planet itself. On earth, on the western part of that great entity called by its inhabitants North America and particularly in and around the fragile and rather delicately constructed young northern city of San Francisco, a number of people realized all too quickly that something of great significance was happening. And if they were sensible and observant they took care to note and remember exactly when it all began, and we have their memories set down for us still. ⌒

> 66 If they were sensible and observant they took care to note and remember exactly when it all began, and we have their memories set down for us still. 99

14

Have You Read?
More by
Simon Winchester

THE MAP THAT CHANGED THE WORLD:
WILLIAM SMITH AND THE BIRTH OF
MODERN GEOLOGY

In 1793, a canal digger named William Smith
made a startling discovery. He found that by
tracing the placement of fossils, which he
uncovered in his excavations, one could follow
layers of rocks as they dipped and rose and
fell—clear across England and, indeed, clear
across the world—making it possible, for the
first time ever, to draw a chart of the hidden
underside of the Earth. Determined to expose
what he realized was the landscape's secret
fourth dimension, Smith spent twenty-two
years piecing together the fragments of this
unseen universe to create an epochal and
remarkably beautiful hand-painted map. But
instead of receiving accolades and honors, he
ended up in debtors' prison, the victim of
plagiarism, and virtually homeless for more
than ten years. Finally, in 1831, this quiet
genius—now known as the father of modern
geology—received the Geological Society of
London's highest award and King William IV
offered him a lifetime pension. With a keen
eye and thoughtful detail, Simon Winchester
unfolds the poignant sacrifice behind this
world-changing discovery.

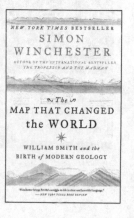

"Winchester brings Smith's struggle to life in
clear and beautiful language."
                    —*New York Times Book Review*

THE PROFESSOR AND THE MADMAN: A TALE OF MURDER, INSANITY, AND THE MAKING OF THE *OXFORD ENGLISH DICTIONARY*

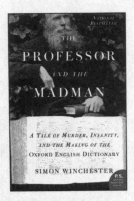

The bestselling tale of madness, genius, and the incredible obsessions of two remarkable men at the heart of the *Oxford English Dictionary*—Professor James Murray, the dictionary's editor, and Dr. W. C. Minor, an inmate at an asylum for the criminally insane who contributed to more than ten thousand definitions.

"An extraordinary tale and Simon Winchester could not have told it better. . . . A splendid book." —*The Economist*

"The linguistic detective story of the decade." —William Safire, *New York Times Magazine*

OUTPOSTS: JOURNEYS TO THE SURVIVING RELICS OF THE BRITISH EMPIRE

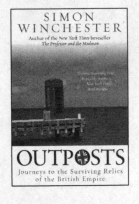

A tour of the last outposts of the British empire—from the chill of the Atlantic to the blue seas of the Caribbean, from the south of Spain and the tip of China to the utterly remote specks in the middle of gale-swept oceans.

"A brilliant and delightful addition to the long and distinguished shelf of British literary odysseys." —Christopher Buckley, *Washington Post Book World*

Don't miss the next book by your favorite author. Sign up now for AuthorTracker by visiting www.AuthorTracker.com.